INTRODUCTION TO MOBILE COMMUNICATIONS

OTHER TELECOMMUNICATIONS BOOKS FROM AUERBACH

Architecting the Telecommunication Evolution: Toward Converged Network Services
Vijay K. Gurbani and Xian-He Sun
ISBN: 0-8493-9567-4

Business Strategies for the Next-Generation Network
Nigel Seel
ISBN: 0-8493-8035-9

Chaos Applications in Telecommunications
Peter Stavroulakis
ISBN: 0-8493-3832-8

Context-Aware Pervasive Systems: Architectures for a New Breed of Applications
Seng Loke
ISBN: 0-8493-7255-0

Fundamentals of DSL Technology
Philip Golden, Herve Dedieu, Krista S Jacobsen
ISBN: 0-8493-1913-7

Introduction to Mobile Communications: Technology,, Services, Markets
Tony Wakefield, Dave McNally, David Bowler, Alan Mayne
ISBN: 1-4200-4653-5

IP Multimedia Subsystem: Service Infrastructure to Converge NGN, 3G and the Internet
Rebecca Copeland
ISBN: 0-8493-9250-0

MPLS for Metropolitan Area Networks
Nam-Kee Tan
ISBN: 0-8493-2212-X

Performance Modeling and Analysis of Bluetooth Networks: Polling, Scheduling, and Traffic Control
Jelena Misic and Vojislav B Misic
ISBN: 0-8493-3157-9

A Practical Guide to Content Delivery Networks
Gilbert Held
ISBN: 0-8493-3649-X

Resource, Mobility, and Security Management in Wireless Networks and Mobile Communications
Yan Zhang, Honglin Hu, and Masayuki Fujise
ISBN: 0-8493-8036-7

Security in Distributed, Grid, Mobile, and Pervasive Computing
Yang Xiao
ISBN: 0-8493-7921-0

TCP Performance over UMTS-HSDPA Systems
Mohamad Assaad and Djamal Zeghlache
ISBN: 0-8493-6838-3

Testing Integrated QoS of VoIP: Packets to Perceptual Voice Quality
Vlatko Lipovac
ISBN: 0-8493-3521-3

The Handbook of Mobile Middleware
Paolo Bellavista and Antonio Corradi
ISBN: 0-8493-3833-6

Traffic Management in IP-Based Communications
Trinh Anh Tuan
ISBN: 0-8493-9577-1

Understanding Broadband over Power Line
Gilbert Held
ISBN: 0-8493-9846-0

Understanding IPTV
Gilbert Held
ISBN: 0-8493-7415-4

WiMAX: A Wireless Technology Revolution
G.S.V. Radha Krishna Rao, G. Radhamani
ISBN: 0-8493-7059-0

WiMAX: Taking Wireless to the MAX
Deepak Pareek
ISBN: 0-8493-7186-4

Wireless Mesh Networking: Architectures, Protocols and Standards
Yan Zhang, Jijun Luo and Honglin Hu
ISBN: 0-8493-7399-9

Wireless Mesh Networks
Gilbert Held
ISBN: 0-8493-2960-4

AUERBACH PUBLICATIONS

www.auerbach-publications.com
To Order Call: 1-800-272-7737 • Fax: 1-800-374-3401
E-mail: orders@crcpress.com

INTRODUCTION TO MOBILE COMMUNICATIONS

Technology, Services, Markets

Tony Wakefield
Dave McNally
David Bowler
Alan Mayne

Auerbach Publications
Taylor & Francis Group
Boca Raton   New York

Auerbach Publications is an imprint of the
Taylor & Francis Group, an informa business

Auerbach Publications
Taylor & Francis Group
6000 Broken Sound Parkway NW, Suite 300
Boca Raton, FL 33487-2742

© 2007 by Informa Telecoms & Media
Auerbach is an imprint of Taylor & Francis Group, an Informa business

No claim to original U.S. Government works
Printed in the United States of America on acid-free paper
10 9 8 7 6 5 4 3 2 1

International Standard Book Number-10: 1-4200-4653-5 (Hardcover)
International Standard Book Number-13: 978-1-4200-4653-3 (Hardcover)

This book contains information obtained from authentic and highly regarded sources. Reprinted material is quoted with permission, and sources are indicated. A wide variety of references are listed. Reasonable efforts have been made to publish reliable data and information, but the author and the publisher cannot assume responsibility for the validity of all materials or for the consequences of their use.

No part of this book may be reprinted, reproduced, transmitted, or utilized in any form by any electronic, mechanical, or other means, now known or hereafter invented, including photocopying, microfilming, and recording, or in any information storage or retrieval system, without written permission from the publishers.

For permission to photocopy or use material electronically from this work, please access www.copyright.com (http://www.copyright.com/) or contact the Copyright Clearance Center, Inc. (CCC) 222 Rosewood Drive, Danvers, MA 01923, 978-750-8400. CCC is a not-for-profit organization that provides licenses and registration for a variety of users. For organizations that have been granted a photocopy license by the CCC, a separate system of payment has been arranged.

Trademark Notice: Product or corporate names may be trademarks or registered trademarks, and are used only for identification and explanation without intent to infringe.

Library of Congress Cataloging-in-Publication Data

Wakefield, Tony.
    Introduction to mobile communications : technology, services, markets / Tony Wakefield.
        p. cm. -- (Informa telecoms & media)
    ISBN 1-4200-4653-5 (alk. paper)
    1. Mobile communication systems. 2. Cellular telephone services industry--Marketing. I. Title.

TK6570.M6W35 2007
621.3845--dc22
                                                                                          2006047941

Visit the Taylor & Francis Web site at
http://www.taylorandfrancis.com

and the Auerbach Web site at
http://www.auerbach-publications.com

Contents

Introduction xi
Acronyms xiii

1   **The Mobile Telecommunications Market** 1
    1.1   Introduction 2
        1.1.1   An Introduction to Telecommunications 2
            The Telecommunications World 2
            The Evolution of Telecommunications 3
            Telecommunications Today 5
            Review Questions 6
        1.1.2   Types of Networks 7
            Introduction to Telecommunications Networks 7
            Fixed Communication Networks 7
            Mobile Communication Networks 11
            Defining Areas of the Network 15
            Review Questions 20
        1.1.3   Services and the User Perspective 21
            Access to Network Services 21
            The Changing Nature of Services 21
            Increasing Service Needs 22
            Telecommunications Services 23
            Network and Product Branding 28
            Review Questions 31
        1.1.4   Review Questions 32
    1.2   The Mobile Market 33
        1.2.1   Segmenting the Market 33
            Mobile Market Sectors 33
            Operators in the Telecommunications Market 33
            Radio Access, Core Networks, and Supporting Systems 44
            OSS, BSS, and MSS 49
            Review Questions 50

		1.2.2 Handset Market 50
		Low-Cost Handsets...51
		Handset Complexity and Features 52
		Examples of Mobile Devices 53
	1.3	Current Mobile Technologies and Markets...............................54
		1.3.1 Current Mobile Technologies 54
		Mobile Generations..54
		GPRS and EDGE 59
		Code Division Multiple Access (CDMA) Technologies 61
		Evolving to 3G Systems..62
		1.3.2 Global Mobile Markets 65
		Market Overview: China 65
		Western Europe..68
		United States...69
		Review Questions..72
		1.3.3 World Market Forecasts 73
		Subscription and Penetration Forecasts 73
		Technology Forecasts 74
		ARPU Forecasts...76
		Review Questions..78
		1.3.4 Looking Ahead..79
		2G: A Changing Market 79
		General Trends in Telecommunications.............................79
		Influential Factors in the Evolution of Mobile Services 80
		Mobile Internet in 2G 81
		Future Mobile Service Needs 81
		Technology for the Future..82
		Review Questions..88
	1.4	Standards and Regulations 89
		1.4.1 Standards and Standards Bodies.................................89
		Benefits of a Standard.. 90
		International Telecommunication Union (ITU) 91
		European Telecommunications Standards Institute
		(ETSI) 91
		Third Generation Partnership Project (3GPP)....................91
		Third Generation Partnership Project 2 (3GPP2)...............92
		UMTS Forum...92
		The GSM Association..93
		The European Conference of Postal and
		Telecommunications Administration (CEPT)...............93
		The Telecommunications Industry Association (TIA).........94
		The American National Standards Institute (ANSI) 94
		The CDMA Development Group..95

The Association of Radio Industries and Businesses
(ARIB) 95
Telecommunication Technology Committee (TTC).......... 96
Telecommunications Technology Association (TTA)........ 96
The Wi-Fi Alliance.. 96
The WiMAX Forum™ 97
Review Questions.. 97
    1.4.2 Telecommunications Regulations.. 98
The Need for Telecom Regulation....................................... 98
Benefits of Regulation to the Consumer............................. 99
Objectives of National Regulatory Authorities 100
Case Study of a National Regulator: Ofcom 101
Regulators of the World 102
Data Protection and User Privacy..................................... 102
Review Questions.. 104

2   Technology Principles: The Basics of Telecommunications 105
  2.1   Introduction 106
    2.1.1 The Basics of Telecommunications 106
Telecommunications: A Definition 106
Review Questions... 111
    2.1.2 Explaining the Principles.. 111
Services and Applications.. 111
Value-Added Services 115
Review Questions... 118
Review Questions...127
Switching 128
Review Questions...133
Signaling and Control 134
Billing 138
Networking Principles...143
Review Questions...144
    2.1.3 Review Questions...145
Practice Questions..146
  2.2   Information Transfer 150
    2.2.1 Representing Information 150
Characteristics of a Waveform...150
Analog Signals... 151
Digital Signals...152
The Frequency Spectrum...154
Propagation, Attenuation, and Noise 155
Review Questions... 157

viii ■ Contents

        2.2.2 Modulation 158
              Modulation Techniques 158
              Example Modulation Scheme:
                  Analog Frequency Modulation 159
              Pulse Code Modulation 160
              Review Questions... 163
        2.2.3 Channels: Organizing the Information 163
              Channel Bandwidth 164
              Channel Errors 165
              Interleaving 167
              Multiplexing Channels 167
              Review Questions... 169
        2.2.4 Example Transmission Systems: Implementing the
            Concepts 170
              Dense Wavelength Division Multiplexing (DWDM) 177
              Analog and Digital Subscriber Loop 178
              Review Questions... 180
        2.2.5 Review Question 181
              Practice Questions.. 182
    2.3 Radio and Cellular Systems 185
        2.3.1 Radio Systems 185
              Example Radio Systems 190
              Review Questions... 192
        2.3.2 The Cellular Concept: Mobile Network Basics 192
              Cell Planning 199
              Review Questions... 203
        2.3.3 GSM, GPRS, and EDGE 204
              Review Questions... 210
        2.3.4 CDMA-Based Systems 211
              Review Questions... 217
        2.3.5 Review Questions 218
              Practice Questions.. 219

3 Mobile Network Infrastructure and Supporting Systems 223
    3.1 Mobile Network 224
        3.1.1 The Overall System 224
              The Requirements 224
              Architecture and Procedures in the GSM Family
                  of Technologies 224
              Review Questions 229
        3.1.2 The User Equipment and Radio Access Network 229
              The GSM Radio Network Elements (BSS) 232

		Review Questions...240

- 3.1.3 Circuit-Switched and Packet-Switched Core Networks 240
  - Core Network Requirements..240
  - The Packet-Switched Domain 244
  - Signaling and Control in the Core Network 245
  - Review Questions...248
- 3.1.4 Section Review Questions 249
- 3.2 Mobile Network Procedures 253
  - 3.2.1 Example GSM Procedures..253
    - GSM Procedures 253
    - Review Questions...259
  - 3.2.2 Making Calls..261
    - Calls to Mobile Networks 261
    - Calls from Mobile Networks 261
  - 3.2.3 Example: GPRS Procedures 261
    - Review Questions.. 266
  - 3.2.4 Procedure Sequences 267
  - 3.2.5 Radio Resources 268
    - Radio Resource Procedures for GSM, GPRS, and EDGE 268
    - Review Questions...272
  - 3.2.6 Section 2 Review Questions 273
- 3.3 Supporting Systems 277
  - 3.3.1 Service Platforms — General.. 277
    - Service Platforms... 277
    - Review Questions.. 286
  - 3.3.2 Messaging Platforms..287
    - Voicemail Platforms 287
    - Review Questions...291
  - 3.3.3 Accessing the Internet 292
    - Methods for Accessing the Internet.................................292
    - Review Questions...295
  - 3.3.4 Intelligent Networks and CAMEL.......................................296
    - Introduction to Intelligent Networks (INs) 296
    - CAMEL (Customized Applications for Mobile Networks Enhanced Logic) 298
    - Review Questions...301
  - 3.3.5 Operational and Business Support Systems (OSS and BSS) and Billing..302
    - Operational Support System (OSS)................................302
    - Review Questions...307
  - 3.3.6 Section 3 Review Questions 308

x ■ Contents

4 Handset, Services, Media, and Content Distribution 313
    4.1   Introduction to Handset Technologies 313
            4.1.1   Handset Basics 313
                      Review Questions..322
            4.1.2   Handset and Standards Bodies..323
                      Review Questions..329
            4.1.3   Types of Handset and the Market 330
                      Review Questions..333
            4.1.4   Section Review Questions 333
    4.2   Handset Components and Architecture...336
            4.2.1   Basic Handset Functions and Processing..........................336
                      Review Questions... 340
            4.2.2   Memory Storage and Requirements.................................. 342
                      Review Questions..347
            4.2.3   Display Technologies..349
                      Review Questions..352
            4.2.4   Handset Sound Capabilities 353
                      Review Questions... 360
            4.2.5   Image and Video Capabilities...361
                      Review Questions..369
            4.2.6   Serial Interfaces 369
                      Review Questions..378
            4.2.7   Section 2 Review Questions 379
    4.3   Mobile Operating Systems...383
            4.3.1   Introduction to Mobile OS..383
            4.3.2   Example Mobile Operating Systems..................................385
                      Review Questions..396
            4.3.3   Section 3 Review Questions 397
    4.4   Services and Security 398
            4.4.1   Mobile Handset Services 398
                      Review Questions... 407
            4.4.2   Security Issues for Mobile Handsets.................................. 407
                      Security Issues for Mobile Handsets............................. 407
                      Review Questions..416
            4.4.3   Section 4 Review Questions 417

Index 419

Introduction

Whether you are technical or non-technical, it is often difficult to make sense of the ever-changing world of telecommunications. Technology seems to advance at an accelerating pace, complicating the business case on the one hand and confusing the customer on the other. The traditionally separate Fixed, Mobile, and Internet sectors are now converging into a single sector with implications for both technology and business. With so much happening, and so many options available, it is increasingly essential for telecommunications professionals to have a clear view of where the industry is heading.

This book provides a solid foundation on which to develop your view of telecommunications. It explores the core requirements of telecommunications — from markets to technology — primarily from the mobile standpoint. It examines how mobile communications is evolving to meet the changing needs of the customer, and how technology is changing to meet those needs. The picture first builds upon first principles to ensure a solid grounding in the basics before delving into more advanced issues.

The material for this book is taken primarily from the four core modules of the Certificate in Mobile Communications Distance Learning program run by the Informa Telecoms Academy. It has been designed by the highly experienced training development team to guide telecommunication professionals and students through the complex and fascinating world of mobile communications — leaving them much better equipped to fulfill roles within their current or prospective companies (be they network operators, vendors, systems integrators or service providers).

Areas covered in this book include the Mobile Telecommunications Market; Technology Principles; Mobile Network Infrastructure and Supporting Systems; and Handsets, Services, Media, and Content Distribution. These areas provide the foundational knowledge required to ensure an excellent grounding in the subject, and the easy format and logical progression make it an ideal resource to read from cover to cover.

The distance learning certificate students would also choose more specific modules covering Radio; Core Network Systems; Business Processes; Marketing,

Branding and Services; Billing and Mediation; and Network Implementation. Those taking the diploma study more advanced topics still. Full details can be found at www.telecomsacademy.com.

Whether as a companion for academic study or as a resource to help you in your current role or ongoing career, this book provides the essential knowledge on which to build and develop your view of the industry and its future direction.

ACRONYMS

1xEV-DO	1x Evolution-Data Optimized
21CN	21st Century Network
3GPP	3rd Generation Partnership Project
3GPP2	3rd Generation Partnership Project 2
A2DP	Advanced Audio Distribution Profile
AAC	Advanced Audio Codec
ACELP	Algebraic Code Excited Linear Prediction
ADSL	Asymmetric Digital Subscriber Line
ARFCN	Absolute Radio Frequency Channel Number
AM	Amplitute Modulation
AMPS	Advanced Mobile Phone System
AMR	Adaptive Multi-Rate codec
ANSI	American National Standards Institute
API	Application Programming Interface
ARIB	Association of Radio Industries and Businesses
ARPU	Average Revenue Per Unit or Average Revenue Per User
ASK	Amplitude shift keying
ATM	Asynchronous Transfer Mode
ATRAC	Adaptive TRansform Acoustic Coding
AuC	Authentication Center
BER	Bit Error Rate
BES	BlackBerry Enterprise Server
BMP	Bitmap
BREW	Binary Runtime Environment for Wireless
BSS	1) Business Support System; 2) Base Station Subsystem
BT	British Telecom
BTS	Base Transceiver Stations
C/I Ratio	Carrier-to-Interference Ratio
c-HTML	Compact HTML
CAMEL	Customized Applications for Mobile network Enhanced Logic
CAP	CAMEL Application Part

CC	Call Control
CCD	Charge-Coupled Device
CCSA	China Communication Standards Association of China
CDC	Connected Device Configuration (J2ME)
CDG	CDMA Development Group
CDMA	Code Division Multiple Access
CDMA2000 1x RTT	Code-Division Multiple Access (CDMA) 1x (single-carrier) Radio Transmission Technology
CEIR	Central Equipment Identity Register
CEPT	European Conference of Postal and Telecommunications (Conférence européenne des administrations des postes et des télécommunications)
CIF	Common Interchange Format
CLDC	Connected Limited Device Configuration (J2ME)
CMOS	Complementary Metal-Oxide Semiconductor
CS	Coding Scheme
CSCF	Call Session Control Function
CTP	Cordless Telephony Profile (Bluetooth)
CVSD	Continuously Variable Slope Delta-modulation
D-AMPS	Digital Advanced Mobile Phone System
dB	decibels
DBS	Direct Broadcast Satellite
DCE	Data Circuit-terminating Equipment
DECT	Digital Enhanced (formerly European) Cordless Telecommunications
DMB	Digital Multimedia Broadcasting
DNP	Dial-up Networking Profile (Bluetooth)
DRAM	Dynamic Random Access Memory
DRM	Digital Rights Management
DSL	Digital Subscriber Line
DSLAM	DSL Access Multiplexer
DSP	Digital Signal Processing
DTE	Data Terminal Equipment
DTMF	Dual-Tone Multi-Frequency
DTT or DTTV	Digital Terrestrial Television
DVB	Digital Video Broadcast
DVB-H	Digital Video Broadcasting-Handheld
DVB-T	Digital Video Broadcasting-Terrestrial
DVD	Digital Versatile Disc (formerly Digital Video Disc)
DWDM	Dense Wavelength Division Multiplexing
E-OTD	Enhanced Observed Time Difference (UMTA)
E2PROM	Electronically Erasable Programmable Read-Only Memory
EDGE	Enhanced Data rate for GSM Evolution

EEPROM	Electronically Erasable Programmable Read-Only Memory
EFR	Enhanced Full-Rate codec
EHF	Extra High Frequencies
EHR	Enhanced Half-Rate codec
EIR	Equipment Identity Register
EMC	ElectroMagnetic Compatibility
EMS	Enhanced Messaging Service
ERTMS	The European Railway Traffic Management System
ESN	Electric Security Number
ETNO	European Telecommunications Network Operators' Association
ETSI	European Telecommunication Standards Institute
EvDO	Evolution Data Optimized
EVDV	Evolution Data Voice
FDD	Frequency Division Duplex
FDM	Frequency Division Multiplexing
FDMA	Frequency Division Multiple Access
FM	Frequency Modulation
FR	1) Frame Relay; 2) Full Rate codec
FSK	Frequency Shift Keying
GAP	Generic Access Profile (Bluetooth)
GCF	Global Certification Forum
GGSN	Gateway GPRS Support Node
GIF	Graphics Interchange Format
GMSC	Gateway Mobile Switching Center
GMSK	Gaussian Minimum Shift Keying
GPRS	General Packet Radio Services
GPS	Global Positioning System
GRX	GPRS Roaming eXchange
GSM	Global System for Mobile Communications
GSMA	The GSM Association
GSN	GPRS Support Node
HSCSD	High-Speed Circuit Switched Data
HDSL	High-speed Digital Subscriber Line
HF	High Frequency
HLR	Home Location Register
HR	Half-Rate codec
HSDPA	High Speed Downlink Packet Access
IETF	Internet Engineering Task Force
IMEI	International Mobile Equipment Identity
IMS	IP Multimedia Subsystem
IMT-2000	International Mobile Telecommunications-2000
IN	Intelligent Network
INAP	Intelligent Network Application Part

IP	1) Internet Protocol: 2) Intercom Profile (Bluetooth)
ISDN	Integrated Services Digital Network
IOT	InterOperability Testing
IrCOMM	Provides COM (serial or parallel) port emulation or connections using IrDA protocol
IrDA	Infrared Data Association
IrMC	Infrared Mobile Communications
ISI	Intersymbol Interference
ISO	International Standards Organization
ITU	International Telecommunication Union
ITU-R	ITU Radiocommunication Sector
ITU-T	ITU Telecommunication Standardization Sector
IWF	InterWorking Function
J2ME	Java 2 Micro Edition
JPEG	Joint Photographic Experts Group
KVM	K Virtual Machine (J2ME)
LA	Location Area
LAC	Location Area Code
LCD	Liquid Crystal Display
LIF	Location Interoperability Forum
LR	Location Register
MGIF	Mobile Gaming Interoperability Forum
MHz	Megahertz
MIDI	Musical Instrument Digital Interface
MIDP	Mobile Information Device Profile (J2ME)
MIMO	Multiple-Input Multiple-Output
MIPS	Million Instructions Per Second
MM	Mobility Management
MMCA	MultiMediaCard Association
MMS	Multimedia Messaging Service
MMSE	Multimedia Messaging Service Environment
MNO	Mobile Network Operator
MPEG	Motiving Pictures Expert Group
MPLS	Multi Protocol Label Switching
MSAN	MultiService Access Node
MSISDN	Mobile Station International ISDN Number
MSC	Mobile Switching Center
MSS	Marketing Support System
MTP	Many-Time Programmable memory
MVNE	Mobile Virtual Network Enabler
MVNO	Mobile Virtual Network Operator
MWIF	Mobile Wireless Internet Forum
nm	nanometer

NMT	Nordic Mobile Telephone
NNG	National Number Group
NTS	Number Translation Services
OEM	Original Equipment Manufacturer
OLED	Organic Light Emitting Device
OMA	Open Mobile Alliance
OPP	Object Push Profile (Bluetooth)
OSA	Open System Architecture
OSS	Operational Support System
OTDOA	Observed Time Difference of Arrival (GSM)
OTP	One-Time-Programmable memory
P-TMSI	Packet-Temporary Mobile Subscriber Identity
P2T	Push-to-Talk
PAMR	Public Access Mobile Radio
PCM	Pulse Code Modulation
PCMCIA	1) Peripheral Component MicroChannel Interconnect Architecture; 2) Personal Computer Memory Card International Association
PCU	Packet Control Unit
PDA	Personal Digital Assistants
PDC	Public Digital Cellular
PDP	Packet Data Protocol
PDH	Plesiochronous Digital Hierarchy
PDN	Packet Data Network
PIM	Personal Information Management
PLMN	Public Land Mobile Network
PM	Phase Modulation
PMR	Professional Mobile Radio (Private Mobile Radio in the UK)
PNG	Protable Network Graphics
PoC	Push to talk over Cellular
POTS	Plain Old Telephone Service
PROM	Programmable Read-Only Memory
PSE	Packet-Switching Exchange
PSK	Phase Shift Keying
PSS	1) Packet Switched Stream; 2) Packet-switched Streaming Service
PSTN	Public Switched Telephone Network
PTT	1) Postal, Telegraph and Telephone; 2) Push To Talk
QCIF	Quarter Common Interchange Format
QoS	Quality of Service
R&TTE	Radio and Telecommunications Terminal Equipment directive
RA	Routing Area
RAM	Random Access Memory
RAN	Radio Access Network

ROM	Read-Only Memory
RNC	Radio Network Controller
RR	Radio Resources
RSN	Robust Security Network (802.11i)
RTT	Radio Transmission Technologies
SBC	SubBand Coding
SCP	1) Service Control Point; 2) Service Control Platform
SDAP	Service Discovery Application Profile (Bluetooth)
SDH	Synchronous Digital Hierarchy
SHF	Super High Frequency
SGSN	Serving GPRS Support Node
SIM	Subscriber Identity Module
SIP	Session Initiation Protocol
SIR	1) Signal to Interference Ratio; 2) Serial InfraRed
SM	Session Management
SME	Small and Medium Enterprise
SMS	Short Message Service
SMSC	Short Message Service Center
SNR	Signal-to-Noise Ratio
SOAP	Simple Object Access Protocol
SONET	Synchronous Optical NETwork
SPP	Serial Port Profile (Bluetooth)
SRAM	Static Random Access Memory
SS7	Signaling System #7
SSP	Service Switching Point
STM	Synchronous Transport Mode
STN	Super-Twisted Nematic display
TACS	Total Access Communications System
TCP/IP	Transmission Control Protocol/Internet Protocol
TD-SCDMA	Time Division Synchronous CDMA
TDD	Time Division Duplex
TDM	Time Division Multiplexing
TDMA	Time Division Multiple Access
TETRA	TErestrial Trunked RAdio
TFT	Thin Film Transistor
TIA	Telecommunications Industry Association
TMSI	Temporary Mobile Subscriber Identity
TN	Twisted Nematic display
UHF	Ultra High Frequency
UICC	Universal Integrated Circuit Card
UM	Unified Messaging
UMTS	Universal Mobile Telecommunications System
URI	Uniform Resource Identifier

USB	Universal Serial Bus
USIM	Universal Subscriber Identity Module
UTRA	Universal Terrestrial Radio Access
UTRAN	UMTS Terrestrial Radio Access Network
UWC-136	Universal Wireless Communications-136
VAS	Value Added Service
VASP	Value Added Service Provider
VDSL	Very high-speed Digital Subscriber Line
VGA	Video Graphics Array
VHE	Virtual Home Environment
VLF	Very Low Frequencies
VM	Virtual Machine
VLR	Visitor Location Register
VoIP	Voice over Internet Protocol
VPN	Virtual Private Network
VLR	Visitor Location Register
W3C	The World Wide Web Consortium
WAN	Wide Area Network
WAP	Wireless Application Protocol
WCDMA	Wideband Code Division Multiple Access
WDM	Wave Division Multiplexing
WiMAX	Worldwide Interoperability for Microwave Access
WLAN	Wireless Local Area Network
WLL	Wireless Local Loop
WML	Wireless Markup Language
WPA	Wi-Fi Protected Access
WRC	World Radio Conference
WSDL	Web Service Definition Language
WWAN	Wireless WAN
X.25	ITU protocol standard for WAN communications
XMF	eXtensible Music Format
XML	Extensible Markup Language

Chapter 1

The Mobile Telecommunications Market

At the end of this chapter, you should be able to:

- Understand the different types of telecommunications networks
- List the kind of services you would expect to be supported by mobile networks
- Explain how the user's experience may be affected by the mobile network
- Appreciate the global scope of GSM and be able to compare this technology with the other mobile networks in use around the world
- List the providers of equipment and services for mobile networks
- Show the extent of coverage offered by the major mobile operators
- Discuss the mobile market trend and services you expect for the future
- Appreciate the need for a standardization process and describe the role of the standards bodies in the development of the mobile telecommunications industry
- Explain the processes of regulation and show how the end user and operators can benefit from these processes

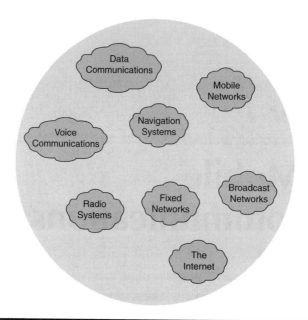

FIGURE 1.1 The telecommunications world.

1.1 Introduction

1.1.1 An Introduction to Telecommunications

The Telecommunications World

At the beginning of the new millennium, telecommunications, in its various forms, is revolutionizing both our society and the world economy. This book shows how technologies have developed, how they work, how they are being utilized today, and how they are likely to come together to satisfy the needs of consumers in the future.

The term "telecommunications," as shown in Figure 1.1, incorporates the following areas:

- Voice communications
- Data communications
- Radio systems
- Navigational systems
- Mobile networks
- Fixed networks
- Broadcast networks

We explore each of these areas to see how both technology push and market pull have acted upon them to create the telecommunications environment we have today.

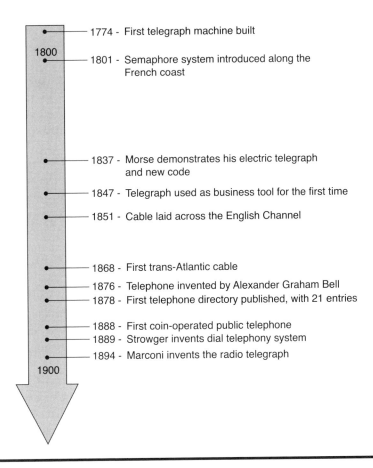

FIGURE 1.2 From Homer to Marconi.

The next few years will see big changes in the telecommunications industry. If all goes according to plan, the consumer will feel the benefits of greater choice; advanced, user-friendly, and more personal services; new billing models; enhanced customer care; and a greater focus on content rather than the underlying technology.

The Evolution of Telecommunications

From Homer to Marconi

Homer, speaking of the fall of Troy in the 11th Century BC, described a chain of beacons that were used to send the news back to Argos. This is one of the earliest examples of communication over a significant distance. (Figure 1.2.) The setting of the story in Greece is significant because the word "telecommunication" derives from the Greek word *tele,* meaning far or distant. As we will see, today the term "telecommunication" is applied to both short- and long-distance communication by electronic means.

The first invention that sought to bring about long-distance communication was the telegraph. The first models utilized 26 pairs of wires, one to transmit each letter of the alphabet. Samuel Morse saw the need to reduce the number of wires to one, which he achieved in 1838. To make his telegraph machine function on a single connection, he devised a coding system that produced an ECG-like line on tickertape. It was only later that one of his assistants produced the now-familiar dots and dashes version of the code that could be sound read by the operators. By 1854 there were 23,000 miles of telegraph cable in use, with the first trans-Atlantic cable link established in 1868.

When Alexander Graham Bell invented the telephone in 1876, not everyone was impressed. The Engineer-in-Chief of the British Post Office commented, "My department is in possession of full knowledge of the details of the invention, and possible use of the telephone is limited."

The early telephony systems employed operators who would manually connect one customer's line to another. On discovering that the local operator was diverting calls to a business rival, a St. Louis undertaker, Amon Strowger, invented dial telephony. As with so many developments within the field of telecommunications, necessity proved to be the mother of invention.

At the end of the 19th Century, Marconi's invention of the radio telegraph was key to the development of future radio-based technologies. Using existing discoveries and inventions, he was able to build a working system that could initially transmit up to a distance of 2.4 kilometers. By 1899 he had refined his system so that British battleships could transmit up to a distance of 121 kilometers.

From Marconi to Telstar

At the turn of the century, Marconi sent the first radio signal across the Atlantic, despite the predictions of his contemporaries that this would fail due to the curvature of the Earth. They had failed to appreciate that the Earth's atmosphere could be used to reflect radio waves.

Television made an appearance in 1926, when Baird produced his electromechanical machine. As is so often the case, a rival system proved to be just around the corner. Farnsworth demonstrated the superior electronic television in 1927 that, thankfully, became the forerunner of today's televisions.

As shown in Figure 1.3, over the next few decades a number of new technologies were developed that proved essential for the development of our modern telecommunications systems. In 1947, the transistor was invented, offering great reductions in the size of telecommunications equipment. The launch of the first communications satellite in 1962 and the development of fiber-optic cable four years later accelerated the pace of development for high bandwidth telecommunications.

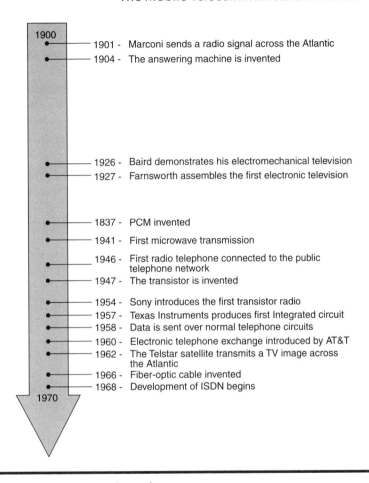

FIGURE 1.3   From Marconi to Telstar.

The Internet and Mobile Communications

The 1970s saw the beginnings of the Internet with the introduction of ARPANET. The original system was designed to encompass many networks from universities and government institutions. However, it took the development of the TCP/IP protocol in 1974 and the subsequent design of the World Wide Web in 1989 by Tim Berners-Lee to make the Internet as easily accessible as it is today.

Telecommunications Today

The world of telecommunications today is a vast and complex place, encompassing mobile, fixed, and Internet connections. In past times, telecommunications and data communications were seen and treated as completely separate fields of technology. It is not possible today to talk of one and neglect the other; voice and

6 ■ *Introduction to Mobile Communications*

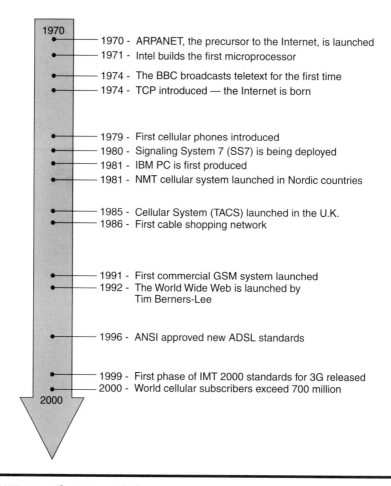

FIGURE 1.4 The Internet influence.

data communications have converged in both mobile and fixed networks. The next significant development, already underway, is the convergence of mobile and fixed networks, providing subscribers with unified communications systems that allow consumers a huge choice in the mode of their communication with another user or service.

Review Questions

Q1. Which of the following is not considered part of the telecommunications world?
   a. Voice communication
   b. Data communication
   c. Personal computers
   d. Radio systems

Q2. The term tele- derives from Greek and means _____.
   a. Distant
   b. Vision
   c. Speak
   d. Data

Q3. The first trans-Atlantic cable was laid in _____.
   a. 1968
   b. 1789
   c. 1868
   d. 1899

Q4. The Internet is made more accessible due to _____.
   a. Expansion of the American ARPANET
   b. Use of TCP/IP
   c. Development of the Web browser
   d. Mobile phone connections

Q5. A major development under way in the world of telecommunications is _____.
   a. Data and voice convergence
   b. The Internet
   c. Fixed and mobile convergence
   d. Satellite communications

1.1.2 Types of Networks

Introduction to Telecommunications Networks

In the world of telecommunications (telecom), there are many ways and means of exchanging information, and the nature of that information can be just as varied. As a result, there are many different kinds of telecommunications networks: fixed and mobile, data and voice, analog and digital. Today, networks such as the Internet allow the transfer of many different types of data, regardless of the underlying technology.

Fixed Communication Networks

The term "fixed" usually refers to the last mile in the communication link. This fixed part connects the user's equipment to the telecom network using some type of physical medium. Copper wires are the most common form of connection but they may also include coaxial cable or, less commonly, fiber optics. In some cases it is possible to provide a fixed connection using a radio connection. These types of systems are often referred to as wireless local loop (WLL) (see Figure 1.5).

8 ■ *Introduction to Mobile Communications*

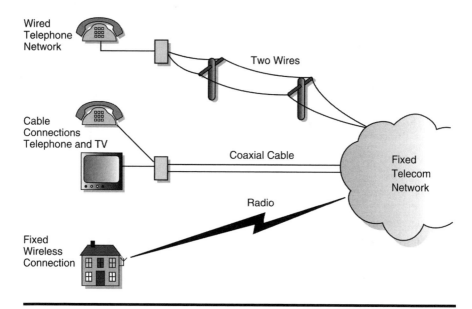

FIGURE 1.5 Types of connections to fixed networks.

Types of Fixed Networks
PSTN > Analog > Voice based
ISDN > Digital > Voice and Data

Radio in the Fixed Network

Fixed networks can also use radio links to provide connections. Microwave point-to-point connections often provide a major portion of the fixed network communications backbone. The copper connection to domestic or commercial premises can also be replaced with radio; this is know as wireless local loop (WLL).

Public Switched Telephony Network (PSTN)

The PSTN (Figure 1.6) is the publicly available dial-up telephone network. It is a complex interconnection of switching centers and end users that offers a voice connection between any two valid subscribers. In addition, international connections to other countries are also possible. In engineering parlance, the PSTN is sometimes referred to as POTS, an acronym for *plain old telephone system*.

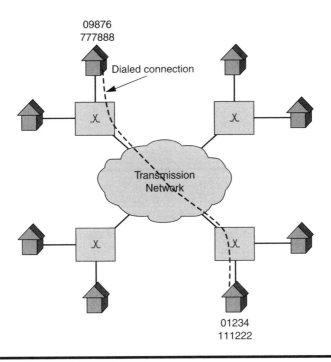

FIGURE 1.6 The public switched telephony network (PSTN).

Connections to the PSTN

As telephone communications evolved, telephone exchanges were used to manually connect users together. A pair of wires ran (typically overhead) from each subscriber to the exchange where these were terminated at a switchboard. An operator then interconnected subscribers in a manual process using short cables. As the number of subscribers increased, the demand for physical space to house the switchboards and operators also increased. This led to massive exchange buildings and armies of staff required to operate the system.

The development of automatic exchanges (or switches) using electromechanical devices simplified the interconnection of subscribers and reduced the amount of space and manpower required. This was taken a stage further by the implementation of digital switches. The only part of the network that remains in its original form is the pair of wires that forms the final connection to the subscriber, commonly known as the local loop.

Dial-up Lines

The sole purpose of the early telephone system was voice communication between two subscribers. As automation was extended into the network, this became a

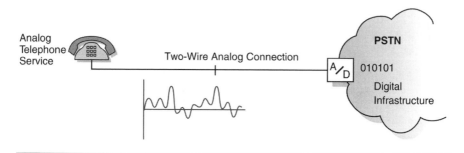

FIGURE 1.7 Analog connection to a digital infrastructure.

dialed service. These dial-up lines were suitable for voice use, but interference from electromechanical switching devices severely limited the speed and quality of data transmission. With the introduction of digital switches, these limitations have been partially overcome and data connection rates up to 56 kbps are achievable.

Analog to Digital

Speech by its very nature is an analog signal, and early systems such as the manual network mentioned above merely carried the signal from end to end. In modern networks, this signal is digitized to preserve audio quality and then carried across the network using digital links and switches (Figure 1.7).

Integrated Services Digital Networks (ISDN)

The modern digital telephony network not only carries voice, but also data at 64 kbps. This service is called an integrated digital network (IDN). As discussed above, there remains a short local loop of analog service over a pair of wires. The next stage is to take the digital channels all the way to the customer so that all services can be integrated on one bearer. This is the Integrated Services Digital Network (ISDN). The ITU-T defines ISDN as:

> A network evolved from the telephony IDN that provides end-to-end digital connectivity to support a wide range of services, including voice and non-voice services, to which users have access by a limited set of standard multipurpose customer interfaces.

Types of ISDN Connections

ISDN is a digitally provided circuit that typically offers three interfaces, as shown in Figure 1.8. These are known as "2B+D" or Basic Rate ISDN. The "2B" refers to

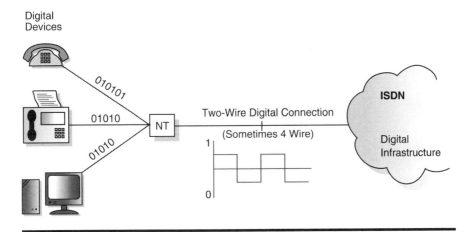

FIGURE 1.8 ISDN Connections.

two 64-kbps circuits that can be used as individual voice and data channels or a combined 128-kbps data channel. "D" is a 16-kbps channel used for signaling.

The basic rate described above is ideal for domestic or small business customers who have a need for a small number of connections at 64 kbps. ISDN offers this connectivity with the added advantage that it allows concurrent voice and data calls.

A faster, primary rate is available and offers 30 "B" channels and one "D" channel. When combined, the total data rate is 2.048 Mbps

Operators of ISDN/PSTN networks offer services to corporate as well as domestic customers. Corporate customer services are more likely to include high-capacity connections and leased line type services.

Mobile Communication Networks

Mobile networks differ from fixed networks in that the "last mile" (or the connection to the user) is based on radio transmission techniques. This wireless service gives the user the added benefit of mobility in the local area or wider area, depending on the nature of the network.

Background to Mobile Radio

Wireless communication, either as part of a fixed network to extend the communications without the need to install physical wires or as the connection to the user, has been a goal ever since the advent of long-distance communications. Providing a wireless connection to the user involves overcoming a large number of problems, not the least of which is the size of the radio equipment that the user would be expected to carry around.

FIGURE 1.9 A modern PMR radio.

Mobile radio systems have been around almost since the development of radio technologies. One of the first installed mobile radio services was in the 1920s by the police in Detroit. Mobile, in this case, meant that the radio equipment was small enough to carry around in a vehicle. Early systems like this tended to be one-way radio. It was not until the technology was pushed forward by World Wars I and II that two-way radio began to emerge. Then the development of commercial radio systems began.

Private Mobile Radio (PMR)

Mobile radio service used to be the preserve of military and commercial organizations. The nature of the service was also very different from the telephone-style service we expect today. However, an important element of modern radio systems reflects the older style of service. Today we refer to these systems as Private or Public Access Mobile Radio (PMR/PAMR) (see Figure 1.9).

Typical uses include emergency services, courier/delivery, and utilities and dispatching services. The nature of this communication is point-to-multipoint or group working (Figure 1.10). Interestingly, the public cellular operators are beginning to offer similar services as *value-added services* on their networks.

Cellular Radio

Advances in technology and radio spectrum management have led to the development of "cellular" radio networks, allowing the networks to support many users over a wide area. Cellular refers only to the method of using the radio spectrum available (Figure 1.11). Many different radio technologies can use cellular deployment of radio channels. However, we often use the term "cellular" to mean a mobile phone network such as GSM or cdmaOne.

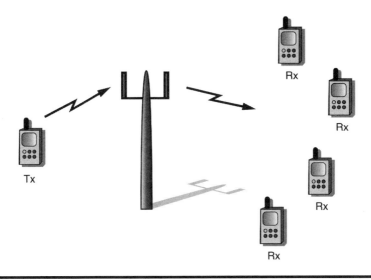

FIGURE 1.10 Typical PMR point-to-multipoint operation.

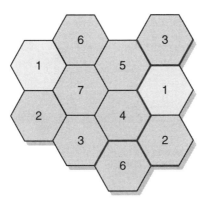

FIGURE 1.11 A cellular network.

Analog Mobile Networks

The first mobile phone networks began to appear in the latter part of the 1970s. These systems were analog and supported a very limited service. Analog networks do not naturally support data transmission. Modems could be purchased to allow users to move data or send faxes, but this was expensive and slow. The handsets were often bulky and had poor battery life (Figure 1.12). However, during the 1980s, many advances in all aspects of mobile phone technology meant that handsets became smaller and more reliable, and battery life also improved.

14 ■ *Introduction to Mobile Communications*

FIGURE 1.12 An early example of a mobile phone.

Analog Mobile Phone Technologies

TACS: Total Access Communications System

AMPS: Advanced Mobile Phone System

NMT: Nordic Mobile Telephone

RadioCom2000

NETZ C

Digital Mobile Networks

During the 1980s, development of other mobile phone systems had begun; this is the second generation of communication technology. Second-generation (2G) mobile phone systems are digital and generally improve the quality of service and

Digital Mobile Phone Technologies

GSM: Global System for Mobile Communications

D-AMPS: Digital AMPS (IS-54, IS136)

cdmaOne (IS-95)

PDC: Public Digital Cellular

increase the number of services that the networks can support. Digital mobile phone networks can be considered ISDN networks because these networks often support both voice and data services.

2.5G Technologies

GPRS: General Packet Radio Service

EDGE: Enhanced Data Rates for Global Evolution

CDMA2000 1x RTT: Code-Division Multiple Access (CDMA) 1x (single-carrier) Radio Transmission Technology

---

More recently, the operators of 2G networks have evolved their systems to support the transmission of data in a more efficient form; this is generally known as *packet data*. The term now used to describe these systems is 2.5G. 2.5G has enabled operators to offer many new and different services to subscribers. Given that more people have access to the Internet and the usage of Internet services continues to grow, the expectations of customers of data services offered by the mobile operators is extremely high. 2.5G enables the operators to partly meet this expectation.

2G networks, which were developed throughout the 1980s and deployed in the 1990s, account for more than 90 percent of the mobile phone users in the world. During the 1990s, so-called 3G networks were developed and are now being deployed all over the world. The main difference between 2G and 3G networks is the level of service a 3G user could expect to receive. 3G supports higher data rates and has been designed primarily with flexible service provision in mind. Many operators have already begun to deploy 3G networks along side their existing 2G and 2.5G networks.

Defining Areas of the Network

In both fixed and mobile networks, it is possible to divide the network into various parts according to their function. As shown in Figure 1.13, some parts of the network deal with connection to the user and others handle the management of the users and the services they wish to consume.

The Access Network

In the fixed network (Figure 1.14), the access network consists of the wires and cable that allow users to be connected to the central part of the network where access to services can be obtained. This is often the largest and most complicated

16 ■ *Introduction to Mobile Communications*

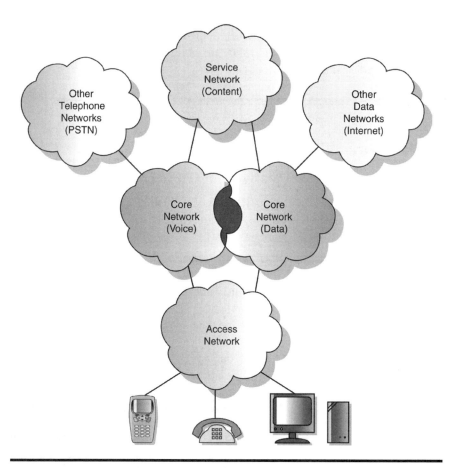

FIGURE 1.13 **The parts of a telecommunications network.**

part because it relies on a pair of copper wires or cable being delivered to every subscriber in the network. In a very large network, this can mean that tens of millions of wires must be connected from the users' premises and run back toward the core network. To reduce the overall complexity, the access network will often contain devices that will allow the information from many wired connections to be sent on a single cable. This reduces the total number of wires in the system.

In some parts of the world, the copper infrastructure is in very poor condition due to lack of maintenance or local conditions. Some operators choose to establish wireless local loop (WLL) style systems, which eliminate traditional copper wires. WLL systems should not be confused with mobile radio systems because they form part of the connection to a user's fixed device.

The access network is able to carry voice information (most likely in analog format) and data where the condition of the line allows. It is becoming more common for domestic premises to have both a speech and a broadband connection with the network service provider.

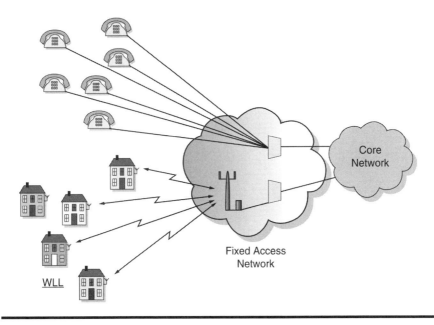

FIGURE 1.14 The fixed access network.

The Radio Access Network

The difference between radio system and fixed system is that the user's terminal is a mobile device. The user then has the freedom to move anywhere within the system radio coverage area and expect to communicate with the core network. Radio access networks (Figure 1.15) are most often associated with cellular phones systems, but PMR and other radio systems such as pagers can be considered mobile radio systems.

The purpose of the radio access network is to allow the users to make connections to the core network anywhere in the coverage area of the network. It is able to do this by placing radio base stations around the area where service is required. The mobile device then uses radio signals to relay the user's information or voice to the core network via the base station.

Satellites can also be part of a radio access network. The satellites are simply radio base stations that orbit the Earth instead of being fixed on the ground.

Core Networks

If the radio access network is there to provide connection to the core network, then the core network exists to provide the service to the user.

The core network (Figure 1.16) is a complex system of switches, interfaces, databases, and transmission systems, all designed to offer the users of the network

18 ■ *Introduction to Mobile Communications*

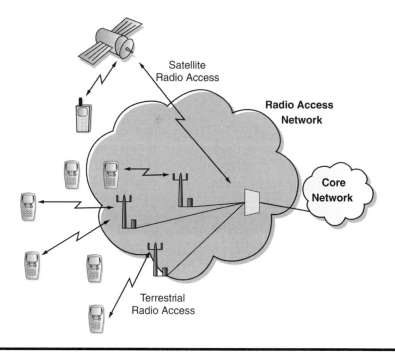

FIGURE 1.15 The radio access network.

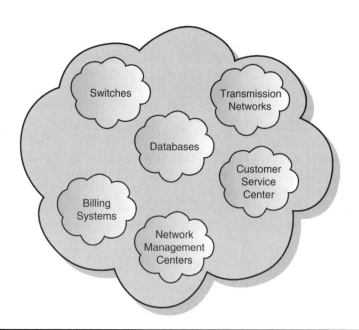

FIGURE 1.16 The core network system.

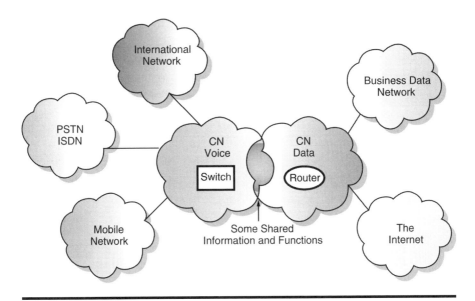

FIGURE 1.17 Voice and data core network (CN) connections.

a service of some kind. The most basic form of service would be a direct-dial speech telephone call, a relatively simple service for the user, but one that still requires a great deal of complexity in the networks.

Today's modern systems (Figure 1.17) often separate the core of the network into two distinct areas: (1) voice service and (2) data services. The core network concerned with voice primarily deals with the routing of telephone calls and those services normally associated with telephony, such as call waiting, call hold, etc. Connections to external telephone systems such as the other public telephone systems or between mobile networks or the international networks are also made from the voice-based core network. The data side of the network deals with data connections and the routing of data to connected data terminals. External connections are also supported, most probably to the public Internet or to local networks associated with business systems.

Other Types of Networks

There are some other types of networks that perhaps do not fit into the precise definitions given above. These are networks that perhaps do not have an access network and do not offer their services directly to members of the public. Their customer is more likely a large business or the fixed and mobile operators themselves. These networks provide long-distance connections within countries or connections between countries (Figure 1.18). They are the national and international carriers.

20 ■ Introduction to Mobile Communications

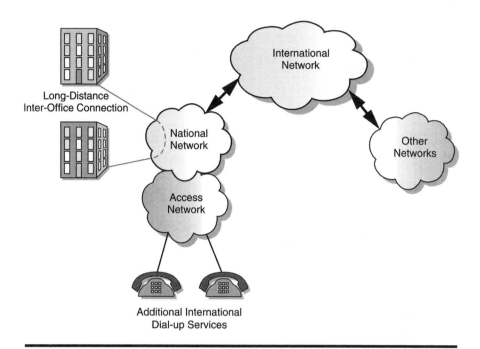

FIGURE 1.18 International and long-distance connections.

It is often the case that a large incumbent fixed operator will offer its service to domestic subscribers via an access network and also provide long-distance connections via its core networks for enterprises requiring fixed point-to-point connections. These types of connections are often referred to as leased lines.

Some of these very large (Tier 1) networks will also own or connect to the international networks to allow calls to be routed between countries around the world.

Review Questions

Q1. The term "fixed" refers to what part of a telecommunications network?
   a. The user equipment and connection to the network
   b. The telephone switch
   c. The service
   d. The bill

Q2. The PSTN is a network designed for carrying analog information.
   a. True
   b. False

Q3. An ISDN basic rate interface consists of:
   a. 2 × 64-kbps bearers
   b. 2 × 64-kbps bearers and 1 × 16-kbps signaling channel
   c. 1 × 64-kbps bearer and 2 × 16-kbps signaling channels
   d. 30 × 64-kbps bearers and 1 × 64-kbps signaling channel

Q4. In mobile communication systems, the term "cellular" refers to:
   a. A cellular phone
   b. The arrangement of radio frequencies
   c. Mobility
   d. A mobile telephone call

Q5. Which of the following enhances the data capability of a 2G mobile network?
   a. ISDN
   b. GPRS
   c. TACS
   d. GSM

Q6. The access network provides a user with _____.
   a. Switching services
   b. Internet connection
   c. Connection to the core network
   d. Voice circuits

1.1.3 Services and the User Perspective

Access to Network Services

The Changing Nature of Services

Relatively unsophisticated technology and networks meant that customers have historically been presented with a fairly narrow range of services from which to choose. The most common service is still that of voice telephony, but over the past few years (especially the past 20), customers have been given the option of an increasing number of valuable and sometimes novel services (Figure 1.19).

Network operators are keen to provide these services in an effort to increase revenue. This is made more important as the margins for voice calls decrease. An additional fact is that networks need to keep existing customers and attract new ones. Different and desirable services are a proven method of achieving this.

With the advent of the Internet and all that it brings in terms of information, together with increasing mobility in the form of cellular networks such as GSM, there is a thirst for new and relevant services. With the most innovation-receptive

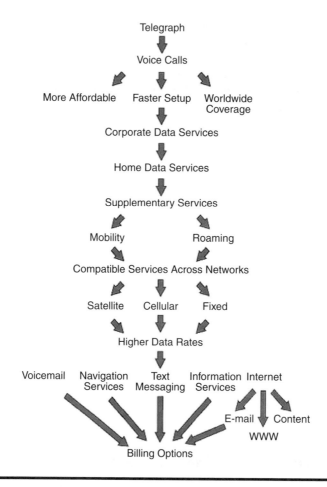

FIGURE 1.19 Changing service requirements.

customers often opting for mobile phones, and the average age of mobile phone users decreasing dramatically over the past few years, the uptake of new services, features, gadgets, and applications is all but assured. New, flexible ways of providing services are now being implemented. The shackles are coming off service creation. The nature of services and their implementation is changing rapidly.

Increasing Service Needs

Users had relatively simple needs in the early days of telecommunications, mainly because they had little knowledge of what was possible as the next step. Simple voice calls completely dominated telecommunications for at least half a century and provided the majority of *traffic* on a network for a full century.

Faster call setup, worldwide coverage, and more affordable services were largely achieved as a result of advances in technology and organization. Data services, however, were provided partly as an advance in technology, but developed in response to customer needs.

As the age of mobility and advanced service offerings approached, the pull of the customers' needs became much more of a factor. At the present time, customers' needs are top priority, evidenced occasionally by media hype about a new service that fails to deliver on its promise (at least initially) due to technology constraints.

The near future looks bright in terms of services, with customer demand on the way to being met by flexible systems with greater bandwidth, more control data, and more innovation than ever before.

Telecommunications Services

Defining a Telecommunications Service

A service offered by a telecommunications company is a product that enables two or more people to exchange information over a long or short distance. The nature and complexity of the service is often reflected in the price the customer has to pay. It was not too long ago that the choice of service was very limited. This was due, in part, to there being little competition in the market, and also to the level of technology deployed. Today, however, in the deregulated and liberalized telecom market, there are many companies competing in the same commercial space, which leads to innovation and development as companies strive to be different. The consumer looking for service in the fixed or mobile arenas is faced with a multitude of apparently different products from many different service providers or telecom companies.

Services in the Fixed Network

Fixed networks traditionally provide wired services, such as direct dial voice. Increasingly, the networks are offering many more advanced services that include data connections.

- *Voice.* Voice is the most basic service offered, and the one the customer is most likely to expect. This is the service that is taken for granted when considering telecommunication networks and, for the operators of the networks, speech is still by far the largest revenue generator.
- *Emergency call.* The once ubiquitous telephone box in the street is to allow all citizens access to the emergency call service. Regulators insist that the emergency services operator is available to every customer of a telecom network and that the service is free of charge. The reason many networks still maintain the telephone box in the street is to allow all citizens access to the emergency call service.

Typical Fixed Network Services

- Voice
- Emergency Call
- International Direct Dial
- Directory Inquiries
- Caller Line Identity
- Ring Back When Free
- Call Waiting
- Alarm Call
- Conference Call
- Call Diversion
- Answer Phone Services
- Call Barring
- Call Blocking
- Broadband Data Connections

- *International direct dial.* Through agreement of international numbering schemes and signaling systems, it is possible to direct dial practically every country in the world. In fairly recent history, international calls had to be booked in advance and the operator would make the connection on behalf of the customer. Undersea cable systems and satellites have made these long-distance connections much easier to achieve and the quality of connection is now very good.
- *Caller line identity.* The presentation of the caller's telephone number to the called party is a common service in mobile networks, but not so often seen in fixed networks. The technology in fixed networks is normally quite old and based on analog transmission. Fixed network service providers will normally expect the customer to subscribe to this service for a small charge.
- *Ring back when free.* If a person's telephone is busy or engaged when a call attempt is made, it is possible to instruct the network to monitor the connection for the busy condition to clear and then alert the calling party so the call can be completed. This is a useful service for customers and for the network because they will charge the subscribers for the ring back service and they are also encouraging the setup of a call for which they will also be able to charge.
- *Call waiting.* Call waiting is a service that alerts a caller to an incoming call attempt by playing an audible tone over the conversation in the current call.
- *Alarm call.* It is possible to book an alarm call from some telecom network providers, for a small charge.
- *Conference call.* A service used by the business community, in which more than two people can participate in a conversation. It is possible to set up the

entire conference from the telephone keypad, without having to resort to operator intervention or more complex mechanisms.
- *Call diversion or forwarding.* It is possible to have all incoming calls forwarded or diverted to a different number, e.g., answering machine. These diversions can be unconditional, where all calls will be forwarded to the new number, or conditional, where the network will make an "intelligent" decision on how to route the call given the current condition of the line (e.g., divert on busy, or divert on unavailable).
- *Answer phone services.* Many people may have their own answering machine at home. However, the fixed networks may also offer a similar service of recording messages left by callers. Access to the answer service is normally via a short four-digit code; then the keypad is used to play, save, and delete the messages. It may also be possible to pick up the messages from another telephone.
- *Call barring.* The network can set up call barring services on behalf of the customer or for account management purposes. A barring service prevents the customer from dialing out on all numbers or certain kinds of numbers. Often, premium rate numbers may be barred for economic reasons.
- *Call blocking.* Nuisance calls can be prevented by reporting the number to the network and setting up the blocking service. Any calls from the blocked number will not be completed; all other calls will be handled normally.
- *Free phone numbers.* Possibly one of the first advanced services to become available from the networks, this is a number prefix, normally X-800, that allows customers to contact businesses free of charge. The number is normally purchased by the business and it is the business that picks up the cost of the connection.
- *Premium rate numbers.* This is another scheme of numbering that allows members of the public to access premium content or services via special prefix numbers (e.g., X976, X845, etc.). These numbers are commonly used to access information services such as weather reports or horoscopes. The recent popularity of "reality" TV programs, in which viewers are urged to interact with the show by voting, has created another excellent way of making money for the telecom networks and the TV production companies.
- *Broadband data connections.* More recently, the telecom networks have begun to offer broadband data services to replace the dial-up modem connection. Broadband data relies on new technology to allow data at up to 2 Mbps to be sent along traditional telephone lines. Asymmetric Digital Subscriber Line (ADSL) and cable modems are examples of broadband data technologies.

Mobile Network Services

Many of the services offered by the mobile networks are the same or similar to those of the fixed networks. The original service was voice. The technology in the early days of mobile networks was not capable of much more than just voice. Technology, both in the networks and in the handsets, has advanced massively over the past 20 years, leading to more complex and sophisticated services. Operators are focusing particularly on data and data services, hoping that it will be a way to sustain growth and revenue in the networks. However, with the complexity of these services and the handsets used to access the service, many subscribers are simply not using them. A major challenge for the near future is for the networks to create services that the population of subscribers actually want and for the handset manufacturers to find a way of presenting the service in a simple and easy to use way.

Typical Mobile Network Services

Basic services:
- Speech
- SMS text messaging
- Caller ID, call waiting, call divert, call barring, etc.
- Answer phone, personal assistant service

Advanced services:
- Dial-up data
- Packet data GPRS
- Download services, MMS, ringtones, wallpaper, icons, etc.
- WAP Web browsing
- Advanced text messaging

- *Basic services.* Because a mobile phone network is principally a voice telephony network, it seems reasonable that some of the services seen in the fixed network transfer to the mobile systems.
- *Short Message Service (SMS).* SMS is a text-based message service that allows subscribers to exchange messages of up to 160 alphanumeric characters. This service was developed as part of the GSM system and has become a very popular service during the past few years. Recent statistics show that some networks are handling more than a billion text messages per month. SMS has evolved from a simple text messaging service to one that allows the transport of other forms of media, including ringtones, pictures, operator logos, wallpapers, and screensavers for mobile devices. Many of the services mentioned here, however, are proprietary to a particular make and model of mobile phone. There are similar services available today that use different methods for transporting the media, which tends to be more compatible over the range of mobile makes and models.

The text messaging service is also finding new applications. SMS short codes, normally five digits, can be used to interact with TV shows and radio stations, subscribe to information services, and to make payments for goods or services.

SMS Applications

1 TV voting via short codes
2 TV and radio show interaction
3 Information service: weather, stocks, horoscopes
4 Payment services: online payments, parking meters

- *Mobile data services.* Since the advent of digital mobile radio, it has been possible — at least in theory — to send and receive data from mobile phones. It has, however, taken many years for data services to become popular. The problem is that, apart from corporate users, general members of the public are not interested in buying or subscribing to a "data service" such as dial-up connections at 9.6 kbps or even more efficient and faster GPRS connections at up to 40 kbps.

  What subscribers want is a useful and engaging service accessible through a simple and reliable interface. The job, then, of mobile network service providers, content providers, and mobile manufacturers is to find such a service and make it available at a price that is appealing. A simple enough statement to make here, but very difficult to execute effectively, given also that in the past, subscribers have had bad experiences with data services.
- *Services that make use of a data connection.* Once the mobile equipment and the network are capable of transferring data at a rate considered reasonable, it is possible to build applications that will exchange information in such a way to offer the user a service that he or she may find useful.
- *Data transfer services.* Business users have a different requirement from normal subscribers. They normally require access to e-mail services, file transfers, and databases. All these types of services are normally run on portable computing equipment such as laptops and PDAs (personal digital assistants). This equipment is often connected to the network using the mobile phone as if it were a modem; in this way, the applications on the laptop are able to exchange information through the data pipe provided by the connection (Figure 1.20).

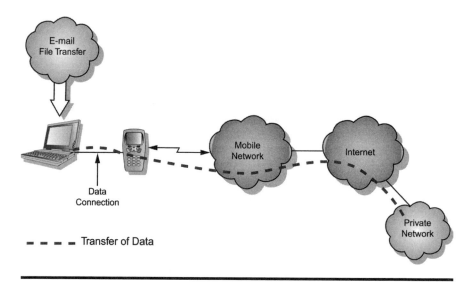

FIGURE 1.20 Mobile as a data connection.

Services on the Handset

The average subscriber would normally consume data services through applications or services provided on the handset itself. The download of ringtones, screensavers, and wallpaper are often-used services. There are a number of different ways in which a subscriber can obtain these services; usually it is a proprietary method defined by the network's operator or handset vendor (Figure 1.21). More recently, standards have begun to emerge for this kind of data service. WAP (Wireless Application Protocol) and MMS (Multimedia Messaging Service) are the most recent.

Network and Product Branding

With growing competition in the deregulated market, it is increasingly important for the operators of telecom networks and product vendors to stand out from the crowd.

Potential customers will have two questions in mind when looking to purchase a new service or product:

1. Is this company a company that we can trust?
2. Will the product or service this company is selling fulfil my need or want?

The company brand will go a long way toward answering these questions, and can reassure the customer that the service or product has the desired quality or value for the money. A brand is an intangible thing that will develop around a company or product regardless of whether or not it is intended. The important thing for a company is to ensure that the branding is the one that it wants. From the

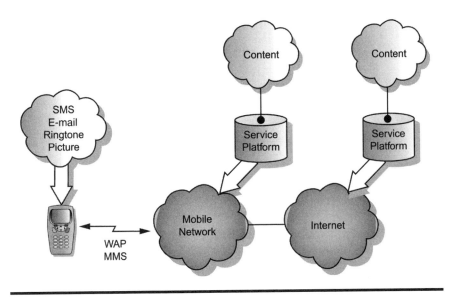

FIGURE 1.21 Mobile handset services.

automotive world one only has to think of two brands — Volvo and Skoda — to see the huge difference branding can make.

A company brand is not the billboards, TV advertisements, and other media exposure. These are simply the way in which information about the brand is propagated. Branding is the enduring, intuitive association a company has in the mind of consumers. The company image or logo is not the only, or indeed the most important, part of the brand.

For example, the customer experience when interacting with a company is an important part of maintaining a brand. Customers will leave a mobile network on the basis of their experience on the phone to the customer service center; every point of contact the company has with its customer will create, or break, a better brand.

Mobile Phone Branding

The mobile operators will generally seek to have the mobile phones they sell for use on their networks branded in some way, as shown in Figure 1.22. This will probably mean that the operator logo will appear on the casing of the device and perhaps even as a "welcome screen" when the phone powers-up. There is often a dedicated button that will take the user to the home page of the operator's WAP Web site, Vodafone Live or T-Zones, for example. It is very common for the operator to lock the handset in such away that it will only accept SIM cards from that operator, making it more difficult for the subscribers to take the handset to a different network. However, the locking of phones in this way has created an underground industry that will unlock the phone for a relatively small fee.

FIGURE 1.22 Network operator branded mobile phones.

FIGURE 1.23 Example screenshot from Vodafone Live.

Network Branding

Network operators will seek to strengthen the brand through consistent presentation of company images or color schemes. It is important, for example, for companies such as Vodafone to ensure that their WAP pages are reproduced accurately and reliably on the all the handsets offered in the Vodafone range. Some network operators will even insist that the menus on the mobile phone also reflect the brand (Figure 1.23).

The Mobile Telecommunications Market ■ 31

FIGURE 1.24 Vodafone Live Web page.

Other customer contact points, such as Web sites, bills, and promotional material, will also carry the branding (Figure 1.24). Remember that the image on its own is not the brand in its entirety; the brand is also the customer's perception of the total service he or she experiences.

Review Questions

Q1. Which of the following does not reflect the changing nature of telecom services over the past few decades?
   a. Faster call setup
   b. Mobility
   c. Less data calls
   d. More services

Q2. Which of the following is not commonly regarded as a fixed network service?
   a. Call waiting
   b. Call hold
   c. Text messaging
   d. Call forwarding

Q3. Which of the following is not considered a mobile network service?
   a. Text messaging
   b. Calling line identity
   c. Broadband Internet access
   d. Conference call

Q4. A data service can be defined as:
   a. The end-to-end transfer of voice in a switched network
   b. The end-to-end exchange of data between applications to describe a service or content
   c. The exchange of signaling to set up a connection
   d. The establishment of a radio connection

Q5. A product brand is:
   a. The logo
   b. The media channels promoting the company image
   c. A combination of company image and consumers' perception of the company
   d. The services provided by the company

1.1.4 Review Questions

The following questions are an opportunity for you to review the information that you have learned in this section. Some questions require you to carry out additional research to discover the answers. The recommended resources for this research are *Mobile Communication International* magazine, *Telecoms Industry Outlook 2005/06* and, of course, the Internet.

Q1. Draw a diagram that shows the types of connections you would expect to find in the PSTN and ISDN.

Q2. What are the main differences between the ISDN and PSTN connects?

Q3. From your own country (or one of your choosing), study the available radio service providers and identify the operators of the following types of radio systems:
   a. Private mobile radio (PMR)
   b. Public access mobile radio (PAMR)
   c. Fixed telecom operator
   d. Mobile phone operator

Q4. Select one of the fixed operators you have identified above and list the services it can offer.

Q5. Choose a mobile operator and find out what services it supports. Compare these services with those offered by a fixed operator.

Q6. Choose a network that has a memorable or popular brand. Define what it is that makes this brand popular.

1.2 The Mobile Market

1.2.1 Segmenting the Market

Mobile Market Sectors

The mobile market is a complex place with many different sectors. These will include the subscriber markets (such as young users and business users) as well as the vendor markets for hardware and software within the mobile industry. Provision of content is an area where there is increasing growth and competition.

Operators in the Telecommunications Market

In the arena of telecommunications, there are many types of operators. Some of these could be considered fixed operators, and some as mobile. There are an increasing number of operators that appear to offer both fixed and mobile services, through the use of partnerships and agreements with operators of different kinds of services. There are also very small regional service providers and very large international carriers that may have no regional presence.

Fixed Telecom Operators

- *Small regional operators.* There are operators of telecom services that can only operate in a very specific region. The branding of the company may reflect this. These companies may own and maintain the wires into domestic and business premises and the network switching and administration hardware. Kingston Communications is an example of such a network, offering service around the North West region of the United Kingdom in Hull.
- *Incumbent telecom operators.* Most countries will have what could be described as an *incumbent* telecom operator. Often, these companies are ex-government run establishments and have represented a market monopoly at some time. More recently, these state monopolies have been broken up and the markets have become more liberalized due to regulator intervention.

  These networks still exist, however, and continue to offer valuable service to many individuals and industries within their respective countries, including:
  - Voice Eircom (Ireland)
  - BT (United Kingdom)

34 ■ *Introduction to Mobile Communications*

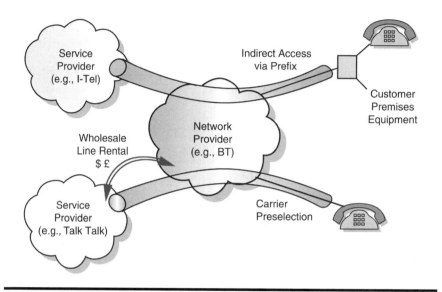

FIGURE 1.25 Service and network providers.

- Swisscom (Switzerland)
- Telia (Sweden)
- KPN (Holland)

All the above are the traditional incumbent operators of these countries.

■ *Service providers and billing-only telcos.* The regulation and liberalization of the telecom market have led to a lot of competition in the domestic markets. The incumbent operators are forced to allow new telecom companies to use the existing connections to offer services to customers. The public telephone networks could now be viewed as consisting of the network provider and the service provider (Figure 1.25).

The *network provider* is the company that owns and maintains the wires, all of or part of the switching infrastructure, and is likely the incumbent operator. They are obliged to offer the service providers a wholesale rental of the fixed connection to subscribers' premises.

The *service provider* is the company from which the individual or corporate buys the service; the service provider will not own or maintain the connection to the subscribers' premises, but it might operate some switching, billing, and customer contact systems.

All service providers are accessible to subscribers via the network providers' connections; this is a condition laid down by the regulator and therefore some means of connecting the subscriber to the service provider system must be in place. This might be an *indirect access*, where the users will dial a prefix before every call, or there will be a piece of equipment installed at the subscriber's home through which calls are made. It is this equipment that will dial the prefix on behalf of the subscriber.

TABLE 1.1 Tier 1 Carriers

AT&T	Singapore Telecom (SingTel)
AOL / AOL Transit Data Network (ATDN)	Qwest
British Telecom (BT)	Sprint Nextel Corporation
Cable & Wireless	Telstra
Telecom NZ	Verizon
Deutsche Telekom	Teleglobe
Global Crossing	Savvis
Level 3	TeliaSonera
NTT / Verio	Verizon Business

The alternative is *carrier pre-selection* (CPS), where the user is automatically directed to a predetermined network by the network provider. This eliminates the need for any customer premises equipment or prefixing codes.

- *Multinational carriers.* There are companies that have been around almost from the beginning of electronic telecommunications. They have grown to be giants in the industry and offer services over a very wide range of markets, from domestic phone services to international cable systems and Internet traffic backbones. These networks (listed in Table 1.1) are often referred to as Tier 1 carriers. Each of these companies will have a presence in more than one country and some own sub-sea telecom cables or satellite systems. Companies such as Verizon, AT&T, and Cable & Wireless have a very large portfolio of international voice and data services.

Mobile Telecom Operators

- *Mobile network operators (MNOs).* An operator of a mobile network is a company or organization that has obtained a license to operate and has invested in all the necessary equipment to provide radio coverage and telephony services for members of the general public. Mobile network operators own, operate, and maintain all the telecom equipment themselves or make use of contractor services for some of this work.

  In the interest of competition, there will usually be more than one operator in a country; this would be a condition set by the government or regulator of the region.

  The functions and responsibilities of an operator include
  - Obtaining a license to operate in a relevant part of the radio spectrum
  - Planning the radio coverage and capacity of the network

TABLE 1.2 Mobile Operators

Operator	Country	System	Subscribers
Vodafone	United Kingdom	GSM9/18	13,503,000
Sonofone	Denmark	GSM9/18	1,423,000
Gibraltar Telecom	Gibraltar	GSM900	20,100
Turkcell	Turkey	GSM900	25,600,000
Swisscom Mobile	Switzerland	GSM9/18	4,002,000
Wind	Italy	GSM1800	12,600,000

- Acquiring the necessary sites to install radio base stations
- Purchasing and installing radio base station, switching, and other telecom equipment
- Provision of mobile telecom services to the general public
- Billing for consumed services
- Customer contact management

There are, of course, many other functions carried out by the mobile operator, and these functions are detailed in later sections.

Table 1.2 Lists examples of mobile phone operators from a number of different countries.

- *Mobile virtual network operators (MVNOs).* It is increasingly common for organizations to offer mobile phone service to the general public where the organization does not own or operate a mobile infrastructure. These entities are called *mobile virtual network operators* (MVNOs). An MVNO is an organization that has made an agreement with one of the existing "real" mobile operators to carry services and products on their behalf. An MVNO will brand the product, services, and handsets as if it were they themselves that were operating a network. This may be a cost-effective way for supermarket chains, music stores, and youth-related industries to move into the mobile service arena. A real operator can also make use of MVNO branding to target niche segments of the market such as youth culture, gaming, etc.

There are several definitions of an MVNO, and Table 1.3 shows some of the areas where differences exist between normal operators, MVNOs, and other providers of mobile service.

According to the Wireless World Forum, there are two classifications of MVNO: (1) discount and (2) lifestyle.

- A discount MVNO provides reduced rate calls to market segments that offer less revenue per customer. Discount MVNOs include Virgin Mobile and EasyMobile. Their strategy is based on cheap, prepaid or postpaid tariffs with little or no data offering beyond SMS.

TABLE 1.3

	Service Provider	Indirect Service Provider	Enhanced Service Provider	Mobile Virtual Network Operator	Mobile Network Provider
Spectrum					Owns spectrum
SIM Card			Branding of host SIM	Issues own SIM	Issues own SIM
Network Infrastructure		Switch and transmission	Depends	Switch/HLR and some transmission	Entire core network and RAN
Pricing		Partially independent pricing	Partially independent pricing	Fully independent pricing	Fully independent pricing
Branding	Some independent branding	Some independent branding	Independent branding	Independent branding	Independent branding

- A lifestyle MVNO focuses on specific niche market demographics that larger operators overlook. MVNOs such as Boost Mobile and AMP'D in the United States, and Hello_MTV and ID&T Mobile in Europe, market solely to young users.

There are three main reasons for mobile operators to allow MVNOs on their networks, depending slightly on the competitive situation in their countries, ARPU levels, and subsidies. These can be defined as:

- *Segmentation-driven strategies.* Mobile operators often find it difficult to succeed in all customer segments. MVNOs are a way to implement a more specific marketing mix, whether alone or with partners and they can help attack specific, targeted segments.
- *Network utilization-driven strategies.* Many mobile operators have capacity, product, and segment needs. An MVNO strategy can generate economies of scale for better network utilization.
- *Product-driven strategies.* MVNOs can help mobile operators target customers with specialized service requirements and get to customer niches that mobile operators cannot reach.

Currently, there are approximately 200 planned or operational MVNOs worldwide. Countries such as the Netherlands, Denmark, United Kingdom, Finland, Belgium, and the United States have the most MVNOs per country, whereas some are just beginning to launch active MVNO business models (e.g., France, the Baltics, and Austria).

Presently, many companies and regulatory bodies are strongly in favor of MVNOs. For example, in 2003, the European Commission issued a recommendation to national telecom regulators (National Regulatory Authority [NRA]) to examine the competitiveness of the market for wholesale access and call origination on public mobile telephone networks. The study resulted in new legislation from NRA in countries such as Ireland and France that forces operators to open up their network to MVNOs. Appendix A lists some of the MVNOs and MVNO activity from three countries.

- *Mobile virtual network enabler (MVNE)* (Figure 1.26). An MVNE provides infrastructure and services to enable MVNOs to offer services and have a relationship with end-user customers but does not have a relationship with those end-user customers.

An MVNE offers infrastructure and related services, ranging from network element provisioning, administration, and operations to operational support system/business support systems (OSS/BSS) support. MVNEs often provide the "middle ground" between MVNOs that do not want to have any control over network elements and those that want complete control.

Some MVNOs want to rely completely on the underlying wireless network infrastructure of the host mobile network operator, whereas other MVNOs want to own or control their own network elements. MVNEs provide the middle ground in the sense that they can provide options to MVNOs for

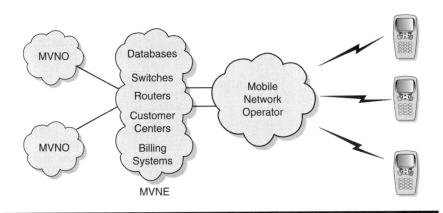

FIGURE 1.26 Mobile virtual network enabler (MVNE).

what they bring in-house versus that for which they rely on the host carrier. For example, a MVNE can provide Home Location Register (HLR), Short Message Service Center (SMSC), Multi Media Service Center (MMSC), as well as more advanced network elements such as GGSN, OSS/BSS, and other systems.

There are many examples of MVNEs around the world; below are just a few with their respective Web addresses:

- Finland   Spinbox: www.spinbox.se/
- France   Transatel: www.transatel.com/
- Germany   Arvato Mobile: www.arvato-mobile.de/
- United Kingdom Martin Dawes: www.martindawessystems.com/
- United States   Visage Mobile: www.visagemobile.com/

Subscriber Markets

Operators are increasingly targeting niche sectors of this market in an effort to increase subscriber average revenue per user (ARPU). They are able to do this by creating services that appeal to some part of the population of mobile subscribers. Important sectors of the subscriber market include young users, corporate users, and industry users.

- *Youth market.* The youth market is one of the fastest growing sectors, and there are many facets of the mobile telecom industry that concentrate on providing services to this age group. Services such as ringtones, screensavers, picture messaging, mobile games, and music are proving to be big money generators for both the network operators and the content providers.

FIGURE 1.27 The "boost" Web site homepage.

Youth Services

Download Services: Music MP3, Games, Videos

Handset Customization: Screensavers, Ring Tones, Pictures, Avatars

Other Services: Greeting Cards, Picture Messages, Chat Rooms

So important is this market that there are *virtual mobile* networks being set up that directly address this market; for example, Boost, an MVNO using the Nextel networks in the United States (Figure 1.27). Boost targets the youth scene with popular services such as the latest chart songs available for download, wallpaper images of rock and rap and other music and movie stars, games for download, and Web-based games. Other services include a cost-effective walkie-talkie mode (push to talk or PTT), operating a little like instant messaging using voice. A Boost credit card is also available that can be used to top up prepaid accounts.

The Mobile Telecommunications Market ■ 41

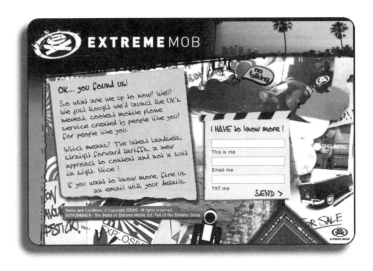

FIGURE 1.28 Screenshot from ExtremeMob.

FIGURE 1.29 Mobile available on the Amp'd Network

Another example is ExtremeMob, whose host network is Vodafone in the United Kingdom (Figure 1.28). ExtremeMob aims its service at the under-24 prepaid user. It offers content based on extreme sports lifestyle, such as skating, blading, MTBing, etc. It claims to be more than just about the sport, however; it is an integral part of the lifestyle of an under-24-year-old. Extreme is a well-established company in the world of extreme sports, with TV channels, a clothing brand, sports drinks, and entertainment-based media production.

Amp'd, another MVNO, aims directly at the youth culture of today with music and video downloads (Figure 1.29). Association with extreme sport culture is also part of this network's appeal, sponsoring personalities from the world of surfing, motocross, skating, etc. This network also has close ties with the music industry.

FIGURE 1.30 BlackBerry PDA mobile.

- *Corporate users.* Operators have traditionally focused on the business user to adopt new services ahead of the general market and sustain a high level of ARPU. The nature of the services used are, of course, business-related tools such as e-mail, fax, and voice. In line with this, the majority of the major mobile telephone manufacturers have business-oriented products that appeal to the traveling business person and executive. The Blackberry device from RIM (Figure 1.30) is a good example of such a device; it allows one to receive and transmit e-mail while away from the office.
  Corporate services include:
  - *Data services:* GPRS, HSCSD, Data Card, WiFi
  - *Voice services:* Fixed Mobile Convergence (FMC), Virtual Private Voice Network
  - *Applications:* E-mail, network access, Internet access, intranet access, fleet management, security
- *Industry users.* Some operators will target users within certain areas of public and private industries. Public utility and maintenance companies will often buy services from national operators to solve their communication needs. This can include integrating existing office-based fixed telecom with mobile services through the use of virtual private networks (VPNs).

  Some sectors of industry may look toward a very different form of mobile communication. Public utilities, emergency services, and haulage firms have traditionally used point-to-multipoint or group communications. Some have purchased and maintained their own radio networks for many years. However, these tend to be aging analog systems that are due for replacement.

FIGURE 1.31 GSM for the railway industry.

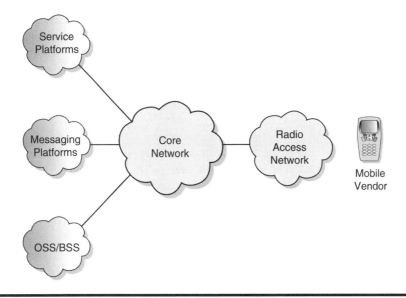

FIGURE 1.32 Key vendor spaces in mobile telecommunications.

The cellular networks do not have the technology to offer these sectors of industry the form of working to which they might be accustomed. The emergency services and some public transport companies all over the world have been investing in other radio systems designed for group communications; TETRA and TETRAPOL are examples of such radio systems.

The European Railway Traffic Management System (ERTMS) uses a version of GSM, modified from the original specification, to provide the special radio communication structure that the railways require. It is sometimes referred to as GSM-R (Figure 1.31).

- *Vendor markets.* The mobile network industry can be divided into the following sectors (Figure 1.32):
  - Handsets
  - Radio access networks (RANs)
  - Core Networks
  - OSS/BSS

- SMS/MMS service platforms
- IP/IMS solutions

Each of these sectors has many suppliers of equipment and solutions, some providing services across several sectors

Handsets

The handset market currently consists of 143 manufacturers and more than 3000 mobile phone models. This market sector is considered very important because without any mobile phones, the user would not be able to access services and the networks would not make any money. It is important for the operators that the users have a high level of choice, particularly when they are launching new services or technologies. A new service such as 3G may only be successful if the network can supply a wide range of mobile devices to satisfy all the consumer requirements.

The handset market is becoming a very complex place. As the network operators begin to target niche consumers with tailored services, those consumers begin to demand equipment that reflects their lifestyle and their image. One of the greatest successes in the mobile market was the development of interchangeable plastic fascias, allowing users to customize their handsets in many different ways. Fashion continues to play the biggest role in mobile phone design and consumer choice.

Radio Access, Core Networks, and Supporting Systems

There are at the present time nearly 2000 mobile networks in operation. They are supplied with equipment by 97 vendors. The range of products offered by the vendors cover all aspects of operating a mobile network, including radio base stations, switches, network transmission equipment, databases, and supporting platforms.

Radio Access Networks

The radio access network (RAN) contains radio base stations, mast antennas, base station controllers, and transmission systems. There are vendors that can provide a "turnkey" solution for RAN and entire networks if required. There are other vendors that specialize in one particular aspect of RAN solutions. As well as equipment vendors in this space, there are service suppliers that can offer site acquisition, mast construction, and radio planning. RAN providers include ADC, Allgon, NEC, AirNet, and ip.access.

Core Network Vendors

The core network is the central part of the network where all the subscriber data is held and the process of switching and routing of data takes place. There are many vendors offering solutions in this area, most of them large and well-established companies that have been serving the telecom industry — both fixed and mobile — for many years. With recent advances in technology and innovation in Internet techniques, companies more commonly associated with IT have been offering telecom products, including:

- Alcatel
- Cisco
- Ericsson
- Fujitsu
- Huawei
- Juniper
- Lucent
- Newport
- Nokia
- Nortel
- Siemens

Messaging Platforms

Most suppliers of network equipment will also supply platforms to support text message and multimedia services. There are, however, a few vendors whose core product is a stand-alone SMS or MMS platform or service. Some companies produce a hardware solution that can be integrated into the operator's switched infrastructure, while other companies produce gateway or software solutions that can support an operator's existing SMS solution.

The following list shows a few of the companies operating in this market:

- 2 ergo ltd
- Beep Marketing
- BulkSMS.com
- Buongiorno UK
- Clickatell
- echovox
- ERA Technology
- Taylor Wessing
- TeleCommunication Systems, Inc. (TCS)

- Sports Insider
- LogicaCMG
- Redback Networks
- Wizcom
- Openwave
- Comverse
- Index Corporation
- mBlox
- Mobile Phone Dating
- MX Telecom
- Netsize
- Segala M Test
- SendMyTxt
- RedRock Software
- Wireless Information Network (WIN)
- Telesoft Technologies
- Telsis
- Dialect Solutions Group
- TynTech
- Tecnomen
- Openwave

Voicemail Platforms

Voicemail is a fundamental service and an important revenue generator for the network operator. Again, the infrastructure vendors are able to offer a turnkey solution for the operators offering the switch products and integrated voicemail. There are also third-party vendors that can offer stand-alone systems. Examples include:

- CTI^2
- Clariti Wireless Messaging
- Comverse
- LogicCMG
- Mailngen
- Openwave
- VoxSurf

IP Multimedia Sub-system

IP Multimedia Sub-System (IMS) is an open, standardized, easily deployed network architecture that enables more flexible control and billing of multimedia services delivered by IP networks using SIP (Session Initiation Protocol). Currently, IMS is heavily promoted by vendors as the "next big thing" for both fixed and mobile operators, promising diverse service opportunities and cost benefits. Some of the vendors providing equipment for the IMS and associated systems include:

- Agere Systems
- Alcatel
- Apertio
- BrookTrout
- Cisco
- Convedia
- Ericsson
- Highdeal
- HP
- IBM
- Lucent
- Marconi
- Motorola
- NetCentrex
- Nortel
- Siemens
- Telcordia
- Ubiquity

Content Providers

There are a number of places that a network or user can obtain content. The network itself can generate the content and deliver it directly to the user via the network operator's own portal. For example, Vodafone Live would be Vodafone's portal where the users can obtain content via their handsets.

An external content provider can generate content and attempt to sell it directly to the operator for inclusion in the operator's own portal (interface b in Figure 1.33). The content could be ringtones, wallpaper, location-based information, etc. The content provider in this case must pass the guidelines and tests set by the operator in order for the content to be accepted.

Alternatively, the content provider can obtain the services of a content aggregator who will advise on the content and format of content and present the data to the operator (interface a in Figure 1.33) as a collection of data from multiple sources. The aggregator will take content from a number of different sources. The operator can then present the information as content within their own portal.

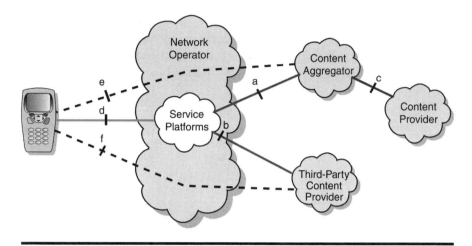

FIGURE 1.33   Content distribution.

TABLE 1.4   Examples of Providers and Aggregators

Type	Name	Comment
Network portal	Vodafone Live	Content portal from Vodafone
Network portal	T-Zones	T-Mobile content portal
Content aggregator	Nellymoser	Music, downloads, games, retail, news channels
Content provider	Billboard Mobile	Top 10 music provider
Content provider	Zingy	General music, mobile gaming

It is possible for a user to bypass the network portal altogether and access information directly from the aggregator of content provider. Enabling technologies such as GPRS and WAP make this type of connection much easier to manage. This opens up many possibilities in terms of the types of service to which the user can gain access. For example, if the user subscribes to the *Financial Times* stocks information via the network portal or uses the FT WAP page directly from the Web on to his mobile device, he finds that there is a cost differential in doing this.

As the number of ways that users can gain access to information increases, the competition between the content providers grows, leading to service innovation, a greater number of services, and a decrease in the cost of services. Table 1.4 provides examples of providers and aggregators.

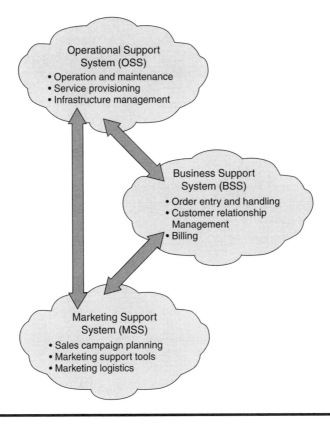

FIGURE 1.34   OSS, BSS, and MSS.

OSS, BSS, and MSS

Support for the operational network must be comprehensive in terms of structure and procedures, and also in terms of systems and software tools. Broadly speaking, the requirement can be split into three main areas (Figure 1.34):

The *operational support system* (OSS) deals with managing the operational network, including infrastructure, maintenance, and fault handling. The requirement is wide ranging, allowing the complete network to be viewed in overview, or detail, with network elements being handled remotely to keep the network running smoothly.

The *business support system* (BSS) allows for effective customer relations, covering areas such as new orders and billing. The requirement is easy to identify, and systems could be dedicated to this role, or be fully integrated within a wider OSS. In either case, information must be available to the customer service representatives, relying on good data interfaces between related functions; e.g., customer billing records being available to the customer service representative.

Finally, the *marketing support system* (MSS) is utilized to facilitate the planning of marketing campaigns, support for sales activities and marketing logistics. It is not uncommon to find that MSS platforms are fully integrated with BSS and may even be referred to as BSS platforms. It should be noted that some within the telecom industry refer collectively to operational, business, and marketing support systems as OSS.

Review Questions

Q1. Which of the following is not considered part of the youth market?
   a. Ringtones download
   b. Music download
   c. Conference calling
   d. Instant messaging

Q2. Which service is most likely to benefit corporate users?
   a. SMS
   b. Video clip download
   c. Picture messaging
   d. Internet access

Q3. Which sector represents the largest market for mobile handset manufacturers?
   a. Entertainment
   b. Business
   c. Fashion
   d. Industrial

1.2.2 Handset Market

Within the handset sector, there are currently 143 manufacturers and approximately 3000 models of mobile phone. The prices of these handsets range from $50 to several thousand dollars, addressing all sectors of the subscriber market.

The manufacturers and network operators must address many different sectors of the market in an effort to satisfy consumer demand. When mobile phone networks first began, there were very few models of mobile phones available. No one would deny that the mobile phone is an important business and day-to-day communication tool. However, it has always been an item of fashion and status as well. Figure 1.35 identifies those areas of the market that manufacturers attempt to address. By far, the largest sector shown here is that of fashion, and fashion can also be considered in other sectors, such as business, sport, gaming, etc.

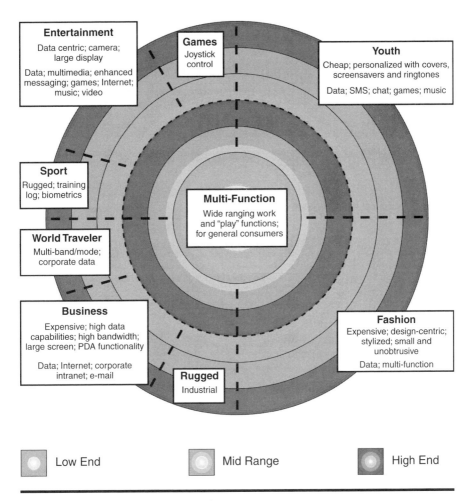

FIGURE 1.35 Mobile handset market segments.

Low-Cost Handsets

Perhaps the most significant development in recent years is the low-cost handset. The GSMA is undertaking a program that it calls the Emerging Handsets Market Program (EHMP), the objective of which is to produce a sustainable low-cost handset below the $30 price point. This product directly addresses the developing markets in Africa, India, parts of the Far East, and other areas of the world. Motorola was the first company to produce a low-cost platform endorsed by the GSMA, called the C114. Features of this and other platforms include long battery life, market-specific design, and durability.

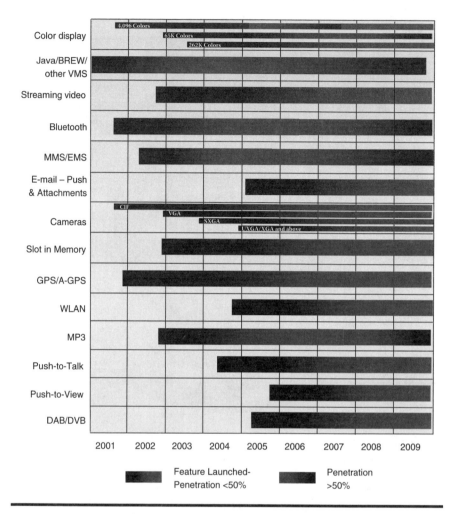

FIGURE 1.36 Mobile handset features.

Handset Complexity and Features

Over the past decade, technology developments have enabled manufacturers to make handsets with an increasing number of features and functions to support multimedia-style services that certain sectors of the subscriber community expect (Figure 1.36).

Monophonic, then polyphonic ring tones, and today MP3 encoded music or sounds may be used to alert the subscriber to incoming calls and text messages. A very competitive element in the handset market is music. Products such as the Apple iPOD and other stand-alone MP3 players have enabled the transfer of technology into mobile devices, with many devices now supporting multi-megabit memory cards or even miniature hard disc drives with capacity up to 4 gigabits.

FIGURE 1.37 Examples of mobile devices: Nokia 1110 (left), Orange SPV M2000 (center), and Vertu Ascent (right).

Amazing as it may sound, some of the earlier analog mobile phones did not support a display; most modern phones of course support at least a gray scale LCD display and most will have a color display. The race is on between the manufacturers to produce displays with the greatest color depth or the highest contrast. Latest technologies make use of organic light emitting diode (OLED) displays. Flexible displays or electronic paper will be seen in the next generation of mobile communication devices.

Examples of Mobile Devices

Figure 1.37 shows three examples of mobile devices from the low, middle, and high end of the market. More expensive phones do not necessarily come with more features, as demonstrated below, although there is a cost associated with exclusivity.

The Nokia 1110 is a very basic model with a monochrome screen and a limited set of features. It can be purchased SIM-free, without a contract, for around U.S.$85 (£50).

The SPV, available on the Orange phone network, is a fairly advanced GSM/GPRS PDA-style device. It combines many advanced features, such as a camera, Bluetooth, Java capability, MMS, and Windows compatibility. It costs £200 with a 12-month contract from Orange.

The Vertu brand is an exclusive mobile phone manufacturer based on Nokia equipment. The Vertu Ascent pictured above costs around U.S.$4000. Other models in the range can cost up to U.S.$40,000. They are designed as jewelry from materials including gold, platinum, silver, sapphire, ruby, leather, and carbon fiber. These models obviously appeal to the super-rich subscriber. A particular feature is a concierge button that puts the user in touch with a Vertu customer care representative who is able to fix all manner of problems for the subscriber, such as booking tickets to a show, arranging for a rental car, etc.

1.3 Current Mobile Technologies and Markets

1.3.1 Current Mobile Technologies

There are at present around 2 billion mobile phone subscribers around the world. The majority use GSM-based systems, but there are many people who use other networks' technologies. This total number of users is expected to grow to 3 billion users before 2010 as the newer 3G technologies are adopted and the developing markets begin to take off.

Mobile Generations

Mobile networks are commonly divided into three "generations," with the third generation (3G), of which UMTS is one such system, currently being deployed in networks around the world.

First-generation (1G) systems were analog systems, designed with the simple aim of making speech services available on the move. They included technologies such as TACS (Total Access Communication System), NMT (Nordic Mobile Telephone), and AMPS (Advanced Mobile Phone System). However, even these simple systems led to annual market growth rates of 30 to 50 percent, leading to around 20 million subscribers by 1990.

However, quality was poor and capability and reliability were low. Thus, as demand grew, the current range of 2G systems was developed to take their place. The most well-known of these systems are GSM (Global System for Mobile Communications), cdmaOne, and the system known in the United States simply as "TDMA," or by its standardization label of "IS-136." These systems were characterized by a move to representing information digitally and brought the following broad changes:

- More consistent and reliable quality of speech
- Increased capacity and spectrum efficiency through more advanced modulation and access schemes
- Easier implementation of advanced voice services, text messaging, fax, plus the addition of basic access to data networks
- Enhanced security and fraud prevention

However, even in the move from 1G to 2G (Figure 1.38), the basic aim was still to optimize for speech services delivered over wide areas (macro cells). 1G and 2G systems are therefore all characterized by circuit switched networks, which are well suited to symmetric, real-time "conversational" services. The term "2.5G" is sometimes used to describe enhancements to second-generation systems that aimed to optimize parts of these systems for data applications using packet switching techniques.

The latest move, to 3G, further advances digital systems with the particular aim of increasing the ability to use data applications on the move (i.e., mobile computing or the wireless office), and to enable "multimedia" services, which may

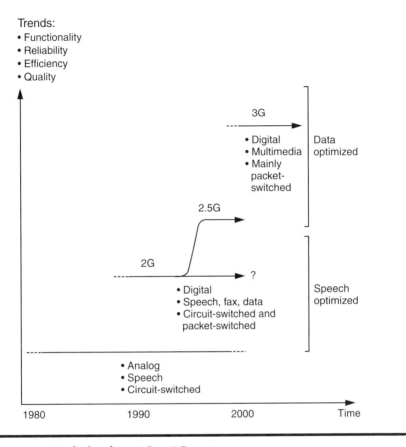

FIGURE 1.38 Evolution from 1G to 3G systems.

mix voice, graphics, video, music, etc. To achieve this, a key change is in increasing the ability of mobile systems to transfer larger quantities of information much faster. (See Table 1.5.)

A factor throughout the evolution of mobile technology (Table 1.6) has been the constant improvement in semiconductor and microwave technologies. While such changes allow for building smaller and more sophisticated mobile equipment, they also result in the expectation of users for more complex, data-intensive applications and services.

The GSM Family of Technologies

GSM is a second-generation digital network, supporting voice and simple data services, including "dial-up" data and text messaging. However, the term "GSM" is also used to describe the more recent advances in the GSM family: GPRS and EDGE.

TABLE 1.5 Mobile Evolution

Generation	Type of Service	Example System
1G	Analog, speech	AMPS, NMT, TACS
2G	Digital, speech, data, fax SMS, supplementary service, roaming	GSM, cdmaOne, PDC, PHS
2.5G	Packet date, WAP, iMODE	GPRS, cdma 1xRTT
2.75G	Faster data	EDGE
3G	Service-oriented, broadband data	UMTS, cdma2000

TABLE 1.6 Mobile Services

Services	2G	2.5G	3G
E-mail	SMS	Text based with small attachments	"Normal" e-email with full attachments
Instant messaging	SMS	Text based	With audio and video
Web browsing	Text based	Text and image, 100-kB page takes approximately 30 seconds to load	Text and image, 100-kB page takes approximately 2 seconds to load
Streaming audit and video	No	Short clips	Yes
VoIP	No	Limited	Yes
File transfers	Very slow	500-kB file; 2-minute load time	500-kB file; 10-second load time
Corporate applications	Very limited	Some	Yes
Location based	No	Limited	Yes

The Development of GSM

During the early 1980s, analog cellular telephone systems were experiencing rapid growth in Europe, particularly in Scandinavia and the United Kingdom, but also in France and Germany. Each country developed its own system, which was incompatible with everyone else's in equipment and operation. This was an undesirable situation because not only was the mobile equipment limited to operation within national boundaries, which in a unified Europe were increasingly unimportant, but there was a very limited market for each type of equipment, so economies of scale, and the subsequent savings, could not be realized.

The Europeans realized this early on, and in 1982 the Conference of European Posts and Telegraphs (CEPT) formed a study group called the Groupe Spéciale Mobile (GSM) to study and develop a pan-European public land mobile system. The proposed system had to meet certain criteria:

- Good subjective speech quality
- Low terminal and service cost
- Support for international roaming
- Ability to support handheld terminals
- Support for range of new services and facilities
- Spectral efficiency
- ISDN compatibility

The developers of GSM chose an unproven (at the time) digital system, as opposed to the then-standard analog cellular systems such as AMPS in the United States and TACS in the United Kingdom. They had faith that advancements in compression algorithms and digital signal processors would allow the fulfillment of the original criteria and the continual improvement of the system in terms of quality and cost.

In 1989, GSM responsibility was transferred to the European Telecommunication Standards Institute (ETSI), and phase I of the GSM specifications was published in 1990. Commercial service started in mid-1991, and by 1993 there were 36 GSM networks in 22 countries, with 25 additional countries having already selected or considering GSM. This is not only a European standard; South Africa, Australia, and many Middle and Far East countries have chosen GSM. By the beginning of 1994, there were 1.3 million subscribers worldwide. The acronym GSM now stands for Global System for Mobile telecommunications.

Significant Events in GSM Development

- 1978: CEPT sets about allocating spectrum in the 900-MHz band
- 1982: Groupe Spéciale Mobile established
- 1984: Three working parties formed developing core network, radio access, and service aspects
- 1985: Decision to specify a digital system taken
- 1987: Initial MoU signed
- 1989: ETSI takes over responsibility for GSM
- 1991: First commercial launch of GSM service in Finland
- 1992: First roaming agreement signed
- 1993: GSM deployed outside Europe
- 1995: First GSM service in the United States
- 1998: 100 million GSM users
- 2001: 500 million GSM users
- 2004: 1 billion GSM users

58 ■ Introduction to Mobile Communications

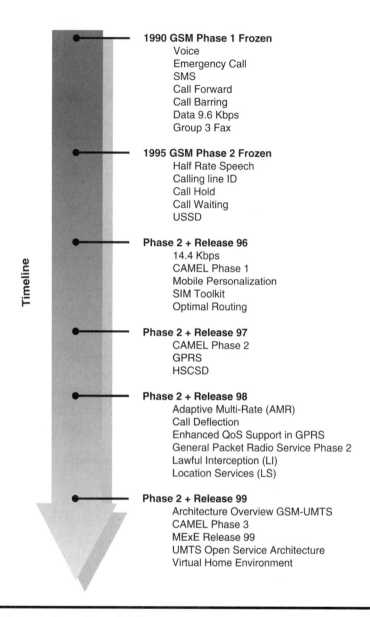

FIGURE 1.39 GSM phase timeline.

Phases of GSM

The development of GSM has never stopped (Figure 1.39). Even in those early days of putting together GSM, the process of development carried the technology forward to support ever more complex mechanisms to improve the quality of operation and to increase the number of services available. In 1990, the first version

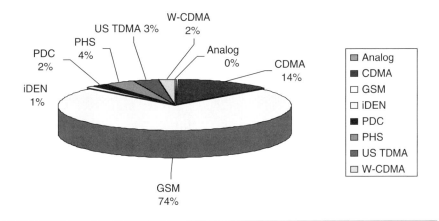

FIGURE 1.40 Comparison of mobile technologies.

or phase was frozen and released. GSM is defined by a series of documents called specifications. The 8000 pages of GSM specifications try to allow flexibility and competitive innovation among suppliers, but provide enough guidelines to guarantee the proper interworking between the components of the system. This is done in part by providing descriptions of the interfaces and functions of each of the functional entities defined in the system.

Phase 1 specifications, as well as defining the vast complexity of GSM, also defined a number of basic services that Phase 1-compatible equipment and networks should be able to provide. Phase 2 enhanced certain areas of GSM operation and defined further services, Phase 2+ brought more technical innovation to GSM and began a series of yearly releases defining services such as GPRS, HSCSD (high-speed circuit switched data), SIM toolkit, and other aspects of GSM technology.

GSM Statistics

GSM is by far the most used technology today, with 75 percent of the world's mobile subscribers using GSM (Figure 1.40). There are more than 1000 networks using GSM technology in over 200 countries. The total number of GSM subscribers is well over 1.5 billion and set to exceed 2 billion by the year 2007.

The GSM family of technologies includes basic 2G GSM, 2.5G GPRS, and EDGE. Most GSM networks that have been deployed support, or are in the process of supporting, GPRS. The choice for a network to deploy EDGE in the network will depend on its strategy for migrating to 3G systems.

GPRS and EDGE

GPRS is classed as 2.5G technology because it takes a step closer to 3G by increasing the data rate available to the user and user applications. Significantly, it uses the

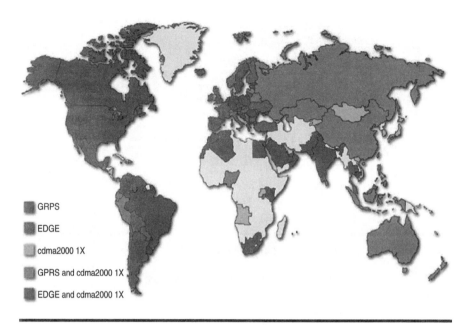

FIGURE 1.41 GPRS deployment around the world. (*Source:* **World Cellular Data Metrics, March 2005.**)

2G radio spectrum more efficiently through the use of packet data techniques. This allows more users to be apparently connected to the network at the same time.

Having an efficient method of connecting users enables network operators to deliver advanced services to their subscribers. When used alongside other technologies such as WAP or XML, this gives the users a "Web-like" experience.

Many network operators around the world have deployed GPRS; the map in Figure 1.41 shows the regions of the world where GPRS can be found.

Operators of networks very rarely sell GPRS as a stand-alone service. What they are able to do, however, is develop services that will use GPRS as the bearer. Picture messaging is an example of a service that most operators are selling, which uses GPRS. Most operators have also established themselves as ISPs (Internet service providers), allowing subscriber access to Internet content.

Current figures suggest that there are approximately 100 million GPRS users (Figure 1.42). This figure should not be confused with the number of GPRS terminals that have been sold, which would be very much higher.

EDGE (Enhanced Data rates for Global Evolution) is an addition to GSM and GPRS technologies that allows the transmission of three times more data across the radio interface. It can be deployed by operators to increase capacity for data services, or in many cases it can be considered a step closer to 3G. EDGE systems may be able to allow support for 3G-like services on a 2G network. This can be seen as an attractive migration route for operators that have not obtained a 3G license, or in those parts of the world where the 3G licenses have not yet been issued.

The Mobile Telecommunications Market ■ 61

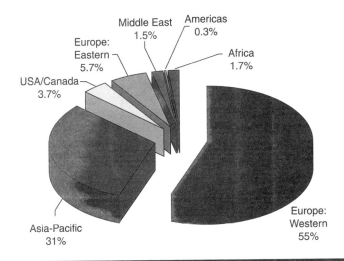

FIGURE 1.42 Distribution of GPRS subscribers (total = 99,786,714).

Code Division Multiple Access (CDMA) Technologies

- *cdmaOne*. The name cdmaOne encompasses a family of CDMA standards also known as IS-95. There are two revisions to the IS-95 standard: revision A and revision B. IS-95 refers to a set of specifications describing all aspects of the operation of an end-to-end wireless system. This system has seen deployments of cellular communication networks and wireless local loop.
- *IS-95A*. The IS-95 specification was first published in 1993 and the first revision (IS-95A) in 1995. This specification forms the basis of many commercial cellular operations around the world, and describes the structure and operation of a 1.25-MHz CDMA radio system. This specification supported basic voice services and "dial-up" data connections at speeds up to 14.4 kbps. The first operational IS-95 system was operated by Hutchison in Hong Kong in 1995.
- *IS-95B*. The second revision, IS-95B, combines a number of standards into one and addresses some interworking issues between IS-95 systems and others. IS-95B can also support higher data rates, with some operators offering 64-kbps connections. It also offers the greater efficiency of packet data, making it a 2.5G system. cdmaOne IS-95B was first deployed in September 1999 in Korea and has since been adopted by operators all over the world.

  There are nearly 300 million cdmaOne subscribers in 400 networks worldwide, the majority of these being in North America and the Asia Pacific regions (Figure 1.43).
- *cdma2000 1 X*. 1 X is an evolution technology bringing further efficiency and higher data rates to existing 2G deployments. This technology can double the voice capacity in a network and support packet data rates of up to 307 kbps.

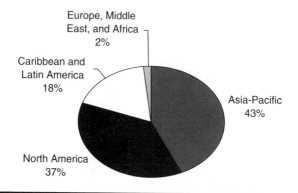

FIGURE 1.43 cdmaOne worldwide distribution. (*Source:* cdg.org.)

The world's first cdma2000 1X commercial system was launched by SK Telecom (Korea) in October 2000. Since then, cdma2000 1X has been deployed in Asia, North and South America, and Europe. There are now 126 operators in 57 countries around the world, with an estimated subscriber base of 200 million; subscriber figures are growing at a rate of 7 million users per month.

Evolving to 3G Systems

Despite the enhancements to 2G networks, it remains the case that many of the "future services" require higher data rates than available even with 2.5G. In addition, the future mix of services remains virtually unknown. Therefore, planning should be flexible, and both circuit and packet switched domains must be supported. The ITU, through its IMT2000 initiative, began the process of trying to describe the required capabilities of a third-generation system. The key feature requirements of a 3G system include:

- Higher data rates, up to 2 Mbps, to enable applications such as large file transfers, mobile video and music, etc.
- Multimedia service support: the ability to multiplex voice, data, video, and other services on a single connection, to be received simultaneously.
- Flexibility: the ability to request *bandwidth on demand,* and variably set data rates to suit applications in progress.
- Efficient delivery of asymmetric services such as Web browsing. This requires the enabling of different bit rates on the uplink and downlink.
- Quality of service (QoS) control, with guarantees over a range from real-time, low-loss services such as speech down to "best-effort" services.
- IP support, to enable efficient interworking with the Internet and other IP-based applications.

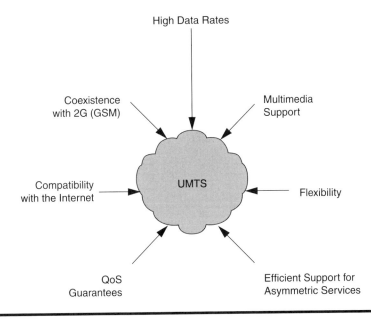

FIGURE 1.44   IMT2000 objectives.

- Coexistence and interworking with existing second-generation networks (GSM) and services, in which operators have already invested a huge amount.
- Ease of global harmonization to ensure that users can gain access to their services wherever they are.

The IMT2000 specification may be seen as a wish list for a future-generation mobile network (see Figure 1.44). Many companies and organizations examined the specifications and produced technical proposals outlining the manner in which they would support the items on the wish list. The IMT committee was to examine each of the proposals and choose just one for further development. This choice would be adopted worldwide as a single global standard for third-generation mobile networks.

In reality, it was not possible to choose a single proposal. There are now three major 3G technologies being developed and deployed around the world: UMTS, cdma2001, and TD-SCDMA (Table 1.7).

- *UMTS.* UMTS (Universal Mobile Telecommunications System) is the 3G proposal from the European community. The ETSI originally started developing UMTS and then handed it to the 3GPP. The 3GPP (Third Generation Partnership Project) is a collaboration of the standards bodies around the world seeking to provide compatibility and interoperation of different systems. UMTS is based on a technology called WCDMA (Wideband Code Division Multiple Access). Figure 1.45 shows worldwide W-CDMA subscriber distribution.

TABLE 1.7 3G Technologies

3G Technology	Development Bodies	Comment
UMTS	3GPP (collaboration of ETSI, ARIB, TTC, CCSA, TTA)	From Europe and Asia Pacific regions
cdma2000	3GPP2 (collaboration of ETSI, ARIB, TTC, CCSA, TTA TIA)	Development of North American standards
TD-SCDMA	CCSA	3G from China

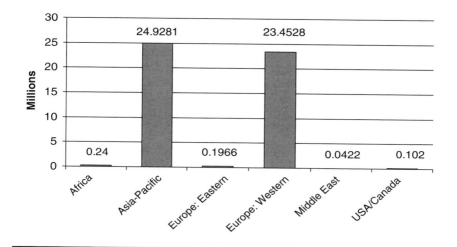

FIGURE 1.45   W-CDMA subscriber distribution.

- *cdma2000.* cdma2000 is a 3G solution from standards bodies and manufacturers in the United States. It is a natural evolution from the existing 2G standard, cdmaOne. The technology is also based on WCDMA, but it is not directly compatible with UMTS. The 3GPP2 standards body is a collaborative effort that seeks to provide interoperability between the various 3G standards from around the world.

  CDMA2000 1xEV-DO is the first step into full 3G technology and delivers peak data speeds of 2.4 Mbps and supports applications such as MP3 transfers and video conferencing. 1xEV-DO Rev A (Revision A) increases efficiency capacity and data rates even further, offering up to 3.1 Mbps.

- *TD-SCDMA.* Time Division Synchronous CDMA (TD-SCDMA) was proposed by the China Wireless Telecommunication Standards group (CWTS) and approved by the ITU in 1999, and technology is being developed by the Chinese Academy of Telecommunications Technology. This is a Chinese solution for 3G networks that is compatible with UMTS.

TABLE 1.8   China Mobile

	2001	2002	2003	2004
Mobile subscriptions (000s)	144.8	206.6	268.7	319.2
Growth (%)	69.8	42.7	30.0	18.8
Net additions (000s)	59.5	61.8	62.1	50.5
Penetration (%)	11.3	16.2	20.9	29.6

1.3.2   Global Mobile Markets

The *global* market for mobile services is a vast and complicated space. Each country will have several mobile operators in competition, leading to complex domestic markets. This lesson contains a brief overview of three of these markets: (1) the Chinese market, (2) the Western Europe, and (3) the U.S. market. Information in this lesson is from the Informa Telecoms and Media Global Mobile Forecast 2005–2010.

Market Overview: China

China's economy grew 9.5 percent year-on-year in 2004. Telecommunications has been one of the priority sectors targeted by the Chinese government since the early 1990s in an effort to speed up China's industrialization process. The telecommunications sector grew 12.6 percent in 2004 (the mobile market alone grew 18.8 percent), faster than the growth of the overall economy. The industry's value-added total accounted for 2.77 percent of China's GDP in 2004.

The Chinese Mobile Market

China became the world's largest mobile market in terms of officially counted subscribers in July 2001. At the end of 2004, there were more than 319 million mobile users, over 7 million more than fixed-line subscribers (312 million), and 50 million new mobile customers were added during the year. Of the two operators, market leader China Mobile enjoys a 64 percent market share but its rival, China Unicom, has a faster rate of growth, with a compound annual growth rate of 40 percent over the past three years.

The number of mobile subscribers surpassed that of fixed-line subscribers in September 2003 (Table 1.8). Mobile capacity had already overtaken fixed capacity on a national scale the year before, with just six provinces having a greater fixed-line than mobile capacity as of 3Q 2002. Despite the large numbers and robust growth, however, market penetration stood at 29.6 percent at the end of 2004, meaning that, unlike most other large mobile markets, there are still many remaining potential users.

China is a vast country in terms of its population and territory. For historical reasons, there exists a huge disparity in its economic development across regions. Large urban centers — Beijing, Shanghai, and Guangzhou — and then, to a lesser extent, on the eastern seaboard linking these three cities are the forerunners in economic development; the mobile penetration rate there is significantly more than 70 percent. However, China's more remote and rural areas, predominantly in China's west, still lack basic access in many places. Therefore, mobile capacity is considerably lower in the west than in the east and south, with the country's industrialized northeast — apart from Beijing, Hebei, and Liaoning — lagging as well.

China's mobile sector has been plagued by price wars over the past few years, despite government pricing rules. As competition stiffens and wireless subscription rates begin to mature, especially in urban areas and high-end segment, operators find their revenue growth slackening. In recent years, the new subscribers have come mainly from the low end. As a result, the overall ARPU has been dropping: for example, the average ARPU at China Mobile has declined steadily since 2000, falling from U.S.$19.08 per month in 2001 to U.S.$12.50 by September 2003 and U.S.$11.10 in 2004.

Operators hope value-added services will help stem the ARPU decline. SMS has been extremely successful in China over the past years. China Mobile and China Unicom reported sending 172.57 billion and 44.22 billion SMS, respectively, in 2004. During the one-week Spring Festival holiday in 2004 alone, 7.8 billion SMS were sent by China Mobile users and more than 2 billion sent by China Unicom. Now all the service providers (SPs) derive the majority of their revenues from SMS.

Other new mobile value-added business grew significantly in 2004. Colour Ring Back Tone (CRBT), known as another "gold mine" following SMS, won more than 20 million customers with a market value reaching nearly RMB1 billion (U.S.$121 million) since it was first launched by China Mobile in May 2003.

WAP service also maintained a rapid rate of growth, increasing at more than 16 percent per month in the domestic market during the first quarter of 2004, due to the improvement of 2.5G networks and the active participation of SPs. By the end of 2004, the number of WAP users grew to 25 million, and the market value climbed by nearly RMB1.2 billion (U.S.$145 million). In addition, entertainment services such as mobile games, pictures, and ringtone downloads; comprehensive information services; and IVR chat services also promised good prospects.

China's Wireless Operators, Networks, and Subscribers

Table 1.9 depicts China's wireless operators, networks, and subscribers.

TABLE 1.9 China's Operators, Networks, and Subscribers

Company	Networks	Subs at End-2004 (millions)	Market Share (%)
China Mobile	GSM, GPRS	204.3	64
China Unicom	GSM cdmaOne CDMA2000 1X	85.9 28.8	27 9
Tibet Telecom	cdmaOne	0.2	0
	Total mobile	319.2	
China Telecom	PAS	43.0	66
China Netcom	PAS	22.5	34
	Total PAS	65.5	
		384.7	

Chinese Mobile Operators

In July 2004, China Mobile (Hong Kong) Limited successfully completed its acquisition of the mobile telecommunications companies in ten provinces from its state-owned parent, China Mobile Communications Corporation, and extended its network coverage to all provinces in mainland China.

Ranking as China's largest telecom operator in terms of revenues, China Mobile posted RMB179.1 billion (U.S.$21.6 billion) in operating revenues in 2003 and RMB203.9 billion (U.S.$ 24.6 billion) in 2004. It also boasts the world's largest mobile subscriber base, with 204.3 million users at the end of 2004, representing an annual growth rate of 23 percent and commanding 64 percent of China's mobile market.

China Unicom (Hong Kong), which owns 30 of China Unicom Telecommunications Corporation's 31 provincial networks (the exception being Guizhou), saw its operating revenue increase by 17.3 percent from 2003 to RMB 79.33 billion (U.S.$9.59 billion) in 2004. By the end of 2004, China Unicom had a total of 114.7 million subscribers — 85.9 million GSM subscribers and 28.8 million CDMA subscribers.

China's largest fixed-line operator, China Telecommunications Corporation (China Telecom), was established as a result of the 2002 industry restructuring, incorporating 21 southern branches of the original China Telecom. Operating revenue in 2004 was RMB161.2 billion (U.S.$19.5 billion), after acquiring the other ten provinces from its parent company in April 2004, up 6.4 percent from 151.5 billion (U.S.$18.3 billion) in 2003.

China Telecom offers PAS-based mobile services (known as "Little Smart" in China. PAS is a personal wireless access system that provides the convenience of a mobile phone with the cost advantages of a fixed-line phone). Its PAS subscriber base increased by 67 percent to 43 million at the end of 2004 — from 25.6 million in 2003 — representing 66 percent share of the national PAS market. China Telecom also offers a CDMA-based limited mobility service, known as "Shihuatong," in Shenzhen.

China Network Communications Corporation (China Netcom) accomplished IPO in New York and Hong Kong in November 2004, incorporating six northern and two southern provinces. It also offers PAS-based mobile services, with 22.5 million PAS subscribers at the end of 2004, more than doubling its customer base during that year. Total revenues in 2004 reached RMB 64.9 billion (U.S.$7.84 billion), up 8 percent from RMB 59.9 billion (U.S.$7.2 billion) in 2003.

Like the current China Telecom, China Netcom was established as a result of the 2002 industry restructuring, incorporating three companies — the original China Netcom, China Jitong, and 10 northern branches of the original China Telecom — into the new China Netcom Group. These companies essentially operated independently until June 2003, when China Netcom purchased China Jitong for RMB 482 million (U.S.$58.2 million). In 2004, China Netcom Group has split itself into three companies to operate future businesses in northern China, southern China, and the global market.

Western Europe

The annual growth rate for the total customer base of western European markets rose in 2003 (9 percent, versus 7 percent in 2002) and maintained this level in 2004. As a result, 2 million more new customers were added in 2004 than in 2003 and penetration for the region has passed 90 percent (Table 1.10). It is not expected that this annual growth rate will be maintained now that the penetration is so high. It is forecast that the growth rate will decline from 2005 onward; despite this, the regional penetration rate is forecast to exceed 100 percent in 2007.

Nine countries in western European have been profiled and forecast. The total customer bases of these nine countries accounted for 86 percent of the regional total at the end of 2004, while their combined populations accounted for 87 percent of the total. In forecasting for the remaining countries, it is assumed that, because the

TABLE 1.10 Western Europe Mobile Market

	2001	2002	2003	2004
Subscriptions (000s)	281,185	301,729	328,896	357,956
Growth (%)	15	7	9	9
Net additions (000s)	37,003	20,544	27,167	29,060
Penetration (%)	71.56	76.65	83.40	90.60

TABLE 1.11 The U.S. Market

	2001	2002	2003	2004
Subscriptions (000s)	126,985	141,303	157,137	179,727
Growth (%)	18	11	11	14
Net additions (000s)	19,498	14,318	15,834	22,590
Penetration (%)	45.46	50.15	55.27	61.05

markets are very similar, their growth, usage of data, and launch of 2.5G/3G would be largely the same as the markets that have been forecast.

As expected, commercial 3G services were launched in many of the region's markets in 2004 and more than 2 percent of the region's subscribers (7.3 million) were using 3G services by the end of the year. This was forecast to rise to 20.9 million (over 5 percent of the customer base) by the end of 2005. By the end of the forecast period, we expect there will be 359 million 3G users — 86 percent of the total mobile users. 2.5G services are also gaining acceptance; in 2004, the number of 2.5G users doubled to 50.7 million, and by end-2006 there are forecast to be 81.8 million 2.5G subs. By 2007, 3G is forecast to overtake 2.5G, having 95.5 million subs compared with 2.5G's 81.9 million subs.

Data services accounted for more than 15 percent of the total revenues in 2004. This figure is set to increase over the forecast period, and in 2010 nearly 30 percent of total revenues are forecast to come from non-voice revenues. However, the increase in enhanced data services does not have the effect of increasing revenues overall — the ARPU is forecast to drop in western Europe over the forecast period. This is likely due to competitive data pricing as well as declining voice prices.

Many of the operators have reappraised their strategy for prepaid and, rather than simply trying to migrate their customers on to contracts, have used prepaid as a way of accessing certain customer segments that have not been targeted until now. As a result, there was more focus placed on prepaid; and in 2004, 43 percent of the total net additions were for prepaid, up from 19 percent in 2003. Even so, the percentage of total customers who use prepaid offerings fell below 60 percent for the first time in several years. It is forecast that the percentage of net additions for prepaid will remain at around 45 percent for the rest of the forecast period, and prepaid's share of the total market will decline to 57 percent by end-2010. Overall, the western European mobile market is forecast to grow by 16 percent over the forecast period (2004–2010), adding 57.3 million customers.

United States

The U.S. mobile market (Table 1.11) has been under-penetrated compared with advanced telecom markets in other regions such as Western Europe and Asia

Pacific. Historically, this was seen as a result of an overly complex and competitive marketplace — numerous small operators servicing small customer bases — and a lack of interoperability between the various mobile operators impeding seamless service offerings. This lack of interoperability was such that roaming agreements were required while still within the United States.

This complexity has been reduced and U.S. operators have undergone a period of consolidation, leading to operators with U.S.-wide mobile coverage. During 2004, there was something of an upturn in the market — the annual growth rate rose and 6.8 million more new customers were added than in 2003 (22.6 million net additions in 2004 vs. 15.8 million in 2003) (Table 1.11).

Cingular Wireless's acquisition of AT&T Wireless, which was completed in October 2004, has triggered a renewed phase of consolidation in the U.S. wireless market that has considerably reduced the number of operators. In 1998, 12 operators accounted for 80 percent of the U.S. wireless market. With the merger of Sprint and Nextel, just four operators account for more than 80 percent of the market: Cingular Wireless, Verizon Wireless, Sprint Nextel and T-Mobile USA.

More mergers and acquisitions among the smaller players in 2005 — Alltel's acquisition of Western Wireless, Alamosa's acquisition of Airgate PCS, and the merger of Horizon PCS and iPCS — will consolidate the market even further.

Cingular Wireless's enlargement has changed the competitive landscape. Cingular and Verizon had more than half of the total U.S. market at the end of 2004: Cingular had a 27 percent share (49.1 million subs) and Verizon had a 24 percent share (43.8 million subs). Even the nationwide operators Sprint PCS and T-Mobile will find it more difficult to compete, while many of the country's numerous smaller regional and local operators are likely to be swallowed up in the long term, unable to compete independently without enjoying the clear advantages that mergers and acquisitions bring: economies of scale, OPEX and CAPEX savings, and the advantage of increased spectrum resources.

The U.S. mobile market embraces several different mobile technologies, including CDMA, TDMA, GSM, and iDEN within the 2G environment (Figure 1.46).

U.S. Operators

Cingular Wireless formed as a joint venture between SBC (60 percent) and BellSouth (40 percent). It is a TDMA/GSM operator and, at the end of 2004, its customer base was approximately two thirds GSM. Cingular currently offers some of the lowest-cost plans in the United States, as well as "rollover" minutes, allowing customers to roll over unused minutes to the next billing period. Cingular completed the acquisition of AT&T Wireless during 2004, making it the biggest wireless operator in the United States. Debt restructuring could have an impact as Cingular absorbs the cost of the acquisition on top of the U.S.$15.1 billion of long-term debt that Cingular's parent (SBC) already carries on its books.

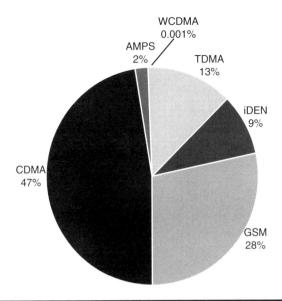

FIGURE 1.46 U.S. subscribers by technology.

AT&T Wireless, like its new owner Cingular, also offered AMPS, TDMA, and GSM. Before the takeover, 37 percent of its customers had moved to GSM. To facilitate the merger with Cingular, AT&T was required to divest itself of holdings in Eurotel Bratislava and Rogers Wireless, and sell customer contracts and assets in 11 states, as well as undertake minimal divestitures of Spectrum and parts of its wireless network.

Sprint is a pure CDMA operator and faces less of a challenge than the two market leaders because it has no legacy analog networks with which to contend. At the end of 2004, its market share was 12 percent, up marginally on its 11 percent share at end-2003.

Sprint has a strategy of reselling network connectivity to affiliates. It has 11 affiliates across the United States, and these wholesale partners and affiliates together contributed U.S.$100 million in revenues to the company in 3Q 2004. Virgin USA, which uses Sprint's network in its role as an affiliate MVNO, became the fastest growing operator in U.S. history in 2003, and its prepaid service offering has given Sprint access to the lucrative under-subscribed youth segment.

T-Mobile USA, a pure GSM operator, is the smallest of the major U.S. operators, with 17.3 million subscribers by year-end 2004. It gained 19 percent of the total net additions in 2004, which gave it a market share of 10 percent — up from 8 percent a year earlier (Table 1.12). Despite its impressive share growth, the operator's position looks less stable within a market characterized by consolidation, and one where economies of scale will likely predetermine future market direction and strategy. Within such an environment, T-Mobile will be hard-pressed to consolidate and grow its position against a backdrop of evolving wireless super-operators.

TABLE 1.12 U.S. Operators

Operator	Subs (000s) end-2002	Market Share (%) end-2002	Subs (000s) end-2003	Market Share (%) end-2003	Subs (000s) end-2004	Market Share (%) end-2004
Cingular Wireless	21,315	15	23,737	15	49,089	27
Verizon Wireless	32,491	23	37,522	24	43,816	24
Sprint	15,458	11	17,806	11	21,507	12
T-Mobile USA	9,900	7	13,133	8	17,314	10
AT&T Wireless	21,241	15	21,596	14	0	0
Others	40,898	29	43,343	28	48,001	27
Total	141,303		157,137		179,727	

Review Questions

Q1. From the Chinese market data, the market reveals that _____.
   a. Instant growth continues to increase
   b. Growth is positive but slowing down
   c. Growth is in decline
   d. Growth has remained the same for the past four years

Q2. Which is the largest Chinese operator?
   a. China Unicom
   b. China Mobile
   c. China Network
   d. China Netcom

Q3. Now that penetration in Western Europe is over 90 percent, market growth is expected to _____
   a. Increase rapidly
   b. Begin to slow
   c. Start to decline
   d. Stay the same

Q4. In the U.S. market, CDMA accounts for _____
   a. 9 Percent of the market
   b. 28 Percent of the market
   c. 98 Percent of the market
   d. 47 Percent of the market

Q5. At the end of 2004, which U.S. operator had the largest market share?
   a. Verizon Wireless
   b. AT&T Wireless
   c. Cingular Wireless
   d. Sprint

1.3.3 World Market Forecasts

Subscription and Penetration Forecasts

The growth rate in 2004 was 24 percent. However, this is not expected to be maintained as many of the world's markets reach saturation. Many people maintain more than one subscription, leading to greater than 100 percent penetration in some of the western European markets (Figure 1.47).

Growth will be below 20 percent in 2005 and will continue to fall, with less than 10 percent in 2007 and as little as 5 percent by 2009. The majority of the growth is still expected to come from the developing regions: China, India, and parts of the Middle East.

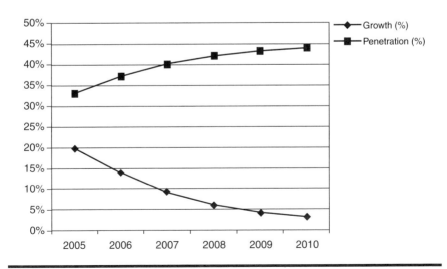

FIGURE 1.47 Growth versus penetration forecast.

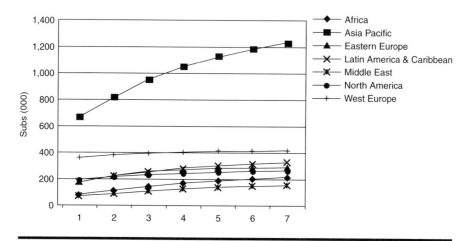

FIGURE 1.48 Regional growth.

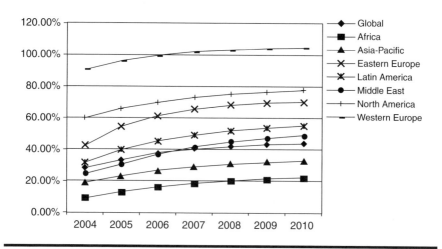

FIGURE 1.49 Regional penetration.

The total growth over the next five years is expected to be 69 percent, in which more than 1.18 billion new subscribers will be added, resulting in almost 3 billion mobile phone users by 2010 (Figures 1.48 and 1.49).

Technology Forecasts

In the different regions around the world, there are different migration or evolution paths that the technology may take, and these have an impact on the forecasts

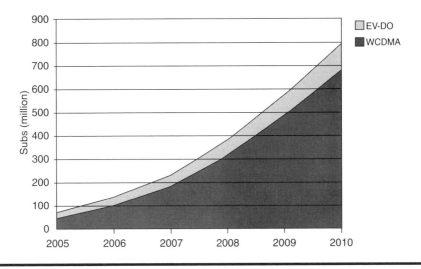

FIGURE 1.50   EV-DO versus WCDMA (UMTS).

of technology uptakes. Because GSM is the main global technology, the evolutionary path through GPRS, and possibly EDGE toward UMTS, sees the greatest numbers. Other regions starting out from 2G technologies, such as cdmaOne or TDMA, have a smaller base in the first instance so the forecast figures for these evolution paths are naturally smaller.

WCDMA in Figure 1.50 represents the deployment of UMTS systems. By the end of 2004, there were 47 systems in place, with more than 60 more networks being rolled out in 2005. By contrast, 20 cdma2000 1 x EV-DO had been established. The adoption of the WCDMA technology is expected to grow at three times the rate of EV-DO. By the end of 2010, EV-DO will only account for 14 percent of 3G subscribers.

It is expected that the established 2G networks will linger for some time to come. Many users in these networks are low-end users, simply making voice calls, even if they possess a more sophisticated device. This fact will slow the take-up of 3G subscriptions and therefore the use of 3G services. 2008 will see a stronger take-up of 3G as people begin to replace their aging 2G handsets, with around 14 percent of the market being 3G. By the end of 2010, 3G will only account for about 27 percent of the total number of mobile subscribers (Figure 1.51). The line marked "Digital" in Figure 1.51 includes GSM, GPRS, EDGE, and cdma2000 1x RTT.

Manufacturers of handsets will continue to produce 2G devices to supply the demand from developing regions. Many of these will support 2.5G. This explains the continuing rise in 2G subscriptions. However, a more regional look at the numbers in Figure 1.52 shows that the developed markets show a decline in 2G subscriptions in the short term.

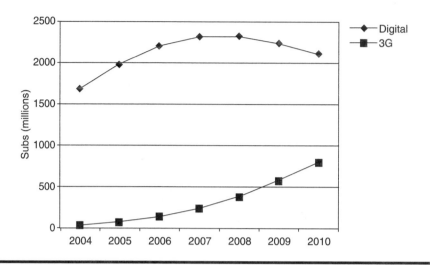

FIGURE 1.51 2G/3G subscriber comparison.

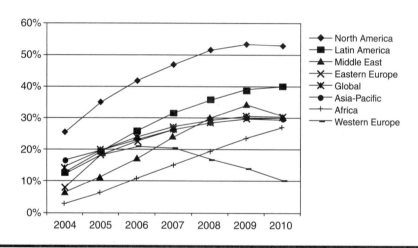

FIGURE 1.52 Regional 2G growth.

ARPU Forecasts

Despite the introduction of 2.5G and 3G networks and services, most operators are still experiencing a declining average revenue per user (ARPU) figure, particularly those in the saturated markets. In addition, many 3G operators are coming to market with aggressively priced voice tariffs, putting further pressure on overall revenues. It is expected that voice ARPU will continue to decline and data ARPU will increase by as much as 30 percent when subscribers begin to use the 2.5G and 3G data services offered by the operators (Figure 1.53).

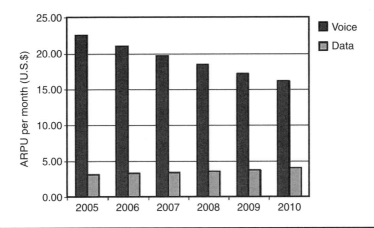

FIGURE 1.53 Data and voice ARPU compared.

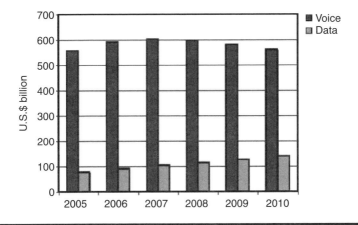

FIGURE 1.54 Global revenues for voice and data.

As shown in Figure 1.54, while revenues from data use will increase by up to 83 percent over the next five years, 80 percent of the operators' revenue will still come from voice. Falling voice revenues and slowing global growth will contribute to falling global revenues by 2010.

Prepaid

Prepaid has proved a mixed blessing for operators. It has stimulated rapid growth in subscriber numbers and driven growth into markets where previously mobile communications would have been unthinkable. However, it has also been a factor in driving down ARPU, because prepaid users are often more concerned with cost, although that is not always the case.

In the recent past, many of the operators in the mature markets have tried to persuade their prepaid users to move to contracts, regarding this as a way of persuading them to spend more money. However, this largely proved unsuccessful — many operators only managed to migrate a fairly minimal percentage to contracts and then discovered that they did not necessarily spend any more if they did. At the same time, it was realized that prepaid users are not necessarily low usage or low-spending users. On the contrary, many prepaid users are from the youth segment, which tends to contain early adopters for new and lucrative enhanced data services.

As a result, many of the operators have reappraised their strategy for prepaid and, rather than simply trying to migrate their customers on to contracts, have used prepaid as a way of accessing certain customer segments that have not been targeted until now. During 2004, 73 percent of the net additions were for prepaid (compared with 51 percent in 2003), and the prepaid share of the market rose to 55 percent. Regional and country markets have shown that as they approach saturation, the proportion of net additions going to prepaid rises. Therefore, it is forecast that a steadily increasing share of net additions will go to prepaid over the forecast period; and by 2009, 64 percent of the total customer base will be using prepaid.

Review Questions

Q1. Over the next five years, most of the global market growth is expected to come from _____.
   a. Western and eastern Europe
   b. The United States and the Americas
   c. China and India
   d. Japan

Q2. The reason for greater than 100 percent penetration on some markets is _____.
   a. Subscribers holding more than one SIM
   b. Subscribers upgrading to 2.5G devices
   c. Increasing 3G take up
   d. More people using SMS services

Q3. By 2010, which technology is expected to account for approximately 85 percent of the world's 3G users?
   a. cdma2000
   b. cdma2000 1 X EV-DO
   c. TD – SCDMA
   d. WCDMA (UMTS)

Q4. Which of the following is not a reason for the slow take-up of 3G subscriptions?
  a. Users primarily want to make voice calls.
  b. Developing markets are mostly 2G and 2.5G.
  c. Poor data services available in 3G.
  d. Long intervals between handset upgrades.

Q5. By 2010, voice and data ARPU are expected to follow the following trend:
  a. Data ARPU increase by 30 percent whilst voice ARPU decline.
  b. Voice ARPU continue to grow whilst data ARPU remains flat.
  c. Data ARPU and voice ARPU climb sharply with new 3G services.
  d. Data ARPU starts to fall and voice ARPU increases.

1.3.4 Looking Ahead

2G: A Changing Market

Although capacity limitations and a need for new spectra are certainly major factors for operators looking to deploy next-generation systems, new revenue opportunities through new services are viewed as the key drivers.

2G, in the most advanced markets at the present time, can be summarized as follows (Figure 1.55). Very little further capacity remains available, resulting in busy signals, dropped calls, and generally poor quality of service in busy areas. There was evidence of a demand for data services shown by the rapid growth in basic SMS (text messaging) traffic, and increased use of GSM for dial-up access to Internet and e-mail while away from the office. Technologies such as WAP were just emerging to further enable mobile Internet services. At the same time, competition was driving profit margins on plain voice subscribers down.

General Trends in Telecommunications

As the 2G mobile market changed, there were also changes taking place in the fixed world. In particular, the Internet has become the *de facto* delivery method for a whole new range of information and entertainment services to consumers. The continued increases in the ability of PCs and modem speeds, and the introduction of faster fixed-line connections (e.g., ISDN, ADSL), means that applications are becoming ever more data-hungry. The result in fixed networks is that they are beginning to carry much more data traffic than voice traffic, and again the revenues from the latter continue to decrease due to competition.

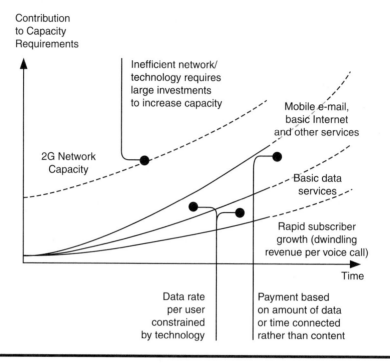

FIGURE 1.55 Mobile market influences.

Influential Factors in the Evolution of Mobile Services

- Rapid growth in the Internet
- Increasingly data-hungry applications
- Data overtakes voice traffic
- Users want mobility
- Prices for voice decrease
- Huge potential for mobile Internet
- Bleak future for voice revenues
- Access needed to desktop applications when mobile

At the same time, as users become accustomed to being able to speak to each other when on the move, and collect their e-mail on the move, the natural desire is for mobility, rather than being tied to fixed terminals and locations for other data applications.

Combining the state of 2G mobile networks with the trends occurring in the fixed and Internet worlds, the obvious market conclusion is that there is huge potential for mobile Internet, a bleak future for simple voice revenues, and an increasing need to be able to access desktop applications from other terminals, including mobile ones.

Mobile Internet in 2G

The first steps to mobile Internet and mobile data using WAP or GSM data cards met with mixed reviews. These reviews most commonly cited slow speed, unreliability, and high cost as the key negatives. Thinking back to the original aim of 1G and 2G mobile networks — optimized to deliver real-time voice — it is not surprising that changes must be made. In 2G networks, access to the Internet involved dialing up for a connection, which could take tens of seconds, and then waiting for data to download at very slow speeds (max. 14.4 kbps in GSM). If the radio signal is broken during this connection, then dial-up would have to be done all over again. All this time, users would pay for the length of the call, regardless of how much or how little data had actually been received. A key aim in defining 3G systems was therefore to allow faster speeds, to avoid the constant need for dial-up, to give better quality, and to allow different charging models (e.g., based on the quantity of data transferred rather than the time used, or maybe even to bill for content). As well as enhancing the user experience, new systems that optimize the transport of data (while preserving traditional voice quality) also contribute significantly to increasing operational efficiency, spectrum usage, and hence cost-savings for the operators.

Future Mobile Service Needs

In addition to evidence from existing applications and services, there have been a number of market studies aimed at trying to identify key opportunities for the future. Clearly, it is difficult to predict with any accuracy which services will and will not succeed, and there will, of course, be entirely new services that are as yet not even thought about. Thus, a prime aim in developing UMTS and other 3G systems is to provide the capability to be flexible and open in providing new services as they are developed, whoever develops them. Some of the likely service groups are shown below. Those in boldface type indicate services that are impossible to offer on 2G systems with any degree of quality, although obviously just about any service will benefit from the increases in data rate (speed) and reliability that 3G will offer.

Potential Future Mobile Services include:

- **Web browsing**: networked games, music downloading, **mobile video clips**
- **Video telephony**: picture mail, **multimedia messaging**, **video conferencing**
- Mobile shopping: mobile banking, consumer information services (timetables, hotels, tourist, etc.), navigation
- Intranet access: corporate database access, corporate business information (stocks, news, sales force etc.)
- Telematics: remote metering and security, machine-to-machine (M2), wireless vending

Location-Based Services

A widespread belief is that the key applications will not be those simply ported from the fixed world, but those that take advantage of the fundamental benefits of the mobile phone. In particular, because the mobile device can change its location, services that are tailored to this location (restaurant bookings, route finding, location information, etc.) are viewed as services with no fixed world competition.

Third-Party Services

There is also plenty of activity from other sectors of industry, such as banks and retailers, who see the mobile phone very much as a "personal trusted device," and hence a key opportunity for transactions and commerce. It is also true of location services such as navigation, that map information, for example, may be held by a digital mapping company. In most cases, operators will not be able to offer such services alone, but will need to work with third-party content and service providers. In many cases, such collaboration will require internetworking, with an access route to third-party application servers and databases located outside the operator's own network. Almost any of the entertainment or service opportunities could fall into this category.

Technology for the Future

Communications technology is constantly evolving. The networks that exist today may be considered reasonably mature second-generation systems. More recently, those networks have begun to roll out 2.5G technology, greatly increasing their data capability and opening the way for new and exciting services. Many networks around the world are well into the rollout of their 3G systems, and recent data shows that subscriber numbers are beginning to increase.

So what is next for networks and subscribers? Is 3G as far as the technology will take us? Already there are modifications and enhancements to the 3G standards that further increase the data capability. Tens of megabits per second is now a real possibility across the cellular networks. 3G technologies will, of course, continue to evolve, improving data rates and service quality. However, there are technological developments taking place elsewhere that may rival the services the cellular networks are providing.

HSDPA

High Speed Downlink Packet Access (HSDPA) is an evolution of existing UMTS radio systems and is viewed as the natural direction in which operators and handset manufacturers will go. As the name implies, this upgrade to existing 3G networks will increase the data rate from the network to the mobiles (i.e., downlink). Increasing the data rate allows network operators to offer more complex content or enables traveling business people to connect their laptops to the Internet via the HSDPA device. Ultimately, higher speed uplink connections will also be supported using HSUPA (High Speed Uplink Packet Access). Data rates for HSDPA are theoretically 10 Mbps, but are more realistically around 1 to 2 Mbps.

WiFi

Intel's Centrino chip set for laptop computers is one of the factors ensuring the success of the IEEE-developed 802.11 wireless technology, more commonly known as WiFi. This technology allows suitably equipped computing equipment to be connected to the Internet, or home and business networks, without wires at multi-megabit data rates. WiFi is a short-range radio system — 20 to 30 meters from the access point in some cases.

WiFi "hotspots" can be found in places such as airports, coffee shops, and hotels. Paying a fee, normally with a credit card, allows users to access the Internet while away from the office. Boingo, BT OpenZone, and The Cloud are examples of companies serving the WiFi market.

WiFi is important to the cellular industry for a couple of reasons. WiFi is a high-speed wireless data connection, which at the present time offers data rates many times that of 3G networks. Thus, it would seem that WiFi could compete directly with 3G. WiFi, however, only supports very short distances. Some operators are enhancing their cellular service by deploying a network of hotspots and allowing subscribers to access the Internet with their laptops and have the access fee charged to their mobile phone accounts. Roaming agreements can be set up between existing hotspot operators and GSM operators. Companies such as MACH, which already brokers normal GSM roaming (Figure 1.56), are helping networks accomplish this.

There is equipment on the market today supporting both GSM and WiFi. A data card for a laptop PC is the natural product to combine GSM/WiFi and possibly 3G, giving the user the maximum possible flexibility (see Figures 1.57, 1.58, and 1.59). Mobile PDA-style devices and some mobile phones also support the WiFi standard. This vastly opens up the market for user applications, particularly if the smart phone devices can support Voice-over-IP (VoIP) applications.

WiMAX

Worldwide Interoperability for Microwave Access (WiMAX) is a solution to the problem of the *last mile* in delivering DSL or cable modem services. Delivering broadband services to premises is always the most difficult part. True broadband services require specific cabling from the network to the customer; this could be

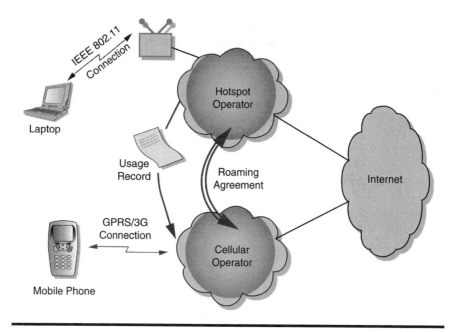

FIGURE 1.56 WiFi hotspot roaming.

FIGURE 1.57 Mobile data card from Orange.

FIGURE 1.58 Nokia 9500 Communicator with WiFi.

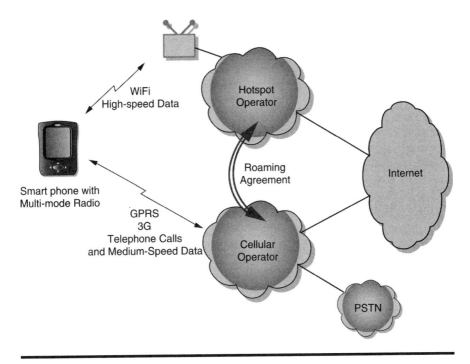

FIGURE 1.59 WiFi-equipped mobile device.

coaxial cable, twisted pair such as cat5, or even fiber in some cases. Installing these connections in a residential area, for example, is very expensive. The solution would seem to be wireless. Many proprietary schemes support this wireless local loop (WLL) or fixed radio access (FRA), some even successfully. A few of these schemes use 802.11b or 802.11g standards to perform this. These existing standards however, were designed for use over short ranges and mostly indoors, so they cannot solve all the problems associated with long-range fixed radio access.

WiMAX is the IEEE solution to this problem. It builds on the success of its previous standards by developing new radio techniques to offer long-range, high data rate services to residential, business, and eventually to mobile users.

WiMAX systems (Figure 1.60) have been deployed primarily in the fixed area, offering wireless networking and high-speed data services at sporting events and bringing Internet connectivity to rural areas. In the future, WiMAX could be deployed in a fashion similar to a cellular radio network and even offer wireless service to mobile device users. WiMAX is capable of supporting up to 70 Mbps under the right conditions, thus rivaling 3G technologies.

WiMAX has in the past been viewed as a threat to the mobile operators, but now it is viewed as more of a complementary technology that could be deployed in certain areas along side traditional 3G services.

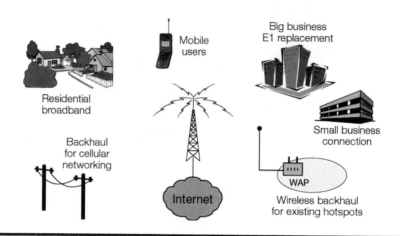

FIGURE 1.60 WiMAX applications.

While the technology is specified by the IEEE, organizations such as the WiMAX Forum exist to ensure maximum interoperability between vendors of WiMAX equipment. The WiMAX Forum has more than 100 members, including Intel, Samsung, British Telecom and Airspan Networks.

Voice-over-IP (VoIP)

In both fixed and mobile networks, voice services have been supported in the same fashion for many decades. However, that is certainly set to change as the Internet grows and becomes a more capable network, able to carry data such as voice in real-time. The collective term for all the technologies that allow voice to transmit over networks such as the Internet is Voice-over-Internet Protocol (VoIP) (Figure 1.61). Many companies have converted their internal voice systems to enable their internal office network data and voice communication to be sent over a common network. This is possible because the quality of the connection can be maintained from end to end. It is much more difficult to maintain quality on the public Internet, but technology is developing very quickly in this area as fixed and mobile operators begin the evolution to next-generation systems.

Domestic broadband users are already taking advantage of the technology by downloading and using freely available software VoIP clients such as Skype, Vonage, MSN Messenger, Sipgate, and SIPphone.com to communicate with other VoIP users at no charge (Figure 1.62). Usually there is no charge to talk to another user within the same systems; that is, a Skype to Skype call is free, even if the users are in different countries. These networks will also offer calls to normal telephone lines for very competitive rates.

An interesting development in this area is mobile smart phones that support a WiFi connection and can also run a VoIP client, thus enabling the user to make voice calls, bypassing the cellular network altogether.

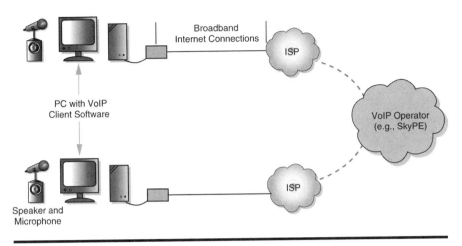

FIGURE 1.61 Voice over the Internet.

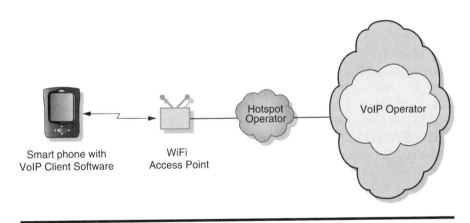

FIGURE 1.62 VoIP mobile telephone call.

BT Bluephone

British Telecom (BT) has launched a fixed–mobile convergent product called the Bluephone. The device will use a Bluetooth radio connection while the user is at home, connecting to the fixed network via a base station rather like a cordless phone. When the user moves away from the house, the mobile will find and use the GSM cellular network. An MVNO agreement between British Telecom and Vodafone allows seamless roaming from the fixed network to the mobile network.

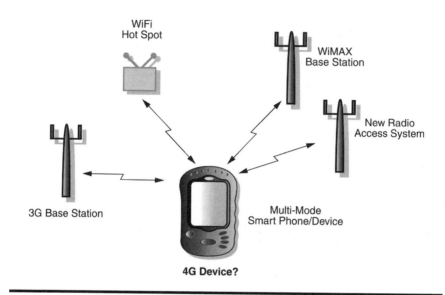

FIGURE 1.63 A 4G mobile device

Already on the market are devices that support multiple radio access systems, cellular, WiFi, 3G, and soon WiMAX. The future mobile communications market is going to become an even more complex place as the operators decide how they want to deliver services to the user.

A fourth-generation (4G) network (Figure 1.63) will be a combination of mobile and portable computing devices that take advantage of many different radio access methods to best suit the application, the user, and the location.

Review Questions

Q1. What evidence is there for increased demand for mobile data services?
   a. More picture messaging
   b. Increase in SMS usage
   c. Reduced number of voice calls
   d. Increasing sales of smart mobile phones

Q2. Which of the following is not considered an influential factor in the evolution of mobile networks?
   a. Rapid growth in Internet traffic
   b. Data connections overtaking voice connections
   c. 100 Percent penetration in some markets
   d. Availability of broadband

Q3. 2G networks support _____.
   a. High-speed packet access connections
   b. Low-speed dial-up connection
   c. ISDN compatible circuit switched connections
   d. Occasional packet access

Q4. UMTS networks will not support _____.
   a. Packet access up to 2 Mbps
   b. QoS negotiation
   c. Coexistence with 2G and 2.5G system
   d. Point-to-point microwave access

Q5. What is likely to be the "killer" application for 3G mobile networks?
   a. Mobile e-mail
   b. WAP Web browsing
   c. Video telephony
   d. Access to any service, any time, any where

1.4 Standards and Regulations

1.4.1 Standards and Standards Bodies

In telecommunications, a *standard* is a document or specification that describes the process by which information is transmitted or received. The standard generally refers to the interface between two entities (e.g., the mobile phone and the base station). A standard also describes the services in a telecommunications network and the manner in which services should be supported.

Standard methods or products normally emerge as a result of commercial success in the marketplace; a classic example always mentioned when discussing standards is the VHS and Betamax videocassette formats. VHS became the industry standard in this case as a result of market success. Letting the market decide on the standards by this process may not always be the best and most efficient way, because there must always be a loser. In the past, the telecom industry has used this method, but in these days of global roaming, there are many more benefits to derive from the industry making agreements about the methods and process of telecommunication.

All of the bodies described below are committees consisting of professionals from manufacturers, network operators, service providers and others, where the technical and non-technical issues of solving telecommunications problems are discussed. In some cases, several solutions may be presented and a vote may take place to decide on the best solution. In this way, all parties can agree and the winning solution will be written up as a technical specification. GSM is a very good example of the success of this method of determining telecom standards.

Benefits of a Standard

For equipment vendors:

- Standards-based, common platforms enable faster innovation and the addition of new components and services.
- Concentrate on specialization (i.e., base stations or customer premises equipment (CPE)); no longer need to create an entire end-to-end solution as in proprietary model.

For consumers:

- More players in the market translates into more choices for receiving broadband access services.
- Quick "trickle-down" effect of cost savings to consumers, translating into lower monthly rates.

For service providers:

- Common platform drives down costs, fosters healthy competition and encourages innovation.
- Enables a relatively low initial CAPEX investment and incremental expenditures that reflect growth.
- No more commitments to a single vendor, a typical by-product of the proprietary technology model.
- Wireless systems significantly reduce operator investment risk.

For component makers:

- Standardization creates a volume opportunity for chip set vendors and silicon suppliers.

The telecom industry has always benefited from the process of standardization. It was in 1865 that the original members of the ITU signed the first international telegraph convention. Following the patent of the telephone and the development of wireless services, the ITU extended its influence with legislation making recommendations on technology and international services.

The ITU has been around influencing telecommunications for nearly 150 years. However, other bodies with a more regional focus have also been established to determine appropriate telecom standards for that region.

The following information outlines only some of the international and regional standards bodies.

International Telecommunication Union (ITU)

The International Telecommunication Union (ITU), headquartered in Geneva, Switzerland, is the United Nations' specialized agency for telecommunications. The main work of the ITU is divided among three sectors, namely:

1. The Radio Communication Sector (ITU-R)
2. The Development Sector (ITU-D)
3. The Telecommunication Standardization Sector (ITU-T)

The ITU-R coordinates matters dealing with radio communication services, radio-frequency spectrum management, and wireless services. The ITU-D focuses on technical assistance to developing countries and countries with economies in transition to allow the development of telecommunications networks and services. The ITU-T ensures the efficient and on-time production of high-quality standards covering all fields of telecommunications on a worldwide basis, as well as defining tariff and accounting principles for international telecommunications services.

European Telecommunications Standards Institute (ETSI)

The European Telecommunications Standards Institute (ETSI) is an independent, non-profit organization, whose mission is to produce telecommunications standards for today and for the future. Based in Sophia Antipolis (France), the ETSI is officially responsible for standardization of Information and Communication Technologies (ICT) within Europe. These technologies include telecommunications, broadcasting, and related areas such as intelligent transportation and medical electronics.

The ETSI unites 688 members from 55 countries inside and outside Europe, including manufacturers, network operators, administrations, service providers, research bodies, and users — in fact, all the key players in the ICT arena.

The ETSI plays a major role in developing a wide range of standards and other technical documentation as Europe's contribution to worldwide ICT standardization. This activity is supplemented by interoperability testing services and other specialisms. The ETSI's prime objective is to support global harmonization by providing a forum in which all the key players can contribute actively. ETSI is officially recognized by the European Commission and the European Free Trade Association (EFTA) secretariat.

Third Generation Partnership Project (3GPP)

The 3GPP (Third Generation Partnership Project) is a collaborative agreement established in December 1998. The collaboration agreement brings together a number of telecommunications standards bodies known as "organizational partners." The current organizational partners include the ARIB, CCSA, ETSI, ATIS, TTA, and TTC.

The establishment of 3GPP was formalized in December 1998 by signing "The 3rd Generation Partnership Project Agreement."

The original scope of 3GPP was to produce globally applicable technical specifications and technical reports for a 3G mobile system based on evolved GSM core networks and the radio access technologies they support (i.e., Universal Terrestrial Radio Access (UTRA), both frequency division duplex (FDD) and time division duplex (TDD) modes). The scope was subsequently amended to include the maintenance and development of the Global System for Mobile communication (GSM) Technical Specifications and Technical Reports, including evolved radio access technologies (e.g., General Packet Radio Service (GPRS) and Enhanced Data rates for GSM Evolution (EDGE)).

Third Generation Partnership Project 2 (3GPP2)

The Third Generation Partnership Project 2 (3GPP2) is a collaborative third-generation (3G) telecommunications specifications-setting project comprising North American and Asian interests developing global specifications for ANSI/TIA/EIA-41 Cellular Radio Telecommunication Intersystem Operations network evolution to 3G and global specifications for the radio transmission technologies (RTT) supported by ANSI/TIA/EIA-41.

3GPP2 was born out of the International Telecommunication Union's (ITU) International Mobile Telecommunications "IMT-2000" initiative, covering high speed, broadband, and Internet Protocol (IP)-based mobile systems featuring network-to-network interconnection, feature/service transparency, global roaming, and seamless services independent of location. IMT-2000 is intended to bring high-quality mobile multimedia telecommunications to a worldwide mass market by achieving the goals of increasing the speed and ease of wireless communications, responding to the problems faced by the increased demand to pass data via telecommunications, and providing "anytime, anywhere" services.

UMTS Forum

The UMTS Forum was set up by a number of telecommunications operators, manufacturers, national governments, and other organizations. Its aim is to define a common strategy and policy for the development and implementation of the future Universal Mobile Telecommunications System, combining personal communications with multimedia services and applications built on existing fixed and mobile infrastructures. It seeks to contribute to the development of a European policy on mobile and personal communications, and provides advice and recommendations to the European Commission, European Radio Communications Office, and European Telecommunications Office.

The GSM Association

The GSM Association (GMSA), founded in 1987, has played a pivotal role in the development of the GSM platform and of the global wireless industry.

> Since its introduction our members and staff have created the landscape of success for global mobile communications via GSM. Ours is a story of international cooperation and collaboration, between people, companies and governments to create the world's first global wireless network.

The GSMA is the global trade association that exists to promote, protect, and enhance the interests of GSM mobile operators throughout the world. At the end of 2004, it consisted of 660 second- and third-generation mobile operators and more than 150 manufacturers and suppliers. The Association's members provide mobile services to nearly 1.3 billion customers across more than 200 countries and territories around the world. The GSMA aims to accelerate the implementation of collectively identified, commercially prioritized operator requirements and to provide leadership in representing the global GSM mobile operator community with one voice on a wide variety of issues nationally, regionally, and globally.

The European Conference of Postal and Telecommunications Administration (CEPT)

CEPT, the European Conference of Postal and Telecommunications Administration, was established in 1959 by 19 countries, which expanded to 26 countries during its first ten years. Original members were the incumbent, monopoly-holding postal and telecommunications administrations. CEPT activities included cooperation on commercial, operational, regulatory, and technical standardization issues.

In 1988, CEPT decided to create the ETSI, European Telecommunications Standards Institute, into which all its telecommunications standardization activities were transferred.

In 1992, the postal and telecommunications operators created their own organizations, Post Europe and ETNO, respectively. In conjunction with the European policy of separating postal and telecommunications operations from policy-making and regulatory functions, CEPT thus became a body of policy makers and regulators. At the same time, central and eastern European countries became eligible for membership in CEPT. With its 45 members, CEPT now covers almost the entire geographical area of Europe.

The role and purpose of CEPT was redefined at its plenary assembly on 5–6 September 1995 in Weimar as follows:

CEPT offers its members the opportunity to:

- Establish a European forum for discussions on sovereign and regulatory issues in the field of post and telecommunications issues.
- Provide mutual assistance among members with regard to the settlement of sovereign and regulatory issues.
- Exert an influence on the goals and priorities in the field of European Post and Telecommunications through common positions.
- Shape, in the field of European posts and telecoms, those areas coming under its responsibilities.
- Carry out its activities at a pan-European level.
- Strengthen and foster more intensively cooperation with Eastern and Central European countries.
- Promote and facilitate relations between European regulators (e.g., through personal contacts)
- Influence, through common positions, developments within the ITU and UPU in accordance with European goals.
- Respond to new circumstances in a nonbureaucratic and cost-effective way and carry out its activities in the time allocated.
- Settle common problems at committee level, through close collaboration between its committees.

This gives its activities more binding force, if required, than in the past, thus creating a single Europe on posts and telecommunications sectors.

The Telecommunications Industry Association (TIA)

The TIA, Telecommunications Industry Association, is the leading U.S. non-profit trade association, and represents providers of communications and information technology products and services for the global marketplace through its core competencies in standards development, domestic and international advocacy, as well as market development and trade promotion programs. The association facilitates the convergence of new communications networks while working for a competitive and innovative market environment. The TIA strives to further members' business opportunities, economic growth and the betterment of humanity through improved communications.

The American National Standards Institute (ANSI)

ANSI, the American National Standards Institute, is a private, non-profit organization that administers and coordinates the U.S. voluntary standardization and conformity assessment system. ANSI's mission is to enhance both the global competitiveness of U.S. business and the U.S. quality of life by promoting and facilitating voluntary consensus standards and conformity assessment systems, and safeguarding their integrity.

The CDMA Development Group

The CDMA Development Group (CDG), founded in December 1993, is an international consortium of companies that have joined together to lead the adoption and evolution of 3G CDMA wireless systems around the world.

The CDG is comprised of CDMA service providers and manufacturers, application developers and content providers. By working together, the members help to ensure interoperability among systems, while expediting the availability of 3G CDMA technology to consumers.

The CDG mission is to lead the rapid evolution and deployment of 3G CDMA-based systems, based on open standards and encompassing all core architectures, to meet the needs of markets around the world.

The CDG and its members work together to:

- Accelerate the definition of requirements for new CDMA features, services and applications.
- Promote industry and public awareness of CDMA capabilities and developments through marketing and public relations activities.
- Foster collaboration and the development of consensus among carriers on critical issues to provide direction and leadership for the industry.
- Define the evolution path for current and next-generation CDMA systems.
- Establish strategic relationships with government ministries, regulatory bodies, and worldwide standards and industry organizations to promote cooperation and consensus on issues facing the CDMA community.
- Serve as the worldwide resource for CDMA-related information.
- Minimize the time-to-market of new CDMA-based products and services.
- Enable global compatibility and interoperability among CDMA systems worldwide.
- Create global economies of scale to make CDMA the preferred choice of operators and end users.

The Association of Radio Industries and Businesses (ARIB)

The ARIB, the Association of Radio Industries and Businesses, was chartered by the Minister of Posts and Telecommunications as a public service corporation on May 15, 1995. Its activities include those previously performed by the Research and Development Centre for Radio Systems (RCR) and the Broadcasting Technology Association (BTA).

The ARIB was established in response to several trends, such as the growing internationalization of telecommunications, the convergence of telecommunications and broadcasting, and the need for promotion of radio-related industries. The ARIB's goal is to advance rapidly the use of radio technology for the benefit of society. This is done by integrating knowledge and experience in various fields of radio use, such as broadcasting and telecommunications, research and development in radio technology, and serving as a standards development organization for radio technology.

Telecommunication Technology Committee (TTC)

The TTC, the Telecommunication Technology Committee, contributes to standardization in the field of telecommunications by establishing protocols and standards for telecommunications networks and terminal equipment, etc., as well as to disseminate those standards. The committee will:

- Develop protocols and standards for telecommunications networks
- Conduct studies and research on protocols and standards for telecommunications networks
- Disseminate protocols and standards for telecommunications networks
- Engage in activities accompanied by the above items
- Engage in other business activities necessary to achieve the purpose of the committee

Telecommunications Technology Association (TTA)

The purpose of TTA, the Telecommunications Technology Association, is to contribute to the advancement of technology and the promotion of information and telecommunications services and industry, as well as the development of national economy, by effectively establishing and providing technical standards that reflect the latest domestic and international technological advances needed for the planning, design, and operation of global end-to-end telecommunications and related information services, in close collaboration with companies, organizations, and groups concerned with information and telecommunications such as network operators, service providers, equipment manufacturers, academia, R&D institutes, etc.

The Wi-Fi Alliance

The Wi-Fi Alliance is a global, non-profit industry association of more than 200 member companies devoted to promoting the growth of wireless local area networks (WLANs). With the aim of enhancing the user experience for mobile wireless devices, Wi-Fi Alliance testing and certification programs ensure the interoperability of WLAN products based on the IEEE 802.11 specification.

Since the introduction of the Wi-Fi Alliance certification program in March 2000, more than 2000 products have been designated as Wi-Fi CERTIFIED™, encouraging the expanded use of Wi-Fi products and services across the consumer and enterprise markets.

The WiMAX Forum™

The WiMAX Forum™ is working to facilitate the deployment of broadband wireless networks based on the IEEE 802.16 standard by helping to ensure the compatibility and interoperability of broadband wireless access equipment. The organization is a non-profit association formed in June of 2001 by equipment and component suppliers to promote the adoption of IEEE 802.16-compliant equipment by operators of broadband wireless access systems.

The WiMAX Forum is comprised of industry leaders who are committed to the open interoperability of all products used for broadband wireless access:

- Support the IEEE 802.16 standard:
  - Propose and promote access profiles for their IEEE 802.16 standard.
- Certify interoperability levels both in network and the cell.
- Achieve global acceptance.
- Promote use of broadband wireless access overall.

Review Questions

Q1. Which of the following bodies provides global coordination on matters of radio spectrum, telecommunications, and technical development in developing countries?
   a. ETSI
   b. ITU
   c. ARIB
   d. TTC

Q2. Which of the following technologies does ETSI no longer develop?
   a. TETRA
   b. DECT
   c. UMTS
   d. TIPHON

Q3. Which of these organizations are not part of the 3GPP ?
   a. ETSI
   b. TTA
   c. ARIB
   d. TIA

Q4. Which of the following organizations was established to promote, protect, and enhance the interests of GSM mobile operators?
   a. UMTS Forum
   b. The GSM Association

c. CEPT
d. WiFi Alliance

Q5. Promoting growth, enhancing user experience, and developing a certification process for WLAN products is the main focus of the _____.
a. WiMAX Forum
b. WiFi Alliance
c. ETSI
d. IEEE

1.4.2 Telecommunications Regulations

The Need for Telecom Regulation

Toward the end of the 20th century, the telecommunications markets underwent enormous change. Many countries had nationally owned and operated networks that were privatized, causing an increase in pro-competitive and deregulatory telecommunications policies. Market-based approaches to the supply of telecommunications services have since been adopted in many places.

This liberalization of the telecom market is motivated by a number of factors:

- Evidence that liberalized markets grow and innovate faster, providing customers with better services
- Attraction of private-sector investment to upgrade existing networks and introduce new services
- Growth of the Internet, leading to many new service providers
- Growth of mobile and other wireless services
- International trade development, and introduction of international telecom service providers

The liberalization of the markets described above has, in turn, led to an increase in the role of regulation and regulators. It seems a little ironic that the increase in market-based service provision and increased competition should increase the amount of regulation, particularly when the operators of these services largely prefer less regulation. However, the role of the regulator is vital when transforming monopolistic telecom markets into competitive ones. Competitive markets are unlikely to emerge without them. Among the key regulatory objectives are:

- Promote universal access to basic telecommunications services.
- Foster competitive markets to promote:
  - Efficient supply of telecommunications service
  - Good quality of service

- Advanced services
- Efficient prices
- In the absence of competitive markets or where markets fail, prevent abuse of market power (e.g., excessive pricing or anti-competitive behavior).
- Create a favorable environment to promote investment and expand telecom networks.
- Promote public confidence in telecom markets through transparent regulatory and licensing processes.
- Protect consumer rights, including privacy.
- Promote increased telecom connectivity for all users through efficient interconnection arrangements.
- Optimize the use of scarce resources (e.g., the radio spectrum, numbers and rights of way).

It could be said that these objectives are also the key requirements of a competitive market-based environment; however, as already stated, this would be unlikely to exist with out some degree of control from independent authorities.

Benefits of Regulation to the Consumer

The process of regulation may yield benefits to the consumer. The definition of "consumer" in this context means the end residential user, small to medium enterprises (SMEs), and large businesses. The benefits experienced by these consumers can be divided into the following areas: choice, price, information, and low switching barriers.

- *Choice.* As the range of services increases in the market, so does the diversity of consumers — from sophisticated consumers such as large businesses that have always been complex users of telecom services, to residential users. Even residential users today are making increasingly sophisticated demands on the suppliers of their services, including flat rate voice, supplementary services, and broadband access. Innovation is a key part of consumer choice, and early adoption or access to new products is seen as an important part of the market. Indeed, competition between operators is vital in promoting innovation and thus bringing new services to the customers. With so many telecommunications products available, consumers may be required to purchase a number of diverse services, mobile phone, fixed phone, data services, TV, etc. It is therefore important that the procurement process remains as simple as possible. Consolidated or bundled services packages are an attractive way of simplifying this; residential users and SMEs tend to favor this form of market.
- *Price.* Low prices are often used as an indication of successful regulation, and it is an important element of the telecom market. Operators and regulators

should not focus too much on lowering prices, however, because other features of the market may suffer as a result (e.g., quality of the service). The cost of delivering a service should be profitable for the provider and sustainable. Allowing the normal forces of competition to govern the market would create a well-functioning wholesale market and lead to sustainable prices for residential and commercial users.

- *Information.* All consumers need access to clear and unambiguous information concerning telecommunications services available. In all cases, it should be possible for the consumer to compare services and prices for that service between telecom service providers. Residential users feel that there is sufficient information available but at times it may be difficult to interpret or make clear comparisons between service providers; this problem may be magnified toward large business users where they may have to make choices between many different services and at a much greater cost.
- *Low switching barriers.* Users must be able to change their providers with a minimum of difficulty and cost in a well-managed competitive telecom market. Innovations such as number portability have made this an easier and more desirable prospect for many users. The fixed line market in the residential and SME markets are least likely to be moved to a different supplier because users perceive moving as a costly and time-consuming exercise.

Objectives of National Regulatory Authorities

Each country around the world will have its own regulatory authority to deal with issues related to telecommunications service. These bodies are often part of the governmental structure of the country or have been spun out of the government to become an independent body. Each country, of course, will have its own regional telecommunications objective, but it can be said that general objectives are those below:

- *Promote competition.* National regulators will promote competition between telecom operators, ensuring that users (including disabled users) derive maximum benefit in terms of choice, price, and quality; they also ensure that there is no distortion or restriction of competition. Competition should also encourage efficient investment in infrastructure and promote innovation, as well as efficient use and management of radio frequencies and numbering resources.
- *Internal market development.* The regulator can develop the internal market by removing obstacles to the provision of the telecommunications service, by encouraging the establishment of international networks, the interoperability of pan-national services and end-to-end connectivity. Cooperation of the regulators with each other and, for example, the European Commission will also ensure consistent regulatory practice. Promotion of the market is also assisted by the elimination of discrimination between operators where similar circumstances occur.

- *Promote the interests of citizens.* The national regulatory authorities ensure that all citizens have access to a *universal service*. When dealing with suppliers, high levels of protection should be available to the citizens via a simple and inexpensive dispute resolution procedure carried out by an independent body. Private and personal data should also be protected. Information relating to tariffs and conditions of use should be promoted in a clear and transparent manner. The needs of specific social groups should also be addressed. The regulator will also ensure that the integrity and security of the public networks are maintained.
- *Management of radio frequencies.* The allocation and assignment of radio frequencies should be based on objective, transparent, nondiscriminatory, and proportionate criteria; this should ensure the effective management of radio frequencies for the provision of radio-based telecommunications services. Harmonization of use of radio frequencies across the community ensures the efficient and effective use of the radio spectrum.
- *Control of numbering and naming.* National regulators should ensure that the management and assignment of national numbering plans provide adequate numbers and number ranges for all publicly available telecommunications services. The national regulator should establish objective, transparent, and nondiscriminatory procedures for numbering resources.
- *Rights of way.* When an operator or competent authority considers an application to install facilities on public or private property, they should act on the basis of transparent and publicly available procedures applied without discrimination and without delay.
- *Co-location and facility sharing.* The national regulators can promote or impose co-location or site sharing, depending on the nature of the installation. The imposition of co-location and facility sharing can occur where there is a need to protect the environment, public health or public security, or to meet the need of country and town planning objectives.

Case Study of a National Regulator: Ofcom

Ofcom, the Office of Communications, is the regulator for the U.K. communications industry, with responsibilities across television, radio, telecommunications, and wireless communications services.

Ofcom's statutory duties under the Communications Act 2003 include:

> The principle duty of Ofcom is to further the interests of citizens in relation to communications matters; and to further the interests of consumers in relevant markets, where appropriate by promoting competition.

Ofcom's specific duties fall into six areas:

1. Ensuring the optimal use of the electromagnetic spectrum
2. Ensuring that a wide range of electronic communications services — including high-speed data services — is available throughout the United Kingdom
3. Ensuring there is a wide range of TV and radio services of high quality and wide appeal
4. Maintaining plurality in the provision of broadcasting
5. Applying adequate protection for audiences against offensive or harmful material
6. Applying adequate protection for audiences against unfairness or the infringement on privacy

Ofcom's Regulatory Principles

- Ofcom will regulate with a clearly articulated and publicly reviewed annual plan, with stated policy objectives.
- Ofcom will intervene where there is a specific statutory duty to work toward a public policy goal that markets alone cannot achieve.
- Ofcom will operate with a bias against intervention, but with a willingness to intervene firmly, promptly, and effectively where required.
- Ofcom will strive to ensure its interventions will be evidence based, proportionate, consistent, accountable, and transparent in both deliberation and outcome.
- Ofcom will always seek the least intrusive regulatory mechanisms to achieve its policy objectives.
- Ofcom will research markets constantly and will aim to remain at the forefront of technological understanding.
- Ofcom will consult widely with all relevant stakeholders and assess the impact of regulatory action before imposing regulation upon a market.

Regulators of the World

Appendix B lists the Telecom Regulators from 84 countries around the world and the Web address for the regulator's site.

Data Protection and User Privacy

In the past few years, technological development has increased the number of ways in which we can communicate electronically. This leads to issues regarding the users' rights to privacy and the manner in which providers of the electronic communications services must handle the users' data.

Within the European Union, directives have been laid out describing the obligation of individuals and companies with respect to users' data and privacy. These are summarized below.

- *Security.* The operators of public communications networks should take appropriate steps to ensure the security of their networks and inform the users of the network should any breach of security occur.
- *Confidentiality of communications.* The regulators should ensure through national legislation the confidentiality of users' communications by prohibiting the tapping, listening, storage, or other kinds of surveillance by persons other than the users, without the consent of the users except where legally authorized to do so (i.e., lawful interception).
- *Traffic data.* Traffic and data related to the users must be erased or made anonymous when it is no longer required to support the communication service; exceptions include data required to support billing services. Where the user gives consent, the user data or traffic may be used for marketing purposes only for the duration of the marketing process. The provider of the service should inform the user of the types of data that are processed and indicate the duration of that process. Access to stored information should only be made available to persons acting under the authority of the provider of the service.
- *Itemized billing.* Subscribers have the right to a non-itemized bill.
- *Presentation and restriction of calling and connected line identification.* Where the calling user identification service is available, the provider of the service must also provide a simple and free method of preventing the identity from being presented. Users should also be able to terminate calls where the identity of the users is presented before the call is completed. There are exceptions in the case of a subscriber or user tracing malicious or nuisance calls and for calls made to emergency services.
- *Location data.* Processing of location data by the provider of a telecommunications service can only be carried out with the consent of the user or subscriber. The service provider must inform the user prior to abating consent of the type of location data that will be transmitted to a third party. The users should also be able to withdraw temporarily or permanently their consent for processing the location data.
- *Automatic call forwarding.* Users should have a simple and free-of-charge means to prevent the automatic forwarding of their calls to a user's terminal.
- *Directories of subscribers.* Subscribers should be informed before they are included in a printed or electronic directory and of any further usage possibilities based on search functions embedded in electronic directories. Subscribers should also be able to determine the presence and relevance of their details in directories and to verify, correct, or withdraw the information.
- *Unsolicited communications.* The use of automated calling systems without human intervention, fax machines, or electronic mail for the purposes of direct marketing is only allowed where the users have given prior consent. However, if a person has obtained the details of a subscriber in the context of a sale of product or service, that person can use that information for direct marketing of similar products and services, providing the user is given a clear opportunity

to object. The sending of electronic mail for the purposes of direct marketing, disguising, or concealing the identity of the sender on whose behalf the communication is made, or without a valid address to which the recipient can send a request that such communications should cease, is prohibited.

Review Questions

Q1. Which one of the following is not a key objective of a regulator?
   a. Foster a competitive market
   b. Cheaper services for everyone
   c. Protect consumer rights and privacy
   d. Optimize the use of scarce resources

Q2. One of the benefits that consumers can expect from a carefully regulated market is _____.
   a. Faster data services
   b. More choice of services and service providers
   c. Cheaper telephones
   d. More efficient customer services

Q3. Data protection and user privacy, according to the European Union, does not include which of the following _____.
   a. Network security
   b. Privacy of user data
   c. Mobile e-mail
   d. WAP Web browsing
   e. Video telephony
   f. Access to any service, any time, anywhere

Chapter 2

Technology Principles
The Basics of Telecommunications

At the end of this chapter, you should be able to:

- State the five basic requirements of every telecommunications network.
- Explain the concepts and requirements in more detail under the headings of Services and Applications; Transmission; Switching; Signaling and Control; Billing and Support Systems (also explain the basics of networking).
- Categorize major telecommunications technologies and techniques in terms of the major headings listed above.
- Identify basic transmission techniques, explaining how information is transferred through a telecommunications system end-to-end, including:
  - Transmission media (metallic, radio, and optical systems)
  - Transmission systems
  - Transmission techniques (including modulation, multiplexing)
- Identify the major requirements, limitations, advantages, and disadvantages of using radio within telecommunications systems — and also identify how it can be used.
- Diagrammatically represent the architecture of a cellular "radio" system, showing where the different elements and transmission techniques are used — including metallic, optical, and radio transmission, as well as switching systems, signaling, and service platforms.
- Within modern cellular systems, describe how the radio interface is divided to provide the required channels and control for the users — difference between 2G, 2.5G, and 3G systems, as well as CDMA-based systems, and non-CDMA systems (i.e., FDMA and TDMA).
- Explain in overview how the overall cellular system operates to provide connections, mobility support, and service control.

2.1 Introduction

2.1.1 The Basics of Telecommunications

Telecommunications: A Definition

The term "telecommunication" is derived from the Greek *tele,* meaning distant, and *communicate,* meaning sharing. From the beginning of time, the need to communicate has been part of man's inherent being. The human race has, throughout the years, communicated using different techniques, dependant on the circumstances and available technology. Early forms of communication included:

- Smoke signals of the early American Indians
- Drums of African tribes
- Semaphore using flags
- Papyrus and paper used to record communication for later use

As time passed, technology advanced and it is now expected that telecommunications is both reliable and efficient. Nothing is more annoying than having to repeat your message to the recipient, whether by phone, facsimile, or e-mail. Other areas that are now paramount to the customer are the availability and quality of additional services that may be requested, and all this has to be at a price that keeps both the customer and the seller happy.

Today, telecommunications can be defined as:

> The reliable and efficient movement of information between two or more points for the purpose of providing services of the required quality and availability, at a price which reflects the needs of the customer and the business.

The Requirements

What is required from a telecommunications system? Before a usable telecommunications system can be used, several key issues should be addressed:

- *Services and applications.* These are the services that the customer has requested, and now the customer has an ever-increasing choice of services. However, as this choice has increased, so has the sophistication of the technology behind them.
- *Transmission.* A suitable medium must be provided for the transmission of the information.
- *Switching.* It is not feasible or cost effective for a transmission medium to run with no breaks from point A to point B. It would require that every device be connected to every other device, and the practicality and economics of this

would severely limit the availability of services to the customer. Thus, there is the requirement that the transmission media (often copper wires) must be connected in some way to create a temporary chain leading from one device to another. Switches provide this function.
- *Signaling.* To set up any call through the network, suitable control or signaling must be used to inform the various units within the network of the requirements of that particular call.
- *Billing.* To have access to this technology, an appropriate price is charged to the customer. This cost depends on the sophistication and value of the ever-increasing services and applications that are available to and accessed by the customer. The system must include appropriate measurements and records to enable accurate billing, and to protect the revenue for the network operator.

Services and Applications

Although customers may not be familiar with (and may not care about) the technology or network on which their services are made available, they are certainly familiar with the services and applications themselves. In fact, they pay only for the services. Hence, services provided by a network operator must be of value to the customer, cost effective to implement, and ultimately attract customers. Available services (Figure 2.1) include:

- Telephony (i.e., a simple phone call)
- Facsimile (i.e., the transmission of written information)
- Video across the network
- Data transfer from and to computers and the World Wide Web
- E-mail across the Internet
- Text to a mobile subscriber (Short Message Service)
- Caller line identification, etc.

Transmission

The transmission medium can take many different forms, such as (1) copper wire, which has been used since the early days of communication; (2) optical fiber, which is a relatively new medium and is increasingly being used; and (3) radio, which also has been used for many years. Each of these media can be bi-directional, that is, allowing information to pass in both directions (Figure 2.2).

Copper may be limited by bandwidth, which limits the amount of information that can be transmitted in a given time, compared to optical fiber, but its cost is low. In contrast, optical fiber is relatively expensive but can carry more data in a given time. Both copper wire and optical fiber cable are more reliable than radio, but radio has many other advantages.

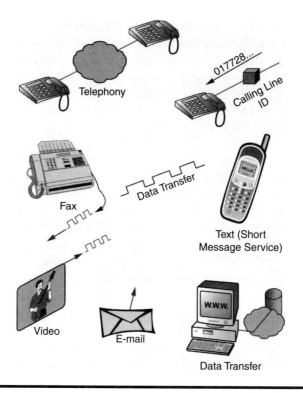

FIGURE 2.1 Services and application examples.

Transmission "pipe"

- Often bi-directional
- Could be copper, radio, or fiber optic
- Capacity, cost, and reliability are important factors
- Informaion can be represented in many different forms

FIGURE 2.2 Transmission requirements.

Radio transmission is a very versatile medium, ideally suited to a mobile environment, but its bandwidth is extremely limited and therefore its total capacity is limited. It can also be adversely affected by atmospheric conditions.

Speed of data transfer, error rates, and other key characteristics determine which transmission medium will be used in particular circumstances. Information can be represented in many different forms. To ensure compatibility across systems, standard *transmission systems and techniques* have been specified in many instances.

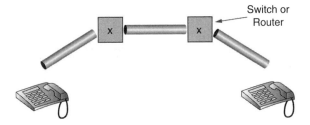

- Allows transmission "pipes" to interconnect only when required
- Points worldwide can now interconnect automatically
- Switching can take different forms (with varying costs)

FIGURE 2.3 Switching.

Switching

To enable the placement of a call from the originator through the network and on to its final destination, that call must pass through several switches or routers (Figure 2.3). Switches or routers allow great flexibility in connecting transmission resources. This gives rise to endless possibilities for end-to-end connections between users. Users may be located nearby or be very remote, but the switches or routers within the network will ensure that the information passes over the required transmission resources to provide end-to-end connectivity.

The number of switches or routers needed in an end-to-end connection will vary, and will depend on many factors, including network topology (the configuration of the various elements), the geographical distance between the user terminals, and the capacity of the switch or router and associated transmission media.

Signaling and Control

To ensure the establishment of end-to-end connections with the required quality of service, control information must be passed between users and the network, and between network elements (Figure 2.4). This control information comes in many different forms, both simple and complex.

The information that must be included in the control information varies from system to system, and also at different points within the same system. Control information is required to provide services of many types, not just end-to-end connections. Control data in telecommunications networks has traditionally been called *signaling*.

Billing

Increasing network sophistication and a corresponding increase in service choices and methods of service provision, have led to an increased requirement for flexible

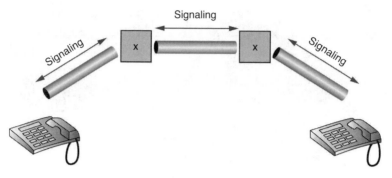

- Allows routing information to be passed from entity to entity as the transmission path is set up or the route established
- Allows the service and the required "quality" to be requested and negotiated
- Takes many forms (simple and complex) – increasing service options = increased signaling requirement

FIGURE 2.4 Signaling and control.

and more complex billing systems. This is confusing for network operators and customers alike, and a great deal of work is underway to provide choices that are achievable, simple to implement, and, very importantly, reflect the requirements of the customer. Service charges that are simple to understand, as well as flexible payment options, are of increasing importance.

Networking Principles

Within a network, various topologies (configurations) can be used to connect the elements together. Shown later in Figure 2.35, these include:

- Mesh
- Bus
- Ring
- Star
- Hierarchic

Different network operators will connect and manage their assets in different ways, and this can lead to incompatibility problems as well as quality issues between networks.

In addition, a single network can incorporate a number of different topologies, depending on which technology is being described. For example, telephone exchanges might be connected together in hierarchical fashion (local exchanges connected to transit exchanges, which are connected to international exchanges), while the underlying transmission network uses fiber-optic connections with a ring topology.

Review Questions

Q1. How does the customer experience telecommunications?
   a. Through transmission and switching
   b. Through signaling
   c. Through networking
   d. Through services and applications

Q2. What is transmission?
   a. The movement of information between two or more places
   b. Control information
   c. Radio signals
   d. Connections and networks

Q3. Switching _____.
   a. Is the process of connecting signaling resources together
   b. Uses digital light pulses
   c. Is the process of connecting transmission resources together
   d. Uses signaling transmission resources

Q4 Signaling is a specific way of describing _____.
   a. Transmission information
   b. Control information
   c. Application data
   d. The connection process

Q5. A network can have _____.
   a. Only one topology
   b. Different topologies within the same overall system
   c. Elements that are not connected together
   d. Only a single point-to-point connection

Q6. The process of calculating and generating bills is known as _____.
   a. Mediation
   b. Charging
   c. Tariffing
   d. Billing

2.1.2 Explaining the Principles

Services and Applications

The Different Categories

In general, services can be split into different categories, depending on their use, demand on system and network resources, and billing options. Services are generally seen as being provided by the network operator or a third party closely related to the network.

These services could include end-to-end data pipes across the network, termed "bearer services"; complete communication packages, which include the means to decode the information into a meaningful format (e.g., such as speech telephony or fax), termed "teleservices"; and "supplementary services", which provide features such as caller line identity.

Applications generally tend to reside in the terminal. This means that their functionality is supported in the end device, and they use bearer services to exchange the information on which much of their functionality depends. An Internet browser application relies on bearer services provided by telecommunications networks to exchange information with the servers hosting the Web pages.

Value-added services refers to services that add value to a network. This is a less formal term that can be used, for example, to describe text messaging or voicemail in mobile or fixed networks. These may be teleservices in their own right, but are not absolute requirements in the network. They do, however, add value for the customer.

The Customer Uses

- Applications
- Teleservices
- Bearer services
- Supplementary services
- Value-added services

Bearer Services and Teleservices

The difference between a bearer service and a teleservice can be explained quite simply. A *bearer service* simply provides the pipe through which information can be moved. Several different applications exist, and each may have different requirements of the bearer service that they use. This could include requirements for error rate, delay, delay variance, and data rate. Hence, in many systems, the bearer services are prescribed formally to ensure application designers have the required means to move information from point to point around the network.

Teleservices go one step further in that both the data pipe is described (in the form of the bearer service), together with the coding scheme used to provide complete end-to-end communication (as opposed to end-to-end data transfer in the case of bearer services). Examples of teleservices include telephony, facsimile (fax), and teletext.

Technology Principles ■ 113

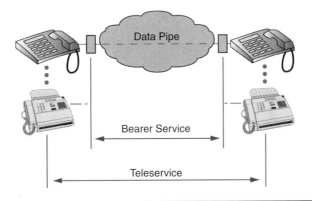

FIGURE 2.5 Bearer services and teleservices are defined for most telecommunications networks and systems.

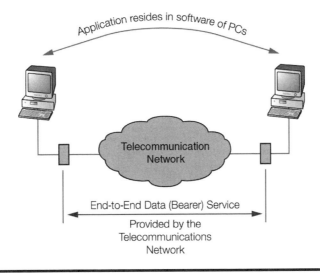

FIGURE 2.6 Applications and the underlying services. Only the underlying service, such as file transfer, need be defined for the telecommunications network concerned. The application, such as videoconferencing, can be defined elsewhere.

Applications

Applications usually reside in the end terminals. This is where the means to access their functionality resides (in the form of a screen for browsing or playing games, or sending e-mails, or perhaps a video camera for capturing visual images). However, information often must be moved around the network to take advantage of the application. To enable this to occur, the application will use bearer services of the required quality (Figure 2.6). The quality is often formally defined to ensure that at least the minimum requirements are met for that application.

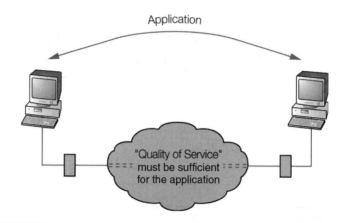

FIGURE 2.7 Quality of Service requirements (QoS). Quality of Service is considered in terms of errors, bit rate, delay, and delay variance.

The Underlying Service Requirements

The *underlying service requirements* are determined by the application. For example, e-mail must be sent error-free, but the exact time it is sent is not a concern as long as it arrives within a few seconds. Speech is more tolerant of errors, but the delay for speech must be minimal to avoid the type of delays that have so often marred satellite interviews on television. The delay must also be constant for speech. For real-time video communications, the delay is again critical, as is the delay tolerance; and in this case, errors must also be minimized.

Quality of Service

Quality of Service (QoS; Figure 2.7) is therefore critical in specifying the bearer requirements for a specific application. There are usually trade-offs to make between quality, cost, and flexibility. Different telecommunications systems are better for certain applications. For example, the Internet in its purest form provides "best-effort" quality of service that is optimized for e-mail but not good enough for speech and real-time video. Many modern systems try to accommodate the need for different QoS requirements and can offer a flexible range of bearer services that can be negotiated by the user or application. These systems include GSM (with relevant enhancements), ISDN, and UMTS (or 3G). IP (Internet Protocol) networks have had new protocols specified to try to ensure that the quality of service issues can be addressed. In terms of managing the packets, for example, voice can be prioritized over data at each node, to minimize delay in the delivery of the voice call. This enhanced quality of service will be essential to enable VoIP (Voice-over-Internet Protocol) to give adequate quality of service to most users.

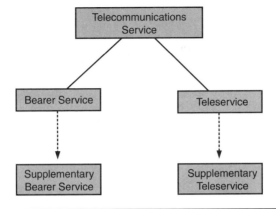

FIGURE 2.8 Supplementary services. Examples include multi-party, call barring, call transfer, caller line identity, called line identity, malicious call identity, ring back when free, call forwarding on busy, call forwarding unconditional, and call forwarding on no answer.

Supplementary Services

Some services cannot be offered in their own right but must be provided in addition to either a teleservice or a bearer service. In effect, they are *supplementary services* (Figure 2.8). An example would be the ability to connect multi-parties into a call. The multi-party supplementary service is in addition to the bearer service that was specified to form the basic connection. Without the bearer connection being specified, the fact that one can connect multi-parties is irrelevant. Hence, this service must be provided as a supplement to the bearer service. Many supplementary services are specified in a range of different systems. They are often formally described in terms of their operation and control (signaling) procedures.

Value-Added Services

Network Provided Value-Added Services

Another way of looking at services is to consider whether they add value for the customer, irrespective of their category — teleservice, bearer service, application, or even supplementary service. In general, they would not form part of the minimum core set of services that need to be provided by a network, but would simply add value to the customer experience.

Examples could include voicemail, text messaging in GSM, or personal assistant services (Figure 2.9). Certainly for text messaging (Short Message Service) in GSM, this is a specified teleservice. It is not needed for basic network operation, but the huge uptake of this service has indicated how valued it is by the customers. At the same time, it provides a valuable revenue source for the network operator.

FIGURE 2.9 Examples of network provided value-added services. Network operators use these services to increase revenue or attract and retain customers.

Many value-added services are provided by the network operator, as shown below and in Figure 2.9.

Third-Party Provided Services

Some services are provided by third parties (Figure 2.10). The network simply acts as a bearer in this case, or in some cases the network components may interact with those of a third party to provide the required service. There is a huge shift toward third-party service providers, not least because of the influence of the Internet. The Internet allows a service provider to advertise and sell services with minimal regulation and potentially low costs.

With telecom networks taking the lead in Internet access, and striving to provide faster and more reliable access, third-party service provision will become a huge factor for both customers and network operators. Third-party service providers are not exclusively limited to the provision of goods, information, and entertainment. Other providers, such as those specializing in card payment methods, may work with a network operator to provide the overall service package to the user; and the user may not be aware of the association.

Technology Principles ■ 117

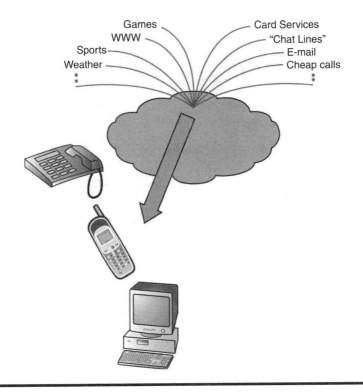

FIGURE 2.10 Third-party provided service examples. The network acts as a data pipe (bearer service). A premium is often paid for these services, with the network operator taking a cut.

The Overall User Experience

Overall, a large (and increasing) number of services are now available to users (Figure 2.11). These services are implemented in many different ways. They form part of the core set of services provided by a network operator, or simply add value for the customer and network operator. From the customer point of view, the network that provides the right mix of services that the user requires at a competitive price, with the required payment options, no matter how they are implemented (as long as the reliability and availability are acceptable), will be a very attractive option.

Service Provisioning

The need to speed up service provision and provide greater flexibility has been recognized for some time. During the 1980s, a great deal of work was undertaken to investigate a concept known as the Intelligent Network (IN). This concept not only

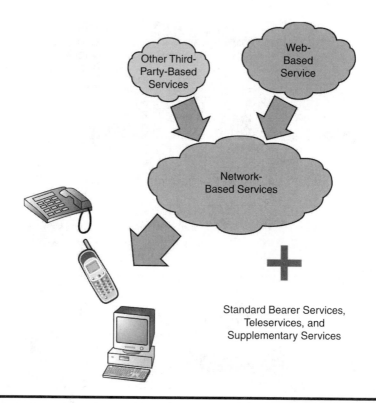

FIGURE 2.11 The overall user experience.

allowed for easier implementation of services on a network, but also made provision for service creation tools. Usually this is in the form of a server used to support a service modeling software tool on which new services can be designed, modified, and initially tested.

The mid-1990s saw IN techniques adopted in many networks, but work has continued to create evermore flexible and functional tools. Many tools allow for end-user control of service parameters. This could even result in users creating their own services and loading them onto the network.

Personalization of services is becoming more widespread, partly due to these service creation techniques (Figure 2.12). However, the tools are usually proprietary, and hence standard service creation across networks does not yet exist.

Review Questions

Q1. Bearer services describes _____.
   a. The complete communication package, including the coding
   b. Any service that adds value to the network

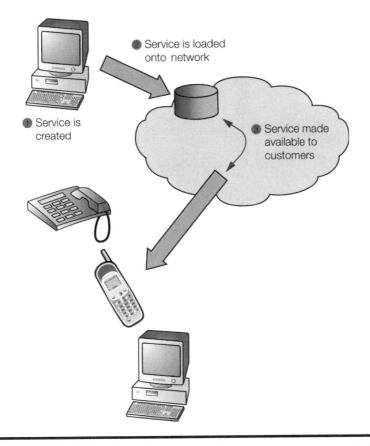

FIGURE 2.12 Service creation tools. These allow the creation and availability of non-standard services on the network. Customers may be able to control and create their own services.

    c. A service that must be supplemental to a bearer or teleservice
    d. How information is moved within specified bearer channels

Q2. Teleservices describes _____.
    a. The complete communication package, including the coding
    b. Any service that adds value to the network
    c. A service that must be supplemental to a bearer or teleservice
    d. How information is moved within specified bearer channels

Q3. Value-added services (VAS) describes _____.
    a. The complete communication package, including the coding
    b. Any service that adds value to the network
    c. A service that must be supplemental to a bearer or teleservice
    d. How information is moved within specified bearer channels

Q4. An application may need to specify _____.
   a. A teleservice over which to transfer information
   b. A bearer service over which to transfer information
   c. A supplementary service over which to transfer information
   d. Specific transmission resources

Q5. Call forwarding is an example of _____.
   a. A teleservice
   b. A bearer service
   c. A supplementary service
   d. An application

Transmission: Media and Systems

Transmission generally consists of two elements: (1) the transmission medium and (2) the transmission system.

The *transmission medium* is the physical means of carrying the information and can be:

- Metallic (usually copper based)
- Radio
- Fiber optic (or optical fiber)
- Infrared (optical line of sight)

The media needs to carry information, and to represent this information requires a variation in the electrical or optical signal. This variation can take many different forms, but is generally referred to as modulation, and can use analog or digital techniques. Modulation is used to vary the electrical or optical signal to represent information on transmission media.

A fundamental aspect of the transmission medium is the frequency at which it is designed to work. This will differ from telecommunications system to telecommunications system and can include multiple frequencies to provide for multiple communication channels.

Radio systems such as GSM can have hundreds of frequencies specified for use within a specified part of the frequency spectrum, whereas copper-based telephony systems might only have a single frequency band specified for use. GSM needs multiple frequencies to allow different frequencies to be used in different geographical parts of the network (to avoid radio interference within the communication channels), whereas copper wire physically separates the communication channels. This illustrates the difference between unbounded and bounded media (respectively).

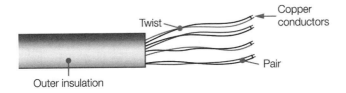

FIGURE 2.13 Copper twisted-pair cable. Twisted-pair cable has good conductivity. The twists reduce noise and crosstalk. It is widely used in the local loop. New technologies, such as xDSL, are available to extend the bandwidth of copper.

Transmission systems are designed to organize the information in a way that allows the equipment at either end of the media to work in unison. They provide a scheme for coding and decoding the information such that one or more communication channels can be identified (on the media, and at the specified frequency), and include extra information so that the equipment can be effectively synchronized and managed. If problems are experienced within the system, this can be notified via specific alarm channels, allowing remedial action to commence.

Copper Twisted-Pair Cable

Although they are the oldest of the media types used in telecommunications, copper cables remain the foundation of most national fixed telecommunications systems, especially within the local loop between the customer's premises and the local telephone exchange (Figure 2.13).

Copper was chosen for its good conductivity, together with its price and flexibility. There are metals that have better electrical conductivity properties, but most are more expensive than copper. The requirement for early telephone circuits was that the media should be capable of carrying low-bandwidth audio signals. The first copper cables were paper insulated. Paper worked well as an insulator, but as the number of cable pairs increased, the cables became extremely stiff due to internal friction. In the 1950s, plastics were introduced as the insulating material, with low-density polyethylene, high-density polyethylene, and polypropylene all being used.

Crosstalk is a phenomenon where signals intended for transmission on a circuit are electrically induced into adjacent circuits, causing interference. This was a particular problem of early cables. To reduce this, twists were introduced along each pair of cables, with up to 25 unique twists being used within a 25-pair cable group, each spaced at a different distance. It was assumed in the 1980s and 1990s that copper was an old technology with a limited future. Telecommunications companies planned ahead to install fiber and coaxial cable systems in the local loop. Eventually, the cost of doing so was judged to be too high in most cases.

Today, new techniques have been developed to transmit higher data rates than had previously been thought possible using copper wires with technologies such as

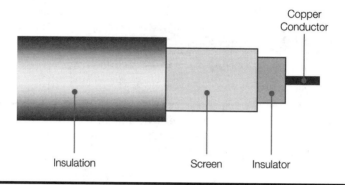

FIGURE 2.14 Coaxial cable. Used in inter-exchange trunks, it has good noise tolerance, low crosstalk, and can carry multiple channels. It is also used for thin-wire Ethernet connections.

Asymmetric Digital Subscriber Line (ADSL). This means that telecommunications companies have the opportunity to bring in revenue from high-speed data services using the existing cable.

Copper Coaxial Cable

A variation of the use of copper is its application in coaxial cables (Figure 2.14). Here, an inner conductor is first covered in an insulating material and then surrounded by a wire mesh or metallic screen. The cable is so named because both the inner conductor and the outer screen share the same axis. Coaxial cables are far more tolerant of electrical noise than traditional copper pairs and are able to transmit higher data rates. The use of coaxial cabling dramatically reduces crosstalk. The main application for coaxial cables was to serve inter-exchange trunk connections where a pair of cables is used for carrying multiple voice channels, one each for the transmit and receive paths. Coaxial cabling is also used for so-called thin-wire Ethernet connections. However, this system has the disadvantage that any failure along the cable route will cause the entire network to fail.

Radio

Radio systems can be used to transmit signals from a few meters to several thousand kilometers and can be used for both point-to-point and broadcasting applications.

Radio signals are a type of electromagnetic radiation, with similar properties to light but with a much shorter wavelength. As with all electromagnetic radiation, radio waves have both an electric and a magnetic field that travel at right angles to each other. As the signal travels outward, it can be compared to a stone being thrown into a pond. Much like the waves on a pond, radio waves get weaker the

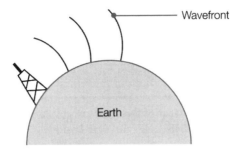

FIGURE 2.15 Ground waves. The wavefront bends close to the Earth's surface, causing it to follow the curvature of the Earth (VLF, LF, and MF bands).

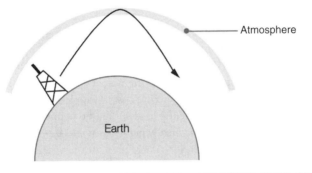

FIGURE 2.16 Sky waves. The radio signal is refracted by the ionized layer in the atmosphere (HF band).

further from the transmitter they are. In radio, this progressive weakening of the signal is referred to as attenuation or path loss.

The frequency of a radio signal will determine how it can be transmitted. Lower-frequency signals, such as those in the *very low frequency* (VLF), *low frequency* (LF), and *medium frequency* (MF) bands that cover frequencies up to about 2 MHz, can propagate using surface or ground waves. As currents are induced in the Earth, this has the effect of slowing down the part of the wave that is closest to the ground, causing a "bending" of the wavefront around the surface of the Earth. Hence, these lower frequencies can be transmitted either by line of sight or by surface waves. The combined effect is termed "ground wave" (Figure 2.15).

Another method of radio propagation is by the use of sky waves (Figure 2.16). Certain frequencies, including those in the *high frequency* (HF) band have the property of being refracted by layers in the atmosphere known as the ionosphere and troposphere. Essentially, when the signal reaches a heavily ionized layer, this layer reflects the signal toward an area of lower ionization. This makes it possible to send HF transmissions many thousands of miles around the globe. It is possible for sky waves to propagate lower frequencies but in most cases these work only under certain atmospheric conditions or at night.

FIGURE 2.17 Line of sight. Most of the signal is sent by direct waves. Line of sight is normally required (VHF, UHF, and SHF bands).

At yet higher frequencies, such as those in the *very high frequency* (VHF), *ultra high frequency* (UHF), and *super high frequency* (SHF) bands, most of the radio signal is transmitted by direct waves. Although there is still some component of ground- and sky-wave propagation, most of the signal relies on line-of-sight transmission (Figure 2.17).

As frequency increases, so path loss increases. This necessitates the use of highly focused, directional antennas for microwave transmission such as that used in satellite communications. One of the main advantages of radio transmission is that it removes the need for expensive cable-laying activities, which can account for 50 percent of telecommunications infrastructure costs. Coupled with the recent progress using the radio spectrum more efficiently has led to radio being heralded as one of the most promising media types for the next generation of telecommunications services.

Optical Fiber

Although first proposed in 1966 by Kao and Hockham, it was only in 1970 that Maurer, Keck, and Schultz designed and produced the first optical fiber that had characteristics that made it suitable for use in telecommunications. Their work enabled the production of fibers that had very little attenuation (loss of signal strength or intensity in the cable), and which kept most of the light traveling through the cable.

An optical fiber has a very thin core of glass or silica surrounded by an outer cladding made of a similar material, but with a lower refractive index. Figure 2.18 illustrates light waves from either an LED or laser entering the inner core of an optical fiber at three different angles, represented by the paths A, B, and C:

- Path A is the ideal, where the critical angle is met. This results in total reflection of the signal in a zigzag fashion within the inner core.
- Path B at first appears to be okay as the signal makes its way along the outer boundary wall of the inner core. However, this leads to spreading of pulses at the receiving end (and consequential data loss) because the signal will be transmitted both along the core wall and internally within the core.
- The example of path C is where the critical angle is exceeded and reflection cannot take place. Consequently, the signal is not transmitted along the fiber. This means that one major shortcoming of fiber is its inability to bend into extreme angles.

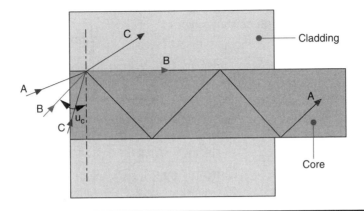

FIGURE 2.18 Propagation in optical fiber. Path A is ideal; path B will cause pulse spreading; and path C fails because reflection does not take place. As a transmission medium, optical fiber has very low attenuation, requires repeaters only every 30 to 40 kilometers, and is immune to electrical noise and crosstalk.

Optical fiber cables have the advantage of being extremely small and able to bear extremely high bandwidths, making them the obvious choice for backbone networks and international circuits.

In traditional copper cable systems, we often use repeaters along the line to compensate for the attenuation in the system. A repeater, as its name suggests, simply reads the signal and repeats it into the line. Because optical fibers suffer from far lower attenuation (data fade) than traditional copper cabling, the need for repeaters is greatly reduced, from about one repeater every 2 kilometers to one repeater in approximately 60 kilometers. This distance between repeaters can increase or decrease, depending on such factors as:

- Laser power
- Quality of fiber used
- Number of splices (the connection of one cable to another)
- Quality of cable splicing

Fiber optic cables are not prone to electrical noise and thus can be used within noisy environments (e.g., near electrical generators, magnetic fields, large motors) where traditional methods of transmission would be ineffective.

Comparison of the Available Media in Terms of Application

Table 2.1 shows the main media types and the typical applications for which they are used. Also shown are some of the key advantages and disadvantages of each media type.

TABLE 2.1 Comparison of Media Types

Media Type	Typical Applications	Advantages	Disadvantages
Copper	Telephone circuits Local area networks ADSL/ISDN connections	Cheap Easy to terminate Can carry power for repeaters, etc. High installed base	Prone to noise Limited bandwidth High initial cost High attenuation Crosstalk
Radio	Radio and TV broadcasting Mobile phone networks Satellite communications Point-to-point radio	No cabling to lay Fast setup Offers mobility	Prone to interference Bandwidth availability and cost Affected by the weather and atmospheric conditions
Fiber	Data backbone networks Inter-exchange trunks WAN interconnections Video on demand Video conferencing	Very high bandwidth Low attenuation Can bear multiple services simultaneously Size Protection Low maintenance cost	Connecting fibers requires skill Prone to mechanical stress High initial cost

Transmission Systems

Transmission systems are complex and involve many different aspects of information transfer and managing that information. They refer fundamentally to the way in which channels can be identified on the transmission medium, rather than the way the application data is coded or any higher layer transport protocols/systems (such as IP technology used to code Internet-type traffic).

Just about in all cases, the application data or transport protocols for any network will require the final coding and synchronization that will allow the information to be carried and identified within one or more specific channels on the transmission medium (often at or around a specified "carrier" frequency). It is this final coding process (and the additional features provided within the coding) that defines the transmission system used.

Example transmission systems include:

- Plesiochronous Digital Hierarchy (PDH)
- Synchronous Digital Hierarchy (SDH)

- Synchronous Optical NETworks (SONET) — almost equivalent to SDH
- Dense Wavelength Division Multiplexing (DWDM)

Many telecommunications systems developed over the past few years include the transmission system as a fundamental part of the overall system. The part of the specification that refers to the transmission is often referred to as Layer 1, and this would often include details about both the medium and the transmission system.

Examples of systems that have Layer 1 (transmission) specified are:

- Integrated Services Digital Network (ISDN)
- Global System for Mobile Communications (GSM) — Radio Interface
- Universal Mobile Telecommunications System (UMTS) — Radio Interface
- Wireless local area network (WLAN), also known as WiFi
- Wireless Interoperability for Microwave Access (WiMAX)

Review Questions

Q1. The three main types of transmission media are _____.
   a. Copper, radio, and light
   b. Radio, light, and fiber optic
   c. Aluminum, copper, and radio
   d. Fiber optic, radio, and copper

Q2. Which is true?
   a. Copper can be referred to as bounded media.
   b. Radio is always bounded media.
   c. Fiber optic is unbounded media.
   d. Copper is unbounded.

Q3. For metallic transmission, the medium _____.
   a. Must be fiber
   b. Can be any metal that conducts electricity
   c. Would usually be copper, but other suitable metals could be used
   d. Is unbounded

Q4. The modulation scheme _____.
   a. Identifies the channel
   b. Specifies which radio frequency is used
   c. Is always digital
   d. Determines how information is represented on the medium (at a particular frequency)

Q5. Transmission systems _____.
   a. Allow identification of one or more channels on the medium
   b. Cannot provide alarms or management features
   c. Only work on metallic systems
   d. Are an integral part of the switching method

Switching

The Principles of Circuit Switching

The Need for Local Switching — To connect telecommunications devices, a transmission path must be set up between them. When the number of devices is low, it may be worthwhile installing a dedicated link between the devices. This type of switching is known as circuit switching because a circuit is established between two devices. Figure 2.19 shows examples of two, three, and five telephones connected together. In the first three examples, the number of links required, if each telephone is connected to every other telephone, is given by the formula:

$$\text{Number of links required} = N \times (N - 1)/2$$

where N is the number of devices.

As each telephone has a connection to every other telephone, there is no possibility of network congestion, but the infrastructure costs would be very high. As an example, even a small village with only 50 telephones to be connected would require 1225 links to provide direct service between all of the subscribers.

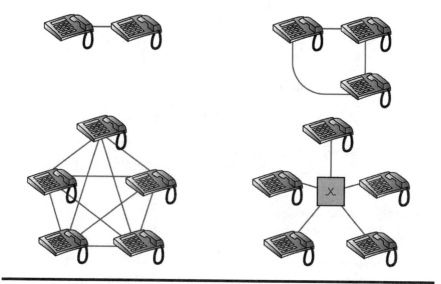

FIGURE 2.19   The need for switching.

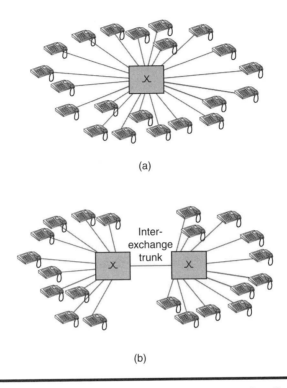

FIGURE 2.20 The need for trunk circuits: (a) increasing subscribers and (b) establishing trunk circuits.

Traditionally, 50 percent of a network's infrastructure costs are attributable to cabling alone. This brings about the need for each telephone circuit to connect to a local exchange, which performs a switching function between the subscribers connected to the exchange. Here it can be seen that the number of transmission links required is greatly reduced. In the example given previously of 50 subscribers in a village, there is now only a requirement for 50 links to service all the telephone circuits. The local exchange is sometimes referred to as the central office or switch.

The Need for Trunk Circuits — As shown in Figure 2.20, as the number of subscribers increases on a local exchange (a), a point may be reached where it is more cost effective to split the subscribers between two or more exchanges (b). This is because the cabling infrastructure costs are high and it may prove more efficient to provide a second exchange, with shorter cables running between the exchange and the subscribers. To connect calls between local exchanges, it is necessary to provide trunk circuits between the exchanges. These trunks must be dimensioned to offer an acceptable level of service availability, but are always equipped with fewer

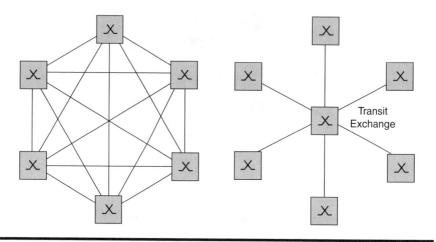

FIGURE 2.21 The need for transit exchanges.

channels than the number of subscribers on each exchange. This is because the probability of every subscriber on one exchange dialing a subscriber on the second exchange at the same time is minimal.

Within the local exchange there must be functionality to switch calls onto the relevant trunk circuits and also to provide the required signaling between the exchanges.

The Need for Transit Switching — Transit exchanges provide interconnections between local exchanges (Figure 2.21). Once again, one of the prime motivators in the provision of transit exchanges is to reduce the amount of cabling infrastructure (trunks) required.

A transit exchange does not normally have any subscribers of its own. It is connected to a number of local exchanges via trunk circuits. To interconnect with the rest of the PSTN, a transit exchange will have connections to other transit exchanges or international exchanges. The trunks that interconnect transit exchanges generally have a higher capacity than those used for inter-exchange trunks. For this reason, fiber-optic transmission is often used, utilizing PDH, SDH, or DWDM transmission systems.

Circuit Switching, Packet Switching, and Message Switching

Circuit Switching — In circuit-switched networks (Figure 2.22), a dedicated path is set up between the two parties. This path remains for the exclusive use of both parties for the duration of the call, and is therefore not available to any other users. This method has traditionally been used within standard telephone exchanges and networks since telephony was first developed.

There is a delay involved in setting up circuit-switched calls because each of the switching nodes has to route the call. However, the actual delays encountered once

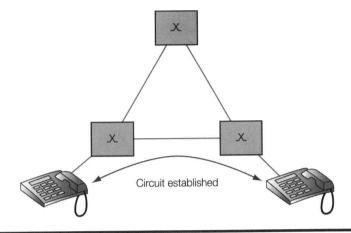

FIGURE 2.22 Circuit switching. In a circuit-switched network, a dedicated part is set up and the circuit remains for the exclusive use of both parties for the duration of the call. Circuit switching, suited to voice and other real-time services, has no significant delays but is an inefficient use of network resources. Examples of circuit-switched services are PSTN and ISDN.

the connection has been established are minimal. This makes circuit switching ideal for voice and other real-time applications.

Circuit switching can be inefficient in its use of network resources. During a voice conversation there are periods when neither party is talking but the connection is still tied up and unavailable for other users. Similarly, *bursty* data, which has gaps between the data, is not efficiently carried over circuit-switching networks.

Charging for circuit-switched services is generally based on the duration of the call. The PSTN and ISDN are examples where circuit switching is employed.

Packet Switching — Packet switching involves dividing the data into packets (or cells or frames) prior to transmission. The length of the packets varies enormously, depending on the technology employed.

Added to each packet is the destination address, together with other control information. The packets are then transmitted across the network. This addressing means there is no requirement to set up a pre-established link. To some extent, each individual packet can be viewed as being able to find its own way to its destination.

In a packet-switched network (Figure 2.23), the resources are shared between many users. This leads to more efficient use of these resources than provided for by circuit-switching techniques. However, with packet switching there is a danger of congestion occurring. The subsequent delays are both variable and unpredictable. Much effort has been put into reducing these delays for applications that require near-real-time transmission such as voice telephony.

When data is divided into packets, it does not follow that all the packets that contain the original data will follow the same route to the destination. This can

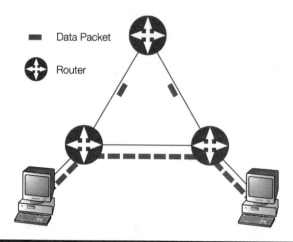

FIGURE 2.23 Packet switching. Packet switching does not require a dedicated circuit. Transmission links are shared between users, making it an efficient use of network resources. It is subject to variable and unpredictable delays, making it better suited for data transmission. Packet-switched services include the Internet, Frame Relay, X.25, and Voice-over-IP (VoIP).

mean that packets arrive out of sequence and must be reassembled in the correct order before delivery to the recipient. Examples of packet-switching technologies include X25, FR (Frame Relay), ATM (Asynchronous Transfer Mode), and IP (Internet Protocol).

Because it is possible to charge for data throughput rather than for the duration of the connection within a packet-switched network, it is more feasible to have permanent online connections than can be provided for by traditional circuit-switched networks. Most commentators see packet switching as the future backbone of new high-bandwidth telecommunications services.

Message Switching — It is not always necessary to establish an end-to-end circuit for the transmission of data. So-called *store and forward* techniques can be applied.

In Figure 2.24, an e-mail message is transmitted between nodes A and D. Because no circuit is established, the message is carried in stages over the links between the nodes. At each stage, the message is stored within the node while the next link is established. This does lead to queuing delays at each stage; but to the applications utilizing message switching, these small delays are unimportant.

Another example of the use of message switching is for SMS text messaging within a GSM network (Figure 2.25.). When a user sends a text message, it is transmitted across the air interface into nodes within the GSM network. The entire message is stored within the node responsible for SMS messaging. Here, the message is stored before it can be forwarded to its destination. Should a node be unable

Technology Principles ■ 133

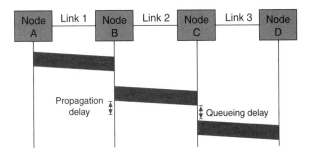

FIGURE 2.24 E-mail via message switching.

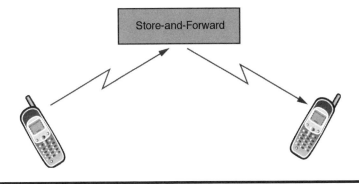

FIGURE 2.25 SMS text messaging.

to forward a message for any reason, it will retry until an expiry time or number of attempts is reached. Depending on the application, a failure message can be transmitted to the originating subscriber to inform them that the message transmission has failed.

Review Questions

Q1. Switched voice connections can be established _____.
   a. Only within an operator's network
   b. Point-to-point between two network elements or telephones
   c. Worldwide
   d. Only within the handset

Q2. Telephony has traditionally been provided by _____.
   a. Circuit-switched connections
   b. Packet-switched connections
   c. Message-switched connections
   d. None of the above

Q3. Internet access is generally more efficient by _____.
   a. Circuit-switched connections
   b. Packet-switched connections
   c. Message-switched connections
   d. None of the above

Q4. Text messaging is generally provided using _____.
   a. Circuit-switched connections
   b. Packet-switched connections
   c. Message-switched connections
   d. None of the above

Q5. For real-time services, packet-switched connections would need _____.
   a. IP technology
   b. To use telephone exchanges
   c. To use secure channels
   d. Effective Quality-of-Service control

Signaling and Control

Control Requirements

To ensure the establishment of end-to-end connections with the required quality of service, control information must be passed between users and the network, and between network elements. This control information comes in many different forms, both simple and complex. The information required in the control information varies from system to system, and also at different points within the same system. Control information is required to provide services of many types, not just end-to-end connections. Control data in telecommunications networks has traditionally been called *signaling* (Figure 2.26).

The Importance of Signaling Systems

Signaling systems are essential to the operation of any telecommunications system. As switching and transmission systems have evolved, signaling systems have had to develop continuously to support the services offered in modern telecommunications networks.

Today, with truly global communications systems in place, signaling plays a critical role. However, most users are unaware of the signal processing that takes place even for a simple local telephone call. This is a long way from the early systems where operators sitting in the local exchange performed most signaling functions.

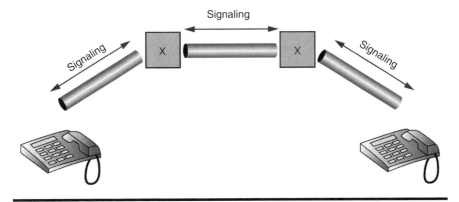

FIGURE 2.26 Signaling and control. Signaling and control allows routing information to pass from entity to entity as the transmission path is set up or the route established. This allows the service and the required quality to be requested and negotiated. Signaling can be categorized as access, network, or user-to-user.

Functions of a Signaling System

A signaling system must be able to perform many functions within the network. One of the most important functions is that of call setup, where the dialed digits are transmitted across the network and the subsequent routing of the call to its destination. Should the called party be engaged, this must be notified to the calling party — who will normally hear this as a tone. When a call is completed, clear down signals must be transmitted to both parties, and the transmission and switching systems released.

When a normal telephone number is received by a telephone exchange, it can use simple look-up tables, held locally, to route the call. However, in the case of freephone or toll-free numbers, the exchange will have to access a regional or central database to route the call. This will be achieved over signaling links, able to interrogate and report back from the databases. This technique forms the basis of Intelligent Network (IN) technology.

Without the ability to accurately and reliably transmit billing information across and between networks, few network operators would stay in business for very long. Another key area for signaling is for operation and maintenance purposes. Should it be necessary to remove an inter-exchange trunk from service, this must be signaled to a remote exchange so that it does not attempt to use the out-of-service circuit. In summary, the functions include:

- Call setup (including transmission requirements)
- Calling party and called party information
- Call supervision
- Advanced call routing (freephone, toll-free, etc.)

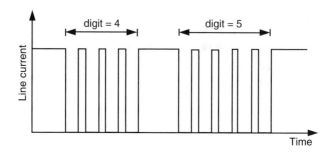

FIGURE 2.27 Loop/disconnect signaling.

- Carrying billing information
- Operation and maintenance
- Look-up service entitlement

Access Signaling

Loop/Disconnect Signaling —The term "loop/disconnect" refers to a type of signaling used within the local loop, whereby the telephone, fax machine, modem, or other device sends signaling information to the local exchange (Figure 2.27). It is so called because the system relies on connecting and disconnecting a loop across the line, allowing an electrical current to flow between the two wires when the loop is made. This current is detectable by circuitry in the telephone exchange.

Although not commonly used anymore for this purpose, loop/disconnect signaling can be used for the transmission of the dialed digits. Each number is transmitted as a corresponding number of breaks or disconnections in the local loop. This form of dialing has a speed of ten pulses per second and consequently it takes a long time to transmit national or international telephone numbers.

While loop/disconnect dialing is not often used, loop/disconnect signaling is still used for the basic signaling requirements of the local exchange, where the telephone signals to the exchange request to make a call (line seizure) and also when the call is cleared down.

DTMF Signaling — Dual-tone multi-frequency (DTMF) signaling (Figure 2.28) is now the preferred signaling method for the transmission of dialed digits. This system works by representing each digit on the keypad with a combination of two frequencies. These frequencies are audible to the caller. DTMF signaling has the advantage of being much faster than loop/disconnect dialing. It also has the advantage that the user, being able to hear the tones, is aware if a button press is missed or accidentally repeated.

freq \ freq	1209 Hz	1336 Hz	1477 Hz
697 Hz	1	2	3
770 Hz	4	5	6
852 Hz	7	8	9
941 Hz	*	0	#

FIGURE 2.28 DTMF signaling.

Other Tones and Signals

Apart from line seizure, clear down, and the sending of digit information, other tones and signals are required to notify both the called and calling parties of the status of the call. A number of different audible tones are transmitted to the caller to let him know the call status (such as engaged or busy tone). The actual tones involved vary from country to country. The called party is made aware of the incoming call by the provision of a ringing signal that will activate a ringing device in the telephone equipment. Again, the cadence of this ring is different between countries.

Local Call Example

Figure 2.29 demonstrates the typical signaling involved in setting up a local call. By taking the telephone off the hook, the calling party (A subscriber) causes a loop to be put across the line, which is a seizure request (1) to the exchange. Upon receiving this signal, the exchange acknowledges by returning dial tone (2) to the A subscriber. Next, the A subscriber dials the number of the called party (B subscriber). This is transmitted as DTMF tones (3) to the exchange, which checks whether the B subscriber's number is available. If it is, then a ringing signal is sent to the B subscriber (4). At the same time, a ringing tone is passed back to the A subscriber (5). Once the B subscriber answers, the off hook signal (6) is sent back to the exchange. The exchange ceases the ringing tone signal that was being transmitted to the A subscriber (7). Once the call is complete, call clear down can be performed in either a forward (8) (9) or a reverse direction (10) (11).

Network Signaling and SS7

Network Signaling — Different applications, features, and services within the modern telecommunications network are supported by various sets of signaling messages. The messages are exchanged between various network elements in a

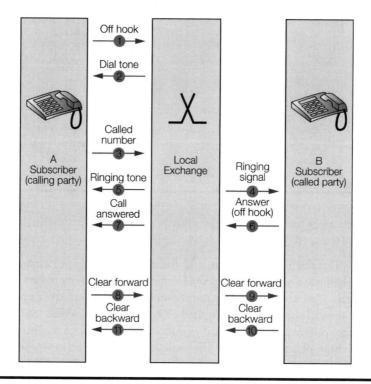

FIGURE 2.29 Signaling for a local call.

standard way, and via a common underlying signaling network called Signaling System Number 7 (SS7).

The message sets, formatting rules, and operation of the signaling entities are all standardized and form part of the SS7 specification. SS7 provides a highly reliable and resilient packet-switched network over which the network control information (in the form of SS7 messages) can be transferred (Figure 2.30). Hence, Call Control, Supplementary Services, and Intelligent Networks are all supported by SS7 in the fixed network, while Mobility Management, Short and Multi-Media Messaging, GPRS (General Packet Radio Service), and other advanced services can be added to the list for mobile networks. It is also an integral part of UMTS (Universal Mobile Telecommunications System).

For security and ease of management, SS7 is only provided within the network and not generally extended into the access network.

Billing

Billing is probably the area that most telecommunications professionals would find closest to their hearts, or at least their pockets. It is the term used to describe the accounting and costing of telephone or data calls made within a network.

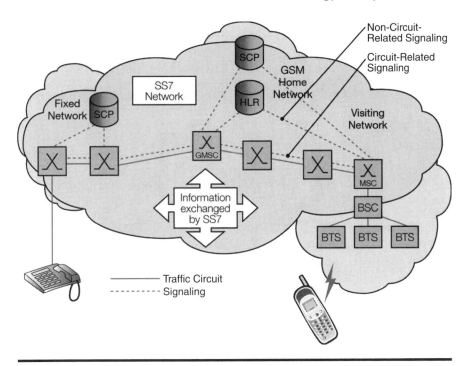

FIGURE 2.30 The SS7 network.

Much focus has traditionally been put into designing the infrastructure of telecommunications networks. Before any work commences on the rollout of networks, plans are drawn up for transmission, switching, and signaling systems. Billing is one area that has not always been fully considered in the planning process. However, with the introduction of the new applications and services, it is increasingly important that the billing model is drawn up before the network is built (Figure 2.31).

What services will customers be willing to pay for? How much would they be willing to pay? Will they pay for content, bandwidth, or duration? Will the network recover its infrastructure and licensing costs?

How Services Are Billed

Several billing methods are available to the customer via the service provider and are consistent with fixed and mobile networks (Figure 2.32). These can be grouped as follows:

- *Postpaid.* At a predetermined time, whether monthly, quarterly, or by direct debit, the service provider sends a bill to the customer detailing the call and service charges for that period. The customer will then have to pay the charges or face service termination and recovery action.

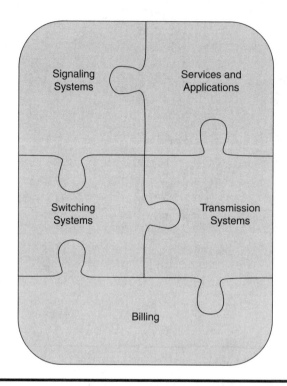

FIGURE 2.31 With the introduction of new applications and services, it is increasingly important to draw up the billing model prior to building the network.

- *Prepaid.* A card or a *top-up* is purchased with a prescribed credit limit. There are various methods of providing this, but most result in credit (held in a central database) being incremented on top-up, and decremented as calls are made or data transferred. The database is checked at call setup to see if enough credit exists to make the call, and for how long it can last.
- *Credit card.* A connection is made to the network operator, and the credit card details given. When the details have been confirmed, a connection is established with the receiving party and the credit card charged accordingly.
- *Free.* Some calls in a network can be free. There are toll-free services found in most networks, primarily used by commercial organizations that ultimately pick up the bill. There are also truly free calls to the subscriber. In countries such as the United States and Canada, free local calls have been a permanent feature of the network. Other operators have followed suit in a limited way, usually offering free off-peak calls or free call minutes allowances. These are often limited to the local exchange and geographic NNG (National Number Group) codes only.
- *Cash (fixed only).* Money is deposited into the phone to enable call completion. When the credit runs out, the connection is terminated. This system

Technology Principles ■ 141

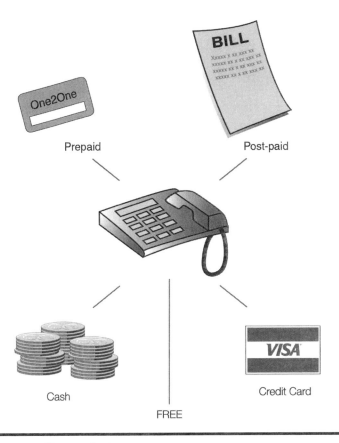

FIGURE 2.32 Billing methods.

only works with pay phones. However, a reverse charge (collect) call can be made via the operator if the receiver is willing to accept the call charges on his account.

What Can Be Billed?

There are a variety of ways in which billing can be achieved in both fixed and mobile networks (Figure 2.33).

- *Time connected.* From the moment that the call is connected, the customer is charged at the relevant rate set by the service provider.
- *Type of service and quality of service.* Depending on the type of service that is requested, whether it is data or voice, the quality of the service can be increased for voice or decreased for data as required. High quality is required for voice to minimize delays and to provide an acceptable experience for the communicating parties. This is then reflected in the cost to the subscriber.

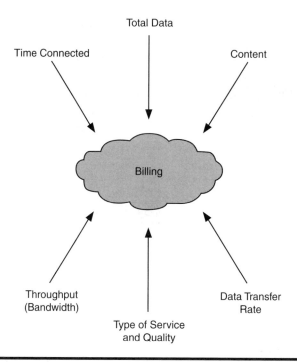

FIGURE 2.33 What can be billed.

- *Total data.* A charge is made for the total amount of data (charge per kilobyte or megabyte) sent or received, regardless of the time connected.
- *Content.* The type and perceived value of the content will dictate the charge levied. The relevant content supplier will possibly pay the delivery overheads to the service provider, enabling a fixed price to be charged to the subscriber and therefore encouraging more custom to the service. This is particularly applicable to media delivery (e.g., pay-per-view services).
- *Throughput (bandwidth).* Where there is spare capacity in the communications channel or system, the redundant channels can be used to enable the data to transmit at a far greater rate. The billing would reflect the fact that additional bandwidth was used to transmit this data.
- *Data transfer rate.* For all services, the charge levied to the subscriber can be fixed to the speed of data delivery. This is particularly used in leased-line applications.

Billing in the Network

A call is initiated from the phone via the local exchange or equivalent exchange in a cellular network (Figure 2.34). The exchange will pass the connection details to

Technology Principles ■ 143

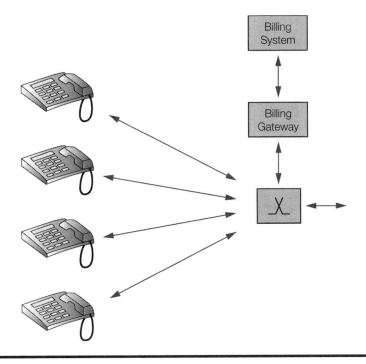

FIGURE 2.34 Billing in the network.

the billing system via the billing gateway. The details of the call will include such details as dialed number, tariff code, and total time connected. This information will be processed, and the local service provider will submit a bill to the subscriber at the predetermined time.

Networking Principles

Within a network, various topologies (configurations) can be used (Figure 2.35). These include:

1. *Mesh.* This connects all nodes in a network and can be applied at different levels (e.g., transit level [between telephone exchanges]).
2. *Bus.* As information is passed along the bus, the address information will identify a particular node or multiple nodes where the information is deposited.
3. *Ring.* This is similar in operation to the bus network.
4. *Star.* Each node is connected to a central hub where switching will connect the relevant nodes. As each smaller star network forms, it can be interconnected to form a multiple star tree.
5. *Hierarchic.* Another method of interconnection is to use a hierarchical structure.

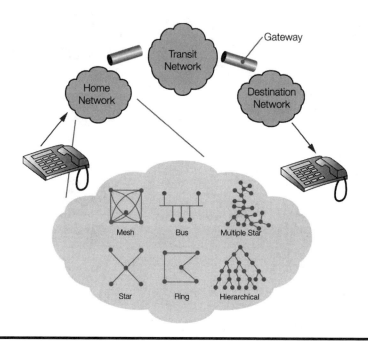

FIGURE 2.35 Network principles.

From the home network, the transmission medium can connect to a transit network via a gateway. The gateway acts as an interface between different networks. The call will then be passed on to the destination's home network, and ultimately on to the required destination. As the call passes through these networks, the quality of the call and compatibility between the networks is always an issue.

Different network providers will manage their assets differently, and this can lead to incompatibility problems as well as quality issues between networks. As additional networks are used, issues arise with regard to billing, such as who to charge for the use of other networks, the customer or the home network provider.

Review Questions

Q1. Signaling systems _____.
    a. Are only found in sophisticated telecommunications networks
    b. Are an integral part of telecommunications networks
    c. Are not needed unless advanced services are used
    d. Are the same in every telecommunications network

Q2. Which is correct?
   a. Core network signaling and access signaling use the same systems.
   b. Access signaling tends to be more complex than core network signaling in fixed networks.
   c. Signaling systems are always complex in all network types.
   d. Access signaling in mobile networks tends to be more complex than access signaling in fixed networks.

Q3. Generally, a network operator _____.
   a. Will support standard services and a range of applications
   b. Will only support standard services
   c. Will support applications and non-standard services
   d. Will never support standard services

Q4. Billing systems _____.
   a. Are all standardized
   b. Tend to be proprietary
   c. Are usually part of the switch or telephone exchange
   d. Can be accessed directly by the customer

Q5. An operator's network would use _____.
   a. Only one network topology
   b. Multiple topologies, depending on what is being described (transmission, switching, etc.)
   c. A hierarchical network for radio systems
   d. Fiber-optic rings for all connections

2.1.3 Review Questions

1. Identify the five major requirements of all telecommunications networks:

   1. ....
   2. ....
   3. ....
   4. ....
   5. ....

2. Categorize each of the following into (Se) Services and Applications, (T) Transmission, (Sw) Switching, (Si) Signaling, (B) Billing/Support, (N) Networking.

   Router
   Copper cable
   Radio

SS7
Full Mesh
Telephone exchange
Tone dialing
Post-paid
Ring tone
MMS
Telephony
Top-ups

3. For a typical GSM/GPRS or cdmaOne system, check off the techniques you think can be found within the overall network (including access/backhaul/core; voice and data).

Copper transmission
Radio transmission
Fiber transmission
Telephone exchanges
Routers
Service platforms
Analog transmission
Digital transmission
Broadband channels
Operational support systems (OSS)
Billing system
Microwave links
Satellite systems
Messaging
Pulse dialing

Practice Questions

Q1. How does the customer experience telecommunications?
  a. Through transmission and switching
  b. Through signaling
  c. Through networking
  d. Through services and applications

Q2. Switching _____.
  a. Is the process of connecting signaling resources together
  b. Uses digital light pulses
  c. Is the process of connecting transmission resources together
  d. Uses signaling transmission resources

Q3. A network can have _____.
   a. Only one topology
   b. Different topologies within the same overall system
   c. Elements that are not connected together
   d. Only a single point-to-point connection

Q4. Bearer services describes _____.
   a. The complete communication package, including the coding
   b. Any service that adds value to the network
   c. A service that must be supplemental to a bearer or teleservice
   d. How information is moved within specified bearer channels

Q5. Teleservices describes _____.
   a. The complete communication package, including the coding
   b. Any service that adds value to the network
   c. A service that must be supplemental to a bearer or teleservice
   d. How information is moved within specified bearer channels

Q6. An application may need to specify _____.
   a. A teleservice over which to transfer information
   b. A bearer service over which to transfer information
   c. A supplementary service over which to transfer information
   d. Specific transmission resources

Q7. General Packet Radio Service is an example of _____.
   a. A teleservice
   b. A bearer service
   c. A supplementary service
   d. An application

Q8. SMS (Short Message Service) is an example of _____.
   a. A teleservice
   b. A bearer service
   c. A supplementary service
   d. An application

Q9. Not all networks provide _____.
   a. Teleservices
   b. Bearer services
   c. Supplementary services
   d. Applications

Q10. The three main types of transmission media are _____.
  a. Copper, radio, and light
  b. Radio, light, and fiber optic
  c. Aluminum, copper, and radio
  d. Fiber optic, radio, and copper

Q11. Which is true?
  a. Copper can be referred to as bounded media.
  b. Radio is always bounded media.
  c. Fiber optic is unbounded media.
  d. Copper is unbounded.

Q12. The modulation scheme _____.
  a. Identifies the channel
  b. Specifies which radio frequency is used
  c. Is always digital
  d. Determines how information is represented on the medium (at a particular frequency)

Q13. Transmission systems _____.
  a. Allow identification of one or more channels on the medium
  b. Cannot provide alarms or management features
  c. Only work on metallic systems
  d. Are an integral part of the switching method

Q14. Switched voice connections can be established _____.
  a. Only within an operator's network
  b. Point-to-point between two network elements or telephones
  c. Worldwide
  d. Only within the handset

Q15. Telephony has traditionally been provided by _____.
  a. Circuit-switched connections
  b. Packet-switched connections
  c. Message-switched connections
  d. None of the above

Q16. Internet access is generally more efficient by _____.
  a. Circuit-switched connections
  b. Packet-switched connections
  c. Message-switched connections
  d. None of the above

Q17. For real-time services, packet-switched connections would need _____.
   a. IP technology
   b. To use telephone exchanges
   c. To use secure channels
   d. Effective Quality-of-Service control

Q18. Signaling systems are _____.
   a. Only found in sophisticated telecommunications networks
   b. An integral part of telecommunications networks
   c. Not needed unless advanced services are used
   d. The same in every telecommunications network

Q19. Which is correct?
   a. Core network signaling and access signaling use the same systems.
   b. Access signaling tends to be more complex than core network signaling in fixed networks.
   c. Signaling systems are always complex in all network types.
   d. Access signaling in mobile networks tends to be more complex than access signaling in fixed network.

Q20. Generally, a network operator _____.
   a. Will support standard services and a range of applications
   b. Will only support standard services
   c. Will support applications and non-standard services
   d. Will never support standard services

Q21. Billing systems _____.
   a. Are all standardized
   b. Tend to be proprietary
   c. Are usually part of the switch or telephone exchange
   d. Can be accessed directly by the customer

Q22. An operator's network would use _____.
   a. Only one network topology
   b. Multiple topologies, depending on what is being described (transmission, switching, etc)
   c. A hierarchical network for radio systems
   d. Fiber-optic rings for all connections

2.2 Information Transfer

2.2.1 Representing Information

Characteristics of a Waveform

To understand the characteristics of a waveform, it is useful to use a sine wave, plotted on a time-domain graph (Figure 2.36). This produces a result similar to that of a device called an oscilloscope, which plots level against time for an input signal. Naturally, real sounds such as the human voice or radio signals are far more complicated than a single sine wave, but the same principles and techniques apply. There are several measures one can use to describe the waveform:

- *Amplitude:* denotes the height or size of the waveform. The amplitude of electromagnetic signals is measured in volts (for voltage) or amps (for current). Note, however, that the most important parameter is power, which is a combination of both voltage and current. Such power is most often measured in decibels (dB).
- *Frequency:* number of cycles per second (e.g., 900 MHz for GSM). It is measured in Hertz (Hz), where 1 Hz is equal to one cycle per second. Often in telecommunications, large numbers are encountered and so the following prefixes are used.

    k kilo × 1000
    M Mega × 1,000,000
    G Giga × 1,000,000,000

As an example, the approximate frequency of the primary GSM band is 900 MHz = 900,000,000 Hz. In practice, the GSM system, like all other systems, requires a range within the frequency band rather than just one "spot" frequency (Figure 2.37).

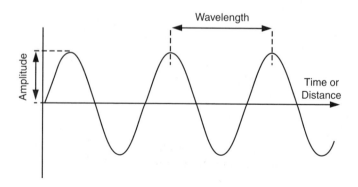

FIGURE 2.36 Waveform characteristics.

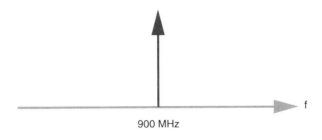

FIGURE 2.37 Representing a frequency in graphical form.

- *Wavelength:* length of one cycle of the waveform (e.g., 33 cm at 900 MHz). The wavelength is measured in meters. There is an inverse relationship between the frequency and the wavelength of a waveform given by:

    Velocity = Frequency × Wavelength
    Velocity of electromagnetic wave = 300,000,000 meters/second

- *Phase:* always a relative value and allows two similar waves (with the same frequency) to be compared in terms of where they are within the "cycle". At any point in time, two signals can be at the "top" of the sine wave, and they are said to be in phase. Alternatively, one signal can be at the "top" and another (with the same frequency) at the "bottom" of the sine wave, and they are said to be in anti-phase. In practice, waves can be any phase relative to each other, and we use values between 0° and ±180° to describe the phase relationship.

Analog Signals

An analog signal (Figure 2.38) is one that can vary between a maximum and minimum limit, which has been determined by the transmission system. The number of levels is infinite within an analog system maximum and minimum. The human voice, when converted to an electrical signal, is a good example of an analog signal.

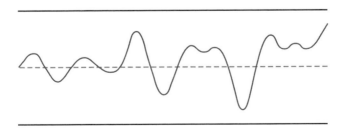

FIGURE 2.38 The analog signal.

- Any amplitude between a maximum and minimum limit
- Infinite number of levels
- Human voice is an example
- Range between lowest and highest frequencies is known as bandwidth

Most signals are not sine waves, but rather are made up of complex waveforms. Through appropriate mathematics, it can be shown that complex waveforms can be represented by many sine waves (of different amplitudes and frequencies) that are combined to give the overall composite signal. The range between the lowest and the highest frequencies contained in the signal is known as the bandwidth.

Although varying from person to person, human voice would be made up of sine waves with frequencies between 50 Hz and 10 kHz, giving a bandwidth of approximately 9.5 kHz, although the stronger frequencies (greater amplitude sine waves) would be located around 800 Hz to 1 kHz.

Digital Signals

A digital signal (Figure 2.39) is one that has discrete values. Although digital signals can often be used to represent any number of discrete values, the binary signal (which is the most commonly used digital signal) has two values set to represent 0 and 1. The actual signal that corresponds to each level will be determined by the transmission medium or system being used. The signals could, for example, be represented by different voltage levels on copper cable, turning the laser *on* or *off* on a fiber-optic cable, or slight changes of frequency in a radio system. Each level or state is known as a *symbol*.

In all cases, a digital signal will change levels (or states); and in modern systems, one commonly finds 2 states (binary), 4 states, 8 states, 16 states, or even 64 states. The number of times the signal selects a new state in a second is known as the *baud rate*.

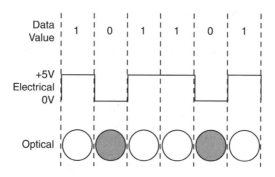

FIGURE 2.39 A digital signal.

Most equipment and systems process information in binary format, using 1s and 0s (bits). If this information needs to be transferred between equipment, it will be carried over the connecting transmission system:

1. If that transmission system uses binary format, the information is simply mapped onto the digital signal, and the baud rate is effectively the same as the bit rate.
2. If the transmission system has more than two levels, then more than one bit at a time can be mapped onto symbols of the transmission system, and the bit rate will be greater than the baud rate.

Transmission systems can therefore be measured in both baud rate and bit rate, although bit rate gives a better measure of performance.

- Made up of discrete values
- Binary signals have two levels
- Measured in terms of baud rate and bit rate

Digital signals have some very significant advantages over analog signals, and some of the more important ones are summarized below. The advantages of a digital signal include:

- *Digital signal noise immunity.* All signals received in telecommunications systems will be received with some amount of electromagnetic noise. The advantage that digital signals have over analog signals in this respect is that noise can be removed from digital signals. Thus digital signals are said to have the property of *noise immunity* (Figure 2.40). This is often described in terms of digital systems providing superior received quality compared to analog systems.

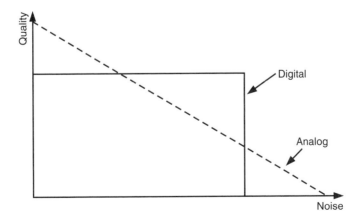

FIGURE 2.40 Digital signal noise immunity.

However, it must be remembered that a digital signal is always a copy of a natural analog signal and that in low-noise environments, the analog waveform is superior to the digital waveform because it represents "the real thing." Cellular radio systems generally suffer from high levels of noise and, in such noisy environments, the advantages of digital transmission become increasingly evident.

- *Confidentiality.* Confidentiality has always been recognized as one of the major advantages of 2G cellular networks. Attempting to make analog transmissions secure is invariably cumbersome and not particularly effective. Digital technology, on the other hand, provides the means of making a transmission extremely secure against accidental or deliberate eavesdropping.
- *Multimedia capability.* The major forms of multimedia (voice, facsimile, image, text, video) tend to look very different in the analog world. Once digitized, these signals may be in a common format (that is, 1s and 0s) and, as such, may conveniently be transported (and stored) over common communication media and storage facilities. This is of increasing importance as the world moves away from simple voice communications toward full multimedia in both fixed and mobile networks.
- *More efficient bandwidth requirements (usually).* Whether an analog or an equivalent digital transmission is more spectrally efficient depends primarily on the nature of the signal being transmitted. With speech, a "raw" digital waveform might occupy a considerably wider bandwidth than an equivalent analog transmission. However, with increasingly sophisticated voice coding and modulation techniques, the bandwidth of a digitally encoded voice can be made similar to, and in some cases less than, its analog equivalent. With other types of multimedia transmissions, the digital signal (due largely to bit reduction techniques) is often the more spectrally efficient. TV provides a good example.

"Use of the frequency spectrum is controlled in order to minimize interference between users, channels or systems / technologies"

The Frequency Spectrum

Whether using cables, radio, or optical systems, the signals will operate at (or, more correctly, around) a specified frequency (Figure 2.41). The overall system can have a number of these specified frequencies available to support multiple users and information channels. The frequencies will usually be specified to minimize interference between different users or channels, and even between different systems or technologies that are using the same transmission medium. Examples include:

Frequency Bands Allocated to:
- Separate users
- Separate channels
- Separate systems/technologies

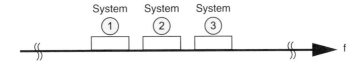

FIGURE 2.41   Using the frequency spectrum.

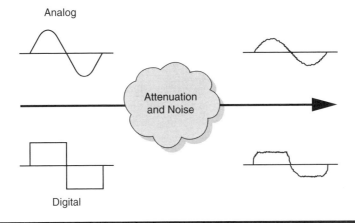

FIGURE 2.42   Attenuation and noise.

- Standard telephony and ADSL Broadband using different frequency bands on the copper cable from the home to the local exchange
- Cellular system frequencies being separated from television broadcast, cordless telephony, civil aviation, or military frequencies by a system of government radio licensing

Propagation, Attenuation, and Noise

Propagation describes how a signal travels through a transmission medium (whether metallic, radio, or optical). The signal energy will generally be confined within bounded media (mainly cables), but radio signals can follow a variety of paths from transmitter to receiver, including direct line of sight, reflected (from buildings or terrain), or refracted (e.g., in layers of the atmosphere).

As shown in Figure 2.42, whichever path a signal takes, it will be subject to:

- Attenuation (loss of power)
- Noise (electrical pickup, interference)

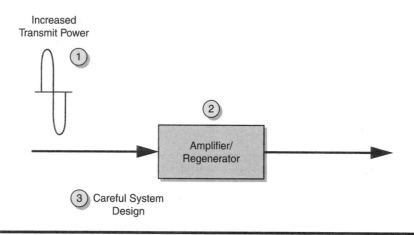

FIGURE 2.43 Options for dealing with attenuation and noise.

Attenuation describes the loss of energy as the signal travels through the medium, resulting in reduced amplitude, while noise will be picked up from other sources of electromagnetic energy, such as nearby cables or magnetic coils.

The noise levels picked up are usually very small, but the cumulative effect over distance, together with the attenuation of the original signal, can quickly degrade the channel to a point where it becomes unintelligible or large data errors occur. High-bandwidth (usually high data rate) channels tend to degrade faster. High-frequency signals also tend to degrade relatively quickly.

Transmission systems are designed to minimize attenuation and provide good immunity to noise. Fiber-optic cable is exceptionally good on both these counts and has become the medium of choice for high data rate (high-bandwidth) channels, especially in the core network.

Even for copper or radio systems, we can still mitigate the problems of attenuation and noise on long transmission paths by careful system design, and also by increasing transmitted power to compensate, or amplifying or regenerating the system at appropriate points in the transmission link before it becomes unintelligible (see Figure 2.43).

Amplifying an analog signal once it has attenuated and also picked up noise would result in a larger-amplitude signal that also retains the amplified noise. Whether this affects the user experience would depend on the proportion of noise in the overall signal and the modulation scheme used.

As long as a digital signal can be regenerated before the noise makes it difficult to recognize whether the bits are 1s or a 0s, the noise can be eliminated, and a *clean* set of data can be retransmitted. Of course, some mistakes will inevitably occur because of the random nature of the noise that is picked up, but these low levels of *bit errors* can be eliminated or minimized by advanced digital processing techniques in the receiver.

In general, the "noise immunity" that digital signals experience compared to analog systems ultimately manifests itself in many ways, including clearer voice channels, low error rate data channels, higher-capacity radio systems (for the same infrastructure costs), or longer transmission paths.

Review Questions

Q1. The frequency describes the _____.
   a. Number of complete electromagnetic sinusoidal variations (cycles) per second
   b. Pitch and amplitude
   c. Number of bits per second carried on the signal
   d. Rate of transmission

Q2. The amplitude of the wave describes the _____.
   a. Instantaneous frequency
   b. Wavelength of the signal
   c. Distance between sinusoidal peaks
   d. P voltage or current and can be used to find the power in a radio signal

Q3. The wavelength is _____.
   a. Proportional to the frequency
   b. Proportional to the amplitude
   c. Always constant
   d. Inversely proportional to the frequency

Q4. Propagation describes _____.
   a. How a signal alters with frequency
   b. How a signal decreases in strength due to wavelength
   c. The process of combining channels
   d. How the energy in the signal travels in the medium

Q5. Attenuation describes _____.
   a. How quickly a signal gains energy after reception
   b. How quickly a signal loses energy after transmission
   c. How fast the signal travels
   d. How fast the signal loses speed after transmission

Q6. The higher the frequency, _____.
   a. The greater the attenuation
   b. The lower the attenuation
   c. The greater the wavelength
   d. The greater the power

2.2.2 Modulation

Modulation modifies a signal to make it suitable for transmission over the selected transmission system.

Modulation Techniques

The principle behind modulation is that a carrier wave, normally at a fixed frequency, is modified by the signal that is to transmit across a medium. This can be achieved by applying various modulation techniques. At the receiving end of the system, demodulation is the technique by which the original signal is extracted from the carrier wave.

This method allows the use of different carrier frequencies to *carry* different information channels. Each carrier frequency is capable of carrying one or more information channels, depending on the system. For example, in GSM, each carrier frequency can carry up to eight voice channels (using a single carrier frequency, but dividing the time into eight regularly repeating *time slots*).

Modulation would be used to carry:

- Analog signals
- Digital signals

The carrier frequency to be modulated is often referred to as the *unmodulated carrier*, and can be represented as a sine wave at a specific frequency.

As the carrier is modulated with the analog or digital signal, the frequency spreads out to cover frequencies on either side of the specified carrier frequency. In general, this spreading depends on how efficient the modulation system actually is (systems vary markedly) and how much information is to be carried.

The amount of frequency spectrum needed to carry the whole (modulated) signal is referred to as the *bandwidth*. Bandwidth required increases with:

- Increasing information
- Less-efficient modulation schemes

Modulation techniques require alteration of one of the major characteristics of the carrier waveform. Modulation techniques are:

- Amplitude modulation (AM)
- Frequency modulation (FM)
- Phase modulation (PM)

Commercial radio stations often use AM or FM.

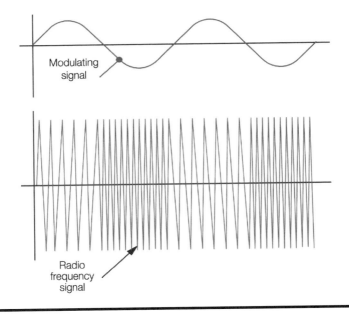

FIGURE 2.44 Frequency modulation. The frequency of the carrier is modulated.

When used to modulate digital signals, the amplitude, frequency, or phase would take discrete values to represent the digital information. In this case, the modulation would be termed *digital modulation*:

- Amplitude shift keying (ASK)
- Frequency shift keying (FSK)
- Phase shift keying (PSK)

GSM radio, for example, would use a form of FSK, whereas UMTS would use a form of PSK.

Example Modulation Scheme: Analog Frequency Modulation

The use of frequency modulation (FM) within the world of telecommunications is very common. Perhaps the best-known use is that within broadcast radio services. FM modulates the frequency of the carrier based upon the level of the modulating signal (Figure 2.44).

- Better signal-to-noise ratio than AM
- High-quality audio transmission
- Unaffected by signal-level variations

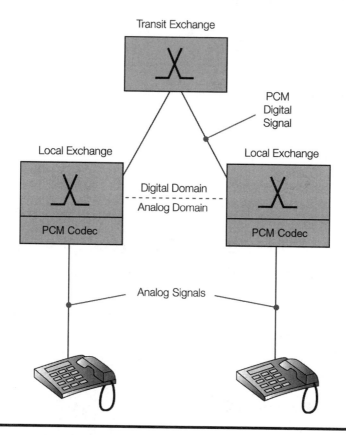

FIGURE 2.45 The need for pulse code modulation.

The main advantage of FM is its ability to be unaffected by signal-level fluctuations. The original signal can be decoded however much the broadcast signal varies in level and will continue to do so down to relatively low levels.

The signal-to-noise ratio is also improved with the use of FM. The process of modulating and demodulating within FM systems is more complex than that of amplitude modulation (AM) and consequentl more costly. However, FM has the advantage of being able to deliver much higher quality transmission.

Pulse Code Modulation

Because digital transmission and switching systems are used within modern telecommunications networks, there is a requirement to convert any analog signals that require transmission over the network into a digital form. Conversely, at the receiving end, the digital signals must be reconverted back to their original, analog form.

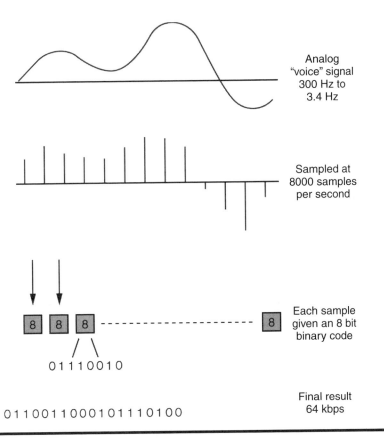

FIGURE 2.46 Pulse code modulation.

Pulse code modulation (PCM) is the process of initially digitizing the analog signal to enable it to be transferred effectively through the network (Figure 2.45). It modifies a signal to make it suitable for transmission over the selected transmission system, and hence the term "modulation" is used to describe it. However, PCM only digitizes the signal, and a second modulation technique would be required to transfer the newly digitized signal over the chosen transmission system.

Once a signal is digitized by PCM, it can be transferred over different transmission systems (and the corresponding modulation techniques) within the network, without having to convert back to analog form as the signal is passed from transmission system to transmission system. In fact, the signal usually (although not always) stays in PCM format until final conversion back to analog ready for presentation to the user.

The devices that perform the analog-to-digital (A-D) and digital-to-analog (D-A) conversions are known as *codecs* (coder/decoder) and are often located within street cabinets or in the local telephone exchange.

There are essentially three processes involved within the production of a PCM signal (Figure 2.46):

1. *Sampling.* The first stage of the PCM process is known as sampling. Here the analog waveform is measured at regular intervals. This frequency, at which the measurements are taken, is known as the sampling rate. The standard sampling rate employed for the A-D conversion of the voice within PCM systems is 8 kHz, or 8000 times per second.
2. *Quantization.* Analog signals have an infinite number of discrete values, between zero and the peak level of the signal, to represent the amplitude. For transmission on a digital network, however, the number of values that represents the amplitude of the signal must be defined. In PCM systems, once the samples of the source analog system have been taken, they must be rounded to the nearest value.

    In standard PCM, we use 256 values (or levels). This number was carefully chosen to provide adequate voice quality, but a reasonably low bandwidth (to allow relatively more channels to be carried over the transmission equipment). The levels are arranged to enable both quiet and loud sounds to be distinguished evenly (i.e., the levels are not linear).
3. *Coding.* Because we use 256 levels, we need 8 bits to represent each level, and the conversion between the level and the 8-bit representation is performed by the coder. There are two main PCM coding formats for this coding:

    μ-law — widely used in North America and Japan (pronounced "mew law")
    A-law — used in the rest of the world

With 8000 samples per second (each requiring 8 bits to represent the sampled level), this means that each channel will require 64 kbps to represent the voice or data. There will usually be a channel required in each direction (duplex).

Each PCM channel requires 64 kbps.

PCM was originally designed to digitize telephone-quality speech. Data can be carried within the PCM channels, as long as the information is initially presented as voice-band tones. This is the case with modem (modulator/demodulator) tones generated by a computer data card in the case of a *dial-up data connection*, or for facsimile (fax) tones.

Review Questions

Q1. Modulation is _____.
   a. The process of adding information to a signal
   b. Combining several channels together
   c. Another name for transmission
   d. Generating a radio signal in the transmitter

Q2. Modulation can be achieved by _____.
   a. Varying the frequency or amplitude of a signal
   b. Increasing the frequency and wavelength
   c. Digitally tuning the analog waveform
   d. Analog methods only

Q3. In the frequency spectrum (frequency domain), modulation will _____.
   a. Be shown as a spread of frequencies about the nominal single carrier (spot) frequency
   b. Require only a single spot frequency to illustrate it
   c. Have no effect
   d. Result in a decrease in the spread of frequencies

Q4. In a digital system, _____.
   a. The signal (waveform) is constantly varying
   b. The signal (waveform) varies only in defined (quantized) steps
   c. The frequency is always constant
   d. The signal (waveform) can be defined as stable

Q5. Which of these is not an advantage of digital signals?
   a. The ability to protect the signal using forward error correction
   b. The ability to encrypt the information
   c. The ability to detect errors in the information
   d. The ability to generate quantization noise

Q6. Digital signals _____.
   a. Allow better error protection than analog signals
   b. Have inferior error protection compared to analog signals
   c. Cannot be protected from errors
   d. Are immune from errors

2.2.3 Channels: Organizing the Information

The term "channel" is used to describe a path through a system or network. It can be used to describe different types of paths, for example, transmission channel, telephony channel, data channel, traffic channel, signaling channel, control channel, radio channel, etc.

Many different scenarios exist, leading to a range of channel types:

- *Direction.* Channels are often provided to support communication, or exchange of data, in both directions; and in this case, the channels are said to be bidirectional. If channels are provided in a single direction only, they are unidirectional channels.
- *Point-to-point or point-to-multipoint.* If the channel is provided between two specified points or users, it is termed point-to-point (e.g., telephony channels). If it is provided between a single point and multiple other points, it is termed point-to-multipoint. In both cases, the channels can be unidirectional or, more commonly, bi-directional.
- *Broadcast.* If a channel is provided between a single origin and many destinations, and it is unidirectional, it is termed "broadcast" (e.g., terrestrial television broadcasts).
- *Common terminology.* For communication channels, we use the following specific terms to describe how the channel operates:
  - *Duplex/full duplex:* allows simultaneous two-way communication.
  - *Half duplex:* transmission in either direction, but not in both directions simultaneously.
  - *Simplex:* the operation of a channel in one direction only, with no reverse capability.

The most common type of channel is probably the telephony channel, which is bi-directional, point-to-point, carrying voice information. It is sometimes referred to as a *voice* channel, *traffic* channel, or *telephony* channel. It is not always clear whether the *channel* being referred to means a single direction of transmission or both directions.

To allow for efficient transfer through a system or network, channels are very often combined using a process called *multiplexing*.

Channel Bandwidth

In general, the greater the amount of information (expressed as data rate) carried, the greater the bandwidth required (Figure 2.47). This can differ markedly from system to system, depending on features such as the modulation system used.

In practice, a system is usually designed with the anticipated channel *bandwidth* requirement as one of the main factors. Increasingly, we need systems with more and more flexibility in bandwidth provision, even changes in channel bandwidth mid-session or mid-call.

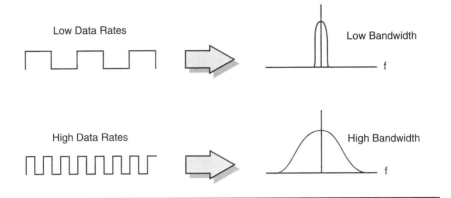

FIGURE 2.47 Relationship between data rates and bandwidth.

Radio systems, especially, are distinguished by:

- Flexibility of bandwidth allocation
- Efficiency relating to:
    - Number of channels
    - Information rate of channels
    - Channel bandwidth required
    - Geographical coverage or length of cable
    - Quality of channels
- Power required

Effectively, the lower the bandwidth required (for a specified channel data rate, coverage area, quality, and transmit power), the better. For newer *multimedia* systems, being able to flexibly alter the channel data rates (and therefore the other related parameters) means that different media services can be offered with similar efficiency — thus making the business case for these services much more attractive to the network operator.

The radio systems based on *cdma* (Code Division Multiple Access) are excellent in relation to the above requirements and provide the basis for many new radio implementations, including 3G cellular networks.

Channel Errors

One of the key benefits of digital technology lies in the ability to remove noise from a received signal (noise immunity). However, the process of recovering *noisy* received information is not without its problems.

To interpret the incoming data as 1s and 0s, the receiver will examine each bit in turn to determine its state. As an example, this decision can be based on comparing the amplitude of the incoming signal with a threshold mid-value. At the instant of bit detection, if the amplitude of the (noisy) signal is above the threshold, then it is interpreted as a logical 1. Below the threshold it is interpreted as a logical 0. Should the amplitude of the noise be significant compared to the amplitude of the signal, then misinterpretation of the incoming data can occur, resulting in *bit-errors*.

Voltage levels are a convenient way of illustrating errors, but different modulation techniques will also generate errors as the receiver differentiates between different frequencies, amplitudes, or phases.

In many transmission systems, errors may be a relatively minor problem, and some systems may experience a bit error rate (BER) of one error per several million bits.

Error Protection

For many systems, the existence of errors is a significant problem as it effectively corrupts the data or information being transferred. There are several different mechanisms available to minimize the final BER, falling into two categories as listed below. Some types of information are less affected by errors than others. For example, voice information and systems can cope with relatively larger errors due to the nature of human speech processing.

1. *Error detection* detects only the presence of errors within received data. Errors can be corrected by requesting a retransmission.
2. *Error correction* is more sophisticated and codes the data in such a way that the receiver is able to correct (an amount of) errors without the requirement for retransmission.

All types of error protection involve adding extra (error protection) bits to the transmitted information. No scheme is foolproof, but more reliable schemes generally require the addition of more error protection bits to the data. Extra bits, of course, increase the bandwidth requirements of the link.

The actual ratio of the transmitted signal arriving at the receiver compared to the "unwanted noise" in the same frequency band is critical in determining the amount of errors, or the amount of error protection required to overcome it (Figure 2.48). The term *signal-to-noise ratio* (SNR) is used to describe this important radio parameter. Poor SNR values cause errors.

Summary of Error Protection

Error detection:

- May request retransmission
- Adds extra bits
- Not perfect
- More bits improve performance

Technology Principles ■ 167

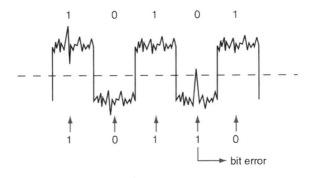

FIGURE 2.48 Example of errors in a received signal.

Error correction:

- No retransmission needed
- Adds extra bits
- Not perfect
- More bits improve performance

Error protection uses valuable capacity

Interleaving

Interleaving involves changing the order of the information plus error protection bits that are to be transmitted. This results in a sudden grouping of errors being more evenly distributed throughout the data stream once the information is put back in the correct order at the receiver. Error protection schemes are far more effective against evenly distributed errors than they are to sudden high groupings of errors. Interleaving therefore increases the effectiveness of the error protection employed.

Multiplexing Channels

Multiplexing is a technique by which several information channels can be combined together in a specified way, allowing them to be separated later. Multiplexing is therefore a very useful technique that provides multiple information channels in a very efficient, cost-effective, and manageable way. The multiplexing technique also forms the basis for the transmission system itself.

> **Multiplexing:** combining together several information channels to enable efficient, manageable, and cost-effective transmission to take place.

Frequency Division Multiplexing (FDM)

Frequency division multiplexing (FDM) is a multiplexing technique used to allow a single, high-bandwidth carrier frequency band to carry multiple information channels. Many different forms of FDM are used in telecommunications, although it is has largely been replaced within the core of a telecommunications network (where copper and fiber-optic transmission is widely used; i.e., bounded media) by time division multiplexing (TDM).

FDM can be found predominantly in radio systems in different forms, including GSM, UMTS, cdmaOne, and microwave links.

Time Division Multiplexing (TDM)

Time division multiplexing (TDM) is a method of multiplexing whereby a timeslot is allocated to each information channel input to the multiplexer.

In theory, any number of signals can be multiplexed together using TDM with the provision that the aggregate speed is at least the sum of the inputs.

An equal amount of data from each input is taken and multiplexed onto the aggregate signal. Data can be collated a bit or byte (8 bits) at a time (bit interleaving, byte interleaving).

In the example in Figure 2.49, each of the four inputs A, B, C, and D is switched on in turn so that a byte of data can be input to the multiplexer. The multiplexer continuously outputs the data from all of the inputs into an aggregate data stream. The arrow (shown) continuously cycles through the inputs, taking data (usually 1 bit at a time, or 8 bits at a time) from the inputs in turn before starting the process again. Each cycle of the data is referred to as a TDM frame.

If four inputs are used, as shown, then each input will have data appearing every fourth slot on the aggregate. The recurring slots are termed "timeslots" or channels. Figure 2.49 illustrates the use of timeslots (or channels) A, B, C, and D.

- Each input is allocated a timeslot.
- Bit or byte interleaving can be used.
- Aggregate speed must be at least the sum of the inputs.
- TDM systems have largely replaced earlier FDM systems in the core network.

Technology Principles ■ 169

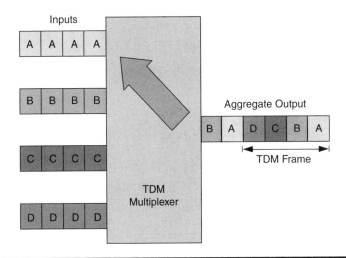

FIGURE 2.49 Time division multiplexing (TDM).

Standard systems have been developed for TDM. The main one used in current networks is the Plesiochronous Digital Hierarchy (PDH). PDH is widely used with:

- Coaxial cable
- Twisted-pair cable
- Microwave radio
- Fiber-optic cable

Review Questions

Q1. The frequency spread required for each channel (after modulation) is known as the _____.
  a. Data rate
  b. Symbol rate
  c. Bandwidth
  d. Baud rate

Q2. In similar systems, higher data rates generally require _____.
  a. Higher bandwidth
  b. Lower bandwidth
  c. Higher frequencies
  d. Lower frequencies

Q3. High-bandwidth channels _____.
   a. Must never be shared
   b. Can be shared by several end users in some systems
   c. Will always be shared by several end users
   d. Will always give less flexibility than several lower-bandwidth channels

Q4. Which is *not* correct? A channel _____.
   a. Is used to describe an identifiable path through a network, or between two pieces of equipment
   b. Is used to create a signal in a transmission system
   c. Provides a system for identifying frequencies
   d. Can be bi-directional

Q5. Duplex describes the following types of channel:
   a. A communications link that does not allow two-way transmission
   b. A communications link that allows two-way transmission
   c. A communications link that allows only one-way transmission
   d. A communications link that will not allow any transmission

Q6. Digital multiplexing _____.
   a. Is a way of carrying multiple digital channels over a single medium
   b. Is a way of altering the frequency or amplitude to represent data on a radio signal
   c. Allows signals to be multiplied and delivered to different endpoints
   d. Is used only rarely

2.2.4 Example Transmission Systems: Implementing the Concepts

> Primary multiplexing systems can be found in the core of the network, or in the access part, depending on the customer's requirements.

Primary Multiplexing

Primary multiplexing takes a number of 64-kbps channels (that are already in Pulse Code Modulation (PCM) format) and combines them into a new, faster aggregate

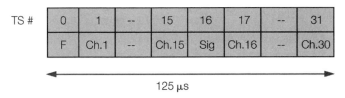

- 32 timeslots (carrying 30 voice/data channels)
- 8 bits per timeslot per frame
- 8000 frames per second
- Aggregate bit rate = 2.048 Mbps
- Timeslot 0 used for frame alignment and alarms
- Timeslot 16 can be used for signaling

FIGURE 2.50 The E1 frame structure (F = Frame).

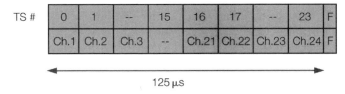

- 24 timeslots (can carry 23 or 24 voice/data channels)
- 8 bits per timeslot per frame
- 8000 frames per second
- Aggregate bit rate = 1.544 Mbps
- Signaling achieved through "bit stealing" or using one of the timeslots
- Framing bit added at end of frame

FIGURE 2.51 The T1 frame structure (F = Frame).

signal. This is achieved by Time Division Multiplexing (TDM). There are two standard primary signals that are used worldwide:

- E1, the European standard, takes 32 × 64-kbps channels (timeslots)
- T1, used in North America and Japan, which takes 24 × 64-kbps channels

These are multiplexed into 2.048-Mbps or 1.544-Mbps signals, respectively.

The 32 timeslots of the E1 primary signal would generally transport 30 voice or data channels. In addition, an extra channel (timeslot 16) is provided for signaling/control and another (timeslot 0) for link management (frame alignment, alarm signals, and error checking).

In a T1 system, signaling is achieved by "stealing" bits from the data and an extra bit follows the 24 timeslots for frame alignment. Figures 2.50 and 2.51 illustrate a single frame of the E1 and T1 multiplex frames, respectively.

E1 or T1 Transmission

Physical transmission of the E1 or T1 link is by:

- *Coaxial cable.* In countries such as the United Kingdom, coaxial cable is generally used for a *wired* E1 link. E1 links carry uni-directional signals only, and therefore a pair of coaxial cables is required for a bi-directional E1 link.
- *Twisted-pair cable.* In other countries (e.g., France), the twisted pair is most common. As above, a pair of twisted pairs is required.
- *Microwave radio.* Increasingly over recent years, the microwave point-to-point radio link has been employed for the transport of E1 or T1 links. Radio frequencies in the range 2 to 50 GHz are used with highly directional dish antennas. Such links often provide a very cost-effective option to the alternative of laying cables.
- *Fiber-optic cable.* Fiber offers several advantages over cable or microwave, not least of which is its tremendously high bandwidth and, therefore, capacity. The installation of fibers has increased over recent years, most notably in the form of SDH (Synchronous Digital Hierarchy), which itself can carry E1 or T1 signals.

PDH and SDH

Digital Hierarchies — The overall structure of modern networks is one of a hierarchy, where higher and higher capacity links are used as one progresses from the access node (local telephone exchange or access router) into the network. In the very core of the network, E1 or T1 links do not provide sufficient transmission efficiencies.

To solve this problem, a hierarchical system was developed to allow E1 or T1 links to further combine into higher and higher data rates. This system is known as the Plesiochronous Digital Hierarchy (PDH) and provided a good solution to the capacity problems.

Unfortunately, PDH was developed before the very high capacity optical cables were introduced, and had a high emphasis on efficiency rather than manageability and resilience. To take full advantage of the higher capacity (mainly optical) infrastructure, the Synchronous Digital Hierarchy (SDH) was developed.

> PDH and SDH are viewed mainly as core network systems, but can sometimes be found in the access network

Newer, even more efficient systems are being developed as the networks become more data-centric.

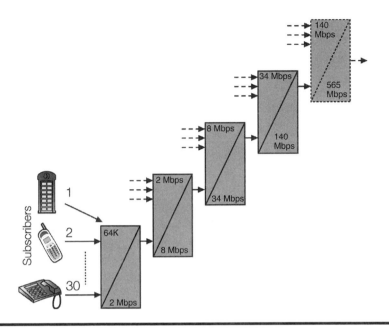

FIGURE 2.52 The Plesiochronous Digital Hierarchy (PDH) — European.

Plesiochronous Digital Hierarchy (PDH)

The PDH system effectively develops the idea of *primary multiplexing* using time division multiplexing (TDM) to generate faster signals. This is done in stages by first combining (multiplexing) E1 or T1 links into what are known as E2 or T2 links, and if required, going even further by combining (multiplexing) E2 or T2 links, etc.

This multiplexing hierarchy is known as the Plesiochronous Digital Hierarchy (PDH). Plesiochronous, meaning "almost synchronous," relates to the inputs that can be of slightly varying speeds relative to each other and the system's ability to cope with the differences.

These groups of signals can be transmitted as an electrical signal over a coaxial cable, as radio signals, or optically via fiber-optic systems. As such, PDH formed the backbone of early optical networks.

The European (and the rest of world) version of PDH is illustrated in Figure 2.52. The aggregate signal can be sent to line at any stage of the hierarchy, using the appropriate transmission medium and modulation techniques.

For comparison, the North American and Japanese PDH systems are shown in Table 2.2.

PDH Network Operation — PDH network equipment is now quite physically small, allowing for its deployment in locations other than a telephone exchange.

TABLE 2.2 North American and Japanese PDH Systems

PDH Level	Europe	North America	Japan	Trans-Atlantic
0	64	64	64	64
1	2,048	1,544	1,544	2,048
2	8,448	6,312	7,876	6,312
3	34,368	44,736	32,064	44,736
4	139,264	274,170	97,728	139,264
5	564,992	—		

Network operators are able to house such equipment in street cabinets; enough equipment to supply 120 telephone lines can be stored in an enclosure measuring 1.5 meter by 1 meter high. This has removed the need for many of the smaller telephone exchange buildings that we used to see in fixed networks.

PDH systems are generally used only for point-to-point communications systems because the signals must be fully demultiplexed to access a single information channel. In addition, proprietary alarm configuration and management means that equipment at either end of a PDH system must be from the same manufacturer. PDH advantages include:

- Equipment small enough for use in street cabinets
- Good for point-to-point connections
- Cost-effective support for access networks

PDH disadvantages include:

- Manufacturer-specific systems
- Multiplexer mountains
- No integrated network management
- Limited management available

Synchronous Digital Hierarchy (SDH)

Although it is a reliable system, PDH has a number of obvious shortcomings. When designing the next generation of transmission systems, consideration was given to overcoming these shortcomings. The Synchronous Digital Hierarchy (SDH) was developed from the American SONET (Synchronous Optical Network) and is designed to provide an effective, well-managed, reliable, and efficient system for use with optical-fiber (high-bandwidth) links. It was developed to be compatible with existing systems and can therefore carry PDH channels as well as other formats.

Although seen as an expensive option compared to the tried and trusted PDH alternative, the advantages of SDH are well recognized, and SDH is now

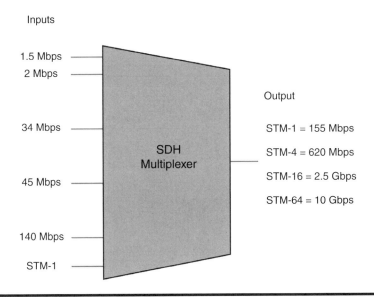

FIGURE 2.53   SDH multiplexing.

the accepted standard for digital transmission around the world. SDH has many advantages over PDH, most notably:

- It is designed to get the best out of high-capacity fiber-optic cables.
- It is compatible with many other accepted standards such as E1 and T1.
- It has built-in network performance monitoring and management facilities.
- It is compatible with both European and American standards.

SDH can multiplex together a variety of different digital signal types, including those that are already multiplexed using PDH, or even SDH (Figure 2.53). These signals are arranged by the system onto a standard frame, called a *synchronous transport module* (STM), ready for transmission. The smallest of these is STM-1, which operates at 155 Mbps. There are larger frames, denoted STM-x. The x merely implies the number of STM-1 equivalents transmitted (systems can employ STM-4, STM-16, STM-64, or even higher). The inputs are known as *tributaries*.

STM-1 is equivalent to 63 × E1 links, or 1890 telephone channels.

The common implementation throughout Europe is a 155.52-Mbps link (carrying many multiplexed channels) in STM-1 (synchronous transfer module) format, which can itself be multiplexed into higher capacity levels (mainly STM-4, STM-16, STM-64). These signals are typically transmitted over optical fiber, although it is possible to send STM-1 over modest distances using coaxial cable or radio.

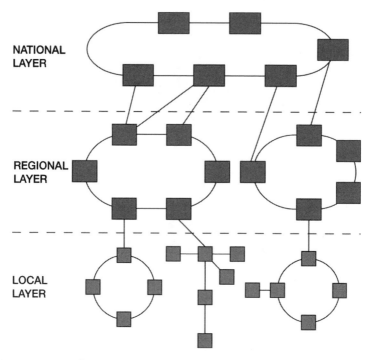

- Inputs at varying bit rates
- Single-stage multiplexer — no mountains!
- Built-in management capability
- Manufacturer interoperability

FIGURE 2.54 Possible SDH ring structures.

SDH Network Operation

Every voice or data channel is identifiable in the STM-x and allows selective demultiplexing. This has the advantage of eliminating the *multiplexer mountains* of PDH and allows new network structures beyond simple point-to-point connections. This also allows some or all of the channels to be effectively protected in case of a network failure. The ability to automatically protect traffic is an inherent feature of SDH.

SDH has inherent management capabilities built into its structure. It is possible to control and configure an entire network remotely. This has given rise to large NOCs (network operation centers) where an operator can monitor, identify, and react to any fault in a network within minutes.

Protection and management systems work best where the fiber optic (or other medium on which SDH is running) is organized in ring structures to provide alternative reconfigurable routes, and therefore more reliable connections for the user (Figure 2.54).

Dense Wavelength Division Multiplexing (DWDM)

Dense wavelength division multiplexing, sometimes referred to as wavelength division multiplexing (WDM), is the optical equivalent of frequency division multiplexing. It is a technique applied to extend the bandwidth of fiber backbone networks.

Essentially, DWDM works by combining and transmitting a number of signals at different wavelengths (colors) on the same optical fiber (Figure 2.55). As an example, it is possible to multiplex a number of 2.5-Gbps STM-16 signals onto a single fiber. If 16 channels are provided in the WDM system, then the capacity of the line is increased from 2.5 Gbps to 40 Gbps, with no need to upgrade the transmission fibers or line equipment.

As DWDM develops, the number of inputs continues to rise and manufacturers are constantly releasing systems with ever-increasing numbers of channels. The resultant bandwidth is consequently increasing.

One of the main advantages of DWDM is that it is both protocol and bit-rate independent, and thus able to carry different types of traffic at various speeds over a fiber-optic channel. This means that a DWDM system can carry data from different network transport systems (ATM, IP, SDH, and Ethernet systems, for example) alongside each other. Installing DWDM systems is one of the quickest and most cost-effective methods for service providers to increase capacity in their networks.

DWDM characteristics:

- Can carry ATM, IP, SDH. and Ethernet
- Uses existing fibers
- Expensive, but quick return on investment
- Ongoing development to increase channels and bandwidth

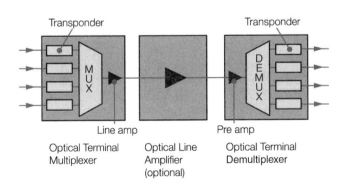

FIGURE 2.55 DWDM system. Transponders convert incoming light to a precise wavelength.

Analog and Digital Subscriber Loop

The Local Loop

The local loop is a pair of copper wires that connects the subscriber to the local telephone exchange.

> The *local loop* describes the pair of copper wires between the customer's premises and the local telephone exchange.

Analog Telephony Connection

The local loop carries bi-directional signals, traditionally in the form of analog (voltage) signals that vary directly with the human voice. The sound energy is turned into electrical energy by the microphone, and converted back into sound at the other end of the connection by the speaker (Figure 2.56a).

Integrated Services Digital Network (ISDN) Connections

The *Basic Rate ISDN* connection uses the existing copper wires to carry digital information to and from the network (Figure 2.56b). Using digital signals brings a number of advantages, allowing ISDN connections to provide:

- Two bi-directional channels (rather than one)
- A mix of voice and/or data channels (rather than one or the other)
- Increased data rates per channel
- Range of advanced services and features
- Clearer speech

Another form of ISDN connection is termed *primary rate,* and this provides connection at the *primary multiplexing* E1 rate, giving up to 30 bi-directional speech or data channels, delivered over copper or radio.

ISDN connections are more expensive than simple analog connections, and the limited value of the extra features and improvements limited the uptake to primarily business users.

If a business wants up to six channels, then basic rate interfaces would usually be provided; but if seven or more channels are required, then primary rate would be used. In terms of the actual network, the ISDN is virtually identical to the PSTN. The options for ISDN access are summarized below:

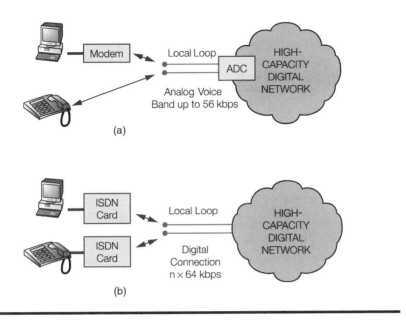

FIGURE 2.56 (a) The analog local loop; and (b) ISDN connections.

- *Basic Rate Access (2B+D):* provides two 64-kpbs traffic channels and a single 16-kbps signaling channel. This connection is most commonly used by domestic subscribers.
- *Primary Rate Access (30B+D):* used by organizations to provide 30 traffic channels at 64 kbps and a single signaling channel (also at 64 kbps).

ISDN is actually the first of the *Digital Subscriber Line* technologies.

Digital Subscriber Line (DSL)

The local loop is extremely useful in the sense that it connects subscribers to a very sophisticated high-capacity digital network. Relatively recent advances have examined ways in which high bit-rate signals can be carried over this local loop. Such advances include HDSL, ADSL, and VDSL (**H**igh Speed, **A**symmetrical, and **V**ery High Speed — Digital Subscriber Line, respectively), each being suitable for different specific scenarios. See Figure 2.57.

ADSL has so far had the biggest impact of these technologies, allowing fixed network operators to offer broadband services to the home using the existing copper wires. The uptake of broadband using this technology continues to be extremely high. It offers up to 8 Mbps of downlink (to the subscriber) capacity, with a reduced rate in the uplink. VDSL can offer up to 52 Mbps in the downlink direction, but over reduced distances compared to ADSL. High bandwidth services such as TV can be offered by these types of systems. It should be noted that all of these

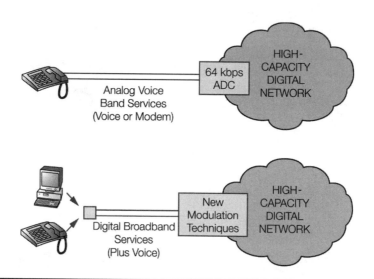

FIGURE 2.57 The local loop.

technologies are limited by distance. A few kilometers is a typical figure quoted between the local exchange and the user.

Symmetric DSL examples include:

- ISDN, 128 kbps (not broadband)
- HDSL, 2 Mbps

Asymmetric DSL examples include:

- ADSL, up to 8 Mbps downlink
- VDSL, up to 52 Mbps downlink

Review Questions

Q1. For most fixed telephone networks, _____.
   a. The basic connection type uses digital copper
   b. The basic connection type uses analog fiber optic
   c. The basic connection uses digital fiber
   d. The basic connection uses analog copper

Q2. Basic Rate ISDN connections use _____.
   a. PDH links
   b. SDH links
   c. Copper links
   d. Fiber-optic links

Q3. ADSL Broadband has been developed to make efficient use of the _____.
   a. Existing core network
   b. Existing copper access network
   c. Existing radio access network
   d. ISDN connections

Q4. PDH _____.
   a. Can only be provided over copper connections
   b. Can be provided in ring architecture only
   c. Uses analog technology
   d. Can be carried over copper, radio, or fiber connections

Q5. SDH would usually be implemented _____.
   a. In a point-to-point topology, _____.
   b. In a ring topology
   c. In one direction only
   d. As a backup to PDH

Q6. DWDM _____.
   a. Uses different PDH levels to generate information
   b. Carries several SDH signals at the same wavelength of light
   c. Uses ADSL and ISDN to format data
   d. Increases the capacity of fiber by providing extra transmission channels at different wavelengths of light

2.2.5 Review Question

1. In the diagrams shown below, mark the amplitude and wavelength.

If the wavelength is 9 cm, what is the frequency (speed of light = 3×10^8 m/s)?

Draw another waveform at two times the frequency of the one illustrated, but the same amplitude. What will the wavelength be?

Which of the two waves will tend to attenuate faster through a transmission medium such as copper?

What is the name given to describe the way the information moves through a transmission medium?

How would you carry information on your carrier? Name one technique.

If the "signal" is used to carry digital information, list three advantages of this scheme over an analog scheme.

Example waveform:

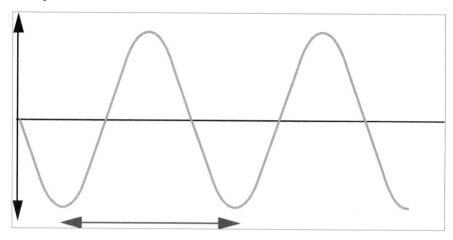

Example waveform at two times the frequency.

Practice Questions

Q1. The correct terms for the generation and "collection" of electrical energy are _____.
    a. Transfer and receipt
    b. Transceive and retrans
    c. Transmit and receive
    d. Transmission and receipt

Q2. Signals *cannot* be represented _____.
   a. Using sinusoidal waveforms
   b. Using rotating vectors
   c. In the frequency domain
   d. By flowcharts

Q3. The wavelength is _____.
   a. Proportional to the frequency
   b. Proportional to the amplitude
   c. Always constant
   d. Inversely proportional to the frequency

Q4. Attenuation describes _____.
   a. How quickly a signal gains energy after reception
   b. How quickly a signal loses energy after transmission
   c. How fast the signal travels
   d. How fast the signal loses speed after transmission

Q5 For similar systems, using a lower frequency _____.
   a. Usually requires less power to transfer the same amount of data over the same distance
   b. Usually requires more power
   c. Always gives a better error performance
   d. Will always give better audio quality

Q6. Modulation can be achieved by _____.
   a. Varying frequency or amplitude of a signal
   b. Increasing the frequency and wavelength
   c. Digitally tuning the analog waveform
   d. Analog methods only

Q7. In the frequency spectrum (frequency domain), modulation will _____.
   a. Be shown as a spread of frequencies about the nominal single carrier (spot) frequency
   b. Require only a single spot frequency to illustrate it
   c. Have no effect
   d. Result in a decrease in the spread of frequencies

Q8. In an analog system, _____.
   a. The signal (waveform) is constantly varying
   b. The signal (waveform) varies only in defined (quantized) steps
   c. The frequency is always constant
   d. The signal (waveform) can be defined as stable

Q9. Digital signals _____.
  a. Allow better error protection than analog signals
  b. Have inferior error protection compared to analog
  c. Cannot be protected from errors
  d. Are immune from errors

Q10. Analog audio signals _____.
  a. Can be better quality than digital on initial transmission
  b. Are always lower quality than digital
  c. Will be better than digital once amplified
  d. Cannot be digitized

Q11. For voice signals, analog-to-digital conversion _____.
  a. Always occurs in the handset/terminal
  b. Is often found in transmission equipment
  c. Would never occur in a connection where the access scheme is analog
  d. Is not necessary

Q12. Higher bandwidth channels usually _____.
  a. Carry higher frequencies
  b. Carry higher data rates
  c. Transmit at a lower modulation rate
  d. Generate less power

Q13. Greater capacity is provided by _____.
  a. Low-bandwidth channels
  b. Lower frequencies
  c. High-bandwidth channels
  d. Higher frequencies

Q14. Simplex describes which type of channel?
  a. A communications link that does not allow two-way transmission
  b. A communications link that allows two-way transmission
  c. A communications link that allows only one-way transmission
  d. A communications link that will not allow any transmission

Q15. Digital multiplexing _____.
  a. Is a way of carrying multiple digital channels over a single medium
  b. Is a way of altering the frequency or amplitude to represent data on a radio signal
  c. Allows signals to be multiplied and delivered to different endpoints
  d. Is used only rarely

Q16. ADSL Broadband was developed to make efficient use of the _____.
   a. Existing core network
   b. Existing copper access network
   c. Existing radio access network
   d. ISDN connections

Q17. PDH is _____.
   a. Primarily a point-to-point transmission system
   b. Designed to work over fiber
   c. An advanced transmission system designed for IP traffic
   d. Faster than SDH

Q18. SDH is _____.
   a. Suitable only for carrying over copper
   b. A less advanced system than PDH
   c. Is a well-managed and resilient transmission system
   d. Has low overheads compared to PDH in order to increase its speed of transmission

Q19. Which is correct?
   a. SDH cannot carry anything other than voice traffic.
   b. SDH can carry only PDH payloads.
   c. SDH is an analog system.
   d. SDH can carry a variety of traffic types, including PDH payloads.

Q20. DWDM is designed to _____.
   a. Make efficient use of copper
   b. Allow several channels to share a radio link
   c. Make efficient use of fiber optic
   d. Allow several channels to share a copper link

2.3 Radio and Cellular Systems

2.3.1 Radio Systems

Electromagnetic Radio Wave

A radio signal consists of magnetic and electrostatic energy. It can be represented as two sinusoidal waveforms, with the magnetic and electrostatic energy perpendicular to each other. In most representations, only the electrical waveform is shown for clarity.

The radio energy travels at approximately the speed of light from the point of transmission in a direction that is perpendicular to the two waveforms.

Electromagnetic energy in the form of radio waves is a commonly understood concept, but it should be noted that "current" within a wire and "flashes of light" within a fiber-optic cable also constitute electromagnetic energy.

Radio Spectrum

The radio spectrum (Figure 2.58) covers a very large scale. For convenience, the spectrum has been divided into a number of radio *bands*, which cover a range from very low frequencies (VLF) up to extra high frequencies (EHF). Each radio band has its own particular characteristics and, therefore, uses. Note that the scale used to nominally specify these bands is logarithmic.

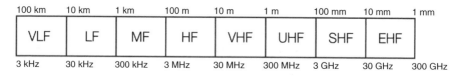

FIGURE 2.58   The radio spectrum. From left to right: increasing frequency, decreasing wavelength, decreasing antenna size, increasing bandwidth, increasing attenuation, line-of-sight propagation above HF.

UHF Band

The Ultra High Frequency (UHF) band (Figure 2.59) is of great importance in the modern telecommunications world because all the leading cellular technologies use this band. The characteristics of this band that have this include:

- *Bandwidth.* The UHF band ranges from 300 to 3000 MHz and therefore has a bandwidth of 2700 MHz. Lower frequency bands have insufficient bandwidth to support the amount of traffic that now travels over cellular systems.
- *Small antennas.* The size of an antenna is related to the wavelength of the signal, with a half wavelength or quarter wavelength antenna being common. At UHF frequencies, the antenna size is small enough to suit pocket-sized mobile devices.
- *Line of sight propagation.* At lower frequency bands (at HF and below), the radio signals are able to *bend* (via various mechanisms) and are therefore almost exclusively used for long-distance, *over-the-horizon* transmissions. The higher bands (VHF and above) are largely characterized by straight-line transmissions that limit the distance to the visible horizon.

    This is actually a major advantage for cellular systems in that (unlike long-distance transmissions) the local line-of-sight transmission allows the highly valuable radio spectrum to be reused many times in different geographical locations. This concept of frequency reuse forms the basis of all modern cellular systems.

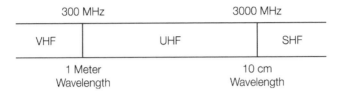

- UHF contains:
- GSM 900, 1800, 1900 MHz
- UMTS 2000 MHz • DECT
- cdmaOne
- cdma2000
- All leading cellular technologies
- UHF properties:
- Wide bandwidth (2,700 MHz)
- Small antennas
- Line of sight - local transmissions
- Frequency reuse
- Attenuation increases significantly above UHF

FIGURE 2.59 The Ultra High Frequency (UHF) band.

- *Limited attenuation of "free space" radio waves.* There is an obvious attraction to higher frequency bands (above UHF) in the search for evermore scarce and valuable radio spectra. Above UHF, however, attenuation of a radio wave increases significantly, meaning that the distance (from mobile device to base station) is reduced. Providing wide area coverage at these higher frequencies would therefore be prohibitively expensive in terms of the number of base stations (and sites) required. (Weather conditions can also begin to affect radio signals at higher SHF frequencies). The spectrum above the UHF band, however, certainly offers wide bandwidth opportunities for indoor or small local area applications.

Radio Propagation

Propagation at UHF — Although *line of sight* indicates an upper limit of the visible horizon, the actual usable distance of a UHF transmission can vary enormously, depending on the transmitted power and the way the signal propagates (travels). The following summarizes these propagation effects (Figure 2.60):

- *Reflection.* Radio waves reflect from surfaces. Generally speaking, the more conductive the surface, the better it will reflect electromagnetic energy. Metal and water are good reflectors. Mountains and buildings can affect cellular transmissions.

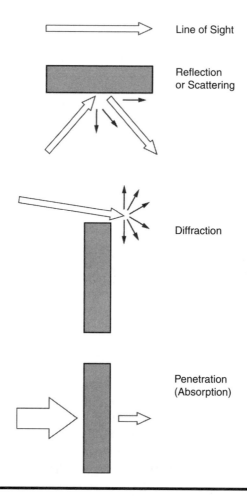

FIGURE 2.60 UHF radio wave propagation characteristics.

- *Refraction.* UHF radio waves will refract (bend) slightly over the visible horizon.
- *Penetration.* Very significantly, radio waves penetrate through solid material such as glass or the walls of buildings. On passing through such obstacles, the radio wave will attenuate (lose power), with overall attenuation being a function of the frequency and also the composition of the material and its thickness.
- *Diffraction.* Diffraction is the "scattering" of radio energy that naturally occurs at the edges of solid objects such as buildings.
- *Multipath propagation.* The above effects combine to cause a phenomenon known as multipath propagation. In many cases, a received signal is a combination of several *radio paths*, either direct or indirect (via a reflection, for example). This causes two distinct problems in cellular systems: fading and distortion.

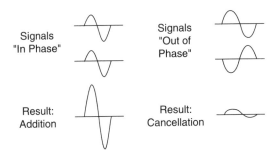

FIGURE 2.61 Fading.

Effects of Multipath Propagation

Fading — When two or more radio waves combine at a receiver, they will add together to form a resultant signal, the amplitude of which can be either greater than, or less than, the amplitude of the individual components — depending on the phase relationship of those components. Two roughly equal and opposite waveforms can cause close to complete cancellation of the received signal. Due to the short wavelength of cellular systems, the distance between a faded (cancellation) and a strong (additive) received signal may be relatively short. In some cases, for a user in the middle of a telephone call, only a few feet may separate a high level from a low level of received signal; see Figure 2.61.

Inter-Symbol Interference

Another result of multipath propagation is *intersymbol interference* (ISI). Rather than the radio signal itself, ISI refers to the distortion of digital information that results from several versions of the same information (time delayed relative to each other) being received. This problem can be minimized by adding extra reference data to the transmitted information to allow the receiver to configure its processing appropriately (as is the case for the GSM radio system).

Errors in Radio Systems

Radio systems are particularly susceptible to errors caused primarily by interference from other sources and systems, and often from transmissions on the same frequency emanating from other radios stations in the same network. Very careful design and optimization of error detection and error correction techniques can minimize the overall effect on the radio channel, and ultimately the user experience or efficiency or cost-effectiveness of the network itself. All digital cellular systems are designed with error correction and detection techniques as a fundamental part of the radio interface scheme.

Example Radio Systems
Introduction

When most people are asked to list applications of radio technologies, broadcast TV and commercial radio services are normally at the top of the list. However, radio is used in many, many applications in many different industry sectors. Some examples include:

- *Analog television.* Analog television has been broadcast for many years. The end user requires a suitable television set and an antenna.
- *Digital terrestrial television (DTT).* Occupying the same frequency range as analog transmission, digital signals are slotted in between existing analog signals and in the main distributed by the same transmitter network. End users can use their existing antennae and only require a compatible TV or a set-top box to decode the digital signals.
- *Cellular telecommunications.* Developments in radio technology have brought about the introduction of cellular telephony systems, with each successive generation's offering better voice quality with more services and applications to cellular subscribers. Radio is used between the user's terminal (telephone or other device) and the cellular base station. Examples include:
  - GSM
  - UMTS
  - cdmaOne
  - cdma2000
- *Satellite systems.* Satellites can extend coverage for communication systems to provide truly global coverage — whether for broadcast (TV or radio stations), telephony, or data services.
- *Direct broadcasting satellite.* Direct broadcast satellite (DBS) is an information delivery system that transmits sound and images in digital form to a television set at a high speed. It is an alternative to cable television and can expand viewing choices up to 500 channels through specialist antennae and set-top boxes.
- *Microwave (point-to-point) links.* Microwave radio links can be used to carry information from one specific point to another, and are extensively used to carry voice and data across both mobile and fixed networks, where the laying of cables is either impractical or the costs involved would be prohibitive.
- *Private mobile radio (PMR).* "Private mobile radio" is the term given to describe a self-provided radio network. This is essentially a network where radio communication is intended for use within an individually licensed, closed user group. Examples of where PMR is used include taxi companies, the emergency services, and the construction industry.
- *Public access mobile radio (PAMR).* Public access mobile radio is basically a PMR radio system with a private operator providing a radio service to a closed user group or a number of closed user groups.

- *TETRA.* TErrestrial Trunked RAdio is the digital radio technology that includes PMR and PAMR, and is similar in operation to GSM but is intended for business users such as taxi companies; fleet operators; site radio; police, ambulance, and fire services; security services; utilities; the military; and public access. A TETRA network is installed, set up, and maintained by a network provider, from whom the air time is then leased by the user.
- *Bluetooth.* Bluetooth is a relatively new standard that uses a short-range radio, enabling wireless connectivity between mobile phones, mobile PCs, hand-held computers, and other peripherals — without the need for cables.
- *WiFi.* Wireless Fidelity (otherwise known as wireless local area network — Wireless LAN — or WLAN) offers a high-speed connection for devices such as laptop computers or smart phones in localized hotspots, at speeds of up to 50 to 100 Mbps.
- *WiMax.* Worldwide Interoperability for Microwave Access (WiMax) offers a high bandwidth, relatively long-distance radio connections, and can be used as a cost-effective solution to deliver broadband services to residential and business customers where cables do not already exist. In addition, it can be used in place of traditional microwave links, including carrying channels from a WiFi (WLAN) installation to the core network.

Wireless Connections

Wireless devices and products can be connected in a number of ways. These include peer-to-peer, ad hoc, and infrastructure-style connections.

- *Peer-to-peer.* Two devices that are directly connected to each other via a wireless circuit can be called a peer-to-peer connection. This would allow the devices to share information between each other. These types of connections are rare.
- *Ad hoc.* Where two or more devices are present, it may be possible for them to connect in ad-hoc mode without the need for voice or data hubs, switches, or routers to control the flow of data. Ad-hoc connections are generally used only over a short distance, and only for a few users, because the radio channels will become congested if too many users are part of the same ad-hoc network.
- *Infrastructure.* A common method of connecting radio devices is via various forms of radio access point. For computers, this could be a wireless router using WiFi technology; and for cellular systems, it would involve a radio station that provides access within a specific geographical area, or cell. These access points or radio stations are often one of many, providing radio coverage over a larger area. All the access points in the system would connect back into the network infrastructure, allowing access to appropriate services and applications, including interconnections to other networks and systems.

Review Questions

Q1. A radio signal travels _____.
   a. At the speed of sound
   b. At the speed of light
   c. Around the world in eight seconds
   d. At 100 meters per second in copper

Q2. Interference is caused by _____.
   a. The bandwidth being too low
   b. Low power
   c. Channels of the same bandwidth
   d. Channels operating at the same (or overlapping) frequencies

Q3. Multipath propagation _____.
   a. Always causes several audio signals to be heard at once
   b. Can reduce the quality of a channel
   c. Is caused by using different sources for the radio signal
   d. Increases the quality of a channel

Q4. In a digital radio system, _____.
   a. Bit errors cannot be detected
   b. Interference will not cause any errors
   c. Errors can be detected but no correction is possible
   d. Errors can be detected and often corrected

Q5. Wireless wide area coverage could be provided by _____.
   a. Bluetooth
   b. WiFi
   c. GSM
   d. Microwave links

Q6. Wireless local area coverage could be provided by _____.
   a. Satellite systems
   b. WiFi
   c. GSM
   d. Microwave links

2.3.2 The Cellular Concept: Mobile Network Basics

The Network — All mobile networks can be divided quite simply into three basic elements:

1. *User terminal.* The user terminal is the mobile element — the subscriber's handset. The terminal is referred to using different terminology, including user equipment, mobile telephone, mobile handset, user terminal, mobile station, or user device.
2. *Radio access network.* The radio access network describes the series of base stations that send and receive the radio signals to and from user terminals. Within the radio access network, there will also be various elements that control these base stations. The radio path between the base station infrastructure and the mobile terminals is often referred to as the *air interface*.

    The base stations are arranged such that each has responsibility for providing a particular geographical area with radio coverage. These regions are known as *cells*, and hence such networks are often described as *cellular* networks.

    In a perfect planning environment, an evenly spaced set of such base stations would result in a series of interconnecting hexagonal cells, although in practice the real shapes of these cells will vary. If a user makes a telephone call and leaves one coverage area (or cell) and enters another one, the call will be *handed over* to ensure that the user experiences a continuous service.
3. *Core network.* The core network is a fixed telecommunications network that interfaces the radio access network with the rest of the telecom world, be that other mobile networks, the Internet, corporate intranets, or the Public Switched Telephone Network (PSTN).

    The core network will include elements that manage the subscriber's information, mobility and access rights; manage the efficient running of the whole network; and the delivery of services through the radio access to the user. Figure 2-62 illustrates the basic elements of a mobile network.

Cellular Principles

The propagation characteristics of UHF provide the capability of frequency reuse. A given amount of radio spectrum (a radio carrier) used at a particular geographical location can be reused at some minimum distance provided that the distance is sufficient to reduce mutual interference to manageable levels. Thus, a cellular system consists of a repeating pattern of cells operating at different frequencies.

The repeating pattern is referred to as a cluster of cells. Within GSM, for example, a cluster might typically consist of a seven- or nine-cell repeating cluster pattern (Figure 2.63). Typically, more complex patterns involving multi-cell sites are often used, particularly in urban environments. Cells operating at the same radio frequency are called co-channel cells.

C/I (Carrier-to-Interference) Ratio

Within a particular cell, there will be interference from nearby co-channel cells (Figure 2.64). The challenge with GSM cell planning is to keep this interference

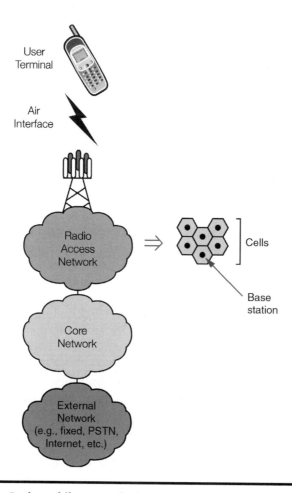

FIGURE 2.62 Basic mobile network elements.

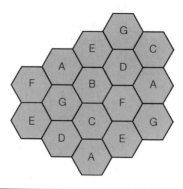

FIGURE 2.63 Seven-cell repeating cluster.

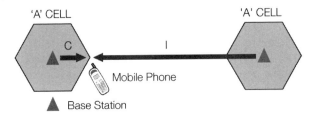

FIGURE 2.64 Cellular principles.

to manageable levels. The ratio between the signal strength of the wanted (carrier) compared to the interference from other cells on the same frequency is known as the C/I (carrier-to-interference) ratio. Cell planning also involves considerations of adjacent channel frequencies in addition to co-channel frequencies, as frequencies that are very close to others can also cause interference.

Example Cellular Systems

Figure 2.65 illustrates the generational evolution of cellular systems:

- 1G (analog) systems include technologies such as TACS (Total Access Communication System) in the United Kingdom, NMT (Nordic Mobile Telephone) in Scandinavia, and AMPS (Advanced Mobile Phone System) in North America.
- 2G (digital) systems include the most well-known cellular standard, GSM (Global System for Mobile communications), deployed initially in Europe and now in most parts of the world, including North America and Asia; and TDMA, the North American digital enhancement to AMPS, deployed in North America and other parts of the world (not Europe).
- 3G systems include:
  - UMTS (Universal Mobile Telecommunication System), which provides an evolution from GSM to 3G and is expected to be the most widely adopted 3G standard
  - cdma2000, the evolution of cdmaOne, which will be deployed mainly in North America and Asia
  - UWC 136, the TDMA 3G evolution, which relies on EDGE (Enhanced Data for Global Evolution) technology for its increased 3G data rates, but will not see widespread deployment
  - In addition, DECT (Digital Enhanced Cordless Telephony) is regarded as a true 3G system by the International Telecommunication Union (ITU)

Cellular Spectrum Use

Cellular systems tend to use frequencies located in a congested part of the spectrum. This is due to the characteristics demanded by all cellular systems — small antenna and propagation that allows adequate (and controllable) area coverage.

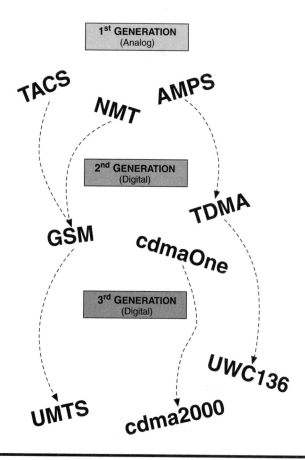

FIGURE 2.65 Evolution of 1G, 2G, and 3G systems.

Because the spectrum is in demand for radio technologies used to support a wide variety of industries and applications, it needs to be tightly controlled and allocated, even across national boundaries. The International Telecommunications Union — Radio (ITU-R) oversees the process, and spectrum is allocated during World Radio Congress (WRC) meetings. The allocation is increasingly on a world or regional basis, and it is then up to national authorities to further allocate the spectrum accordingly.

National authorities either allocate or auction spectra to the highest bidders and, typically, regulation within the country would demand that several network operators each share the overall available bandwidth (Figure 2.66). 3G spectrum auctions raised upward of $30 billion in Germany and the United Kingdom (between five operators). Spectrum can be allocated for use with a specific technology, or in some cases, may be left technology agnostic.

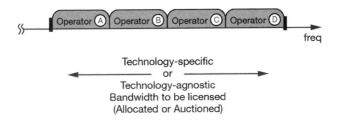

FIGURE 2.66 Spectrum is a scarce resource. Typical cellular frequency bands: 400 MHz, 800 MHz, 900 MHz, 1800 MHz, 1900 MHz, and 2 GHz.

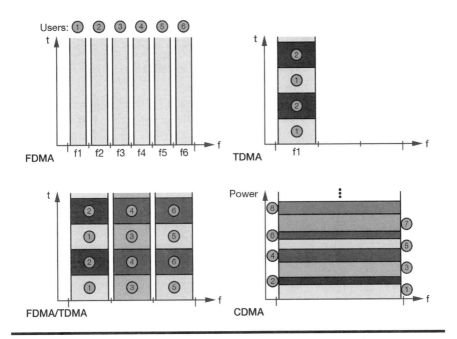

FIGURE 2.67 Radio access schemes.

Radio Access Schemes

In transmitting different signals within a given frequency allocation, there are two basic ways to divide this spectrum (see Figure 2.67):

1. *FDMA.* Frequency Division Multiple Access schemes divide a spectrum allocation into smaller frequency segments, allocating each signal a different frequency. Simple first-generation systems used this method.

2. *TDMA.* Time Division Multiple Access allows signal transmission on the same frequencies, but not at the same time — each signal is given its own timeslot within this frequency band.

   GSM uses a combination of both FDMA and TDMA. Network operators are allocated a portion of spectrum, which is divided into radio carrier frequencies spaced 200 kHz apart (FDMA). Each carrier frequency band is then divided into eight separate timeslots (TDMA).

3. *CDMA.* The third type of access scheme, CDMA (Code Division Multiple Access), allows all signals to share the same frequency and time domains. To distinguish signals at the receiver, unique codes are used on each signal. A common analogy made between the TDMA and CDMA schemes, which are the basis of 2G cellular systems, is as follows.

Imagine a crowded room. In a TDMA system, everyone in the room is speaking the same language. Therefore, to hear someone speaking on the other side of the room, it is necessary for everyone else to stop speaking. Each person could therefore be allocated a recurring timeslot during which they can speak, with multiple conversations supported by allocating a different timeslot to each person.

In CDMA, everyone in the room is speaking a different language. So, even when other people in the room are speaking at the same time, it is still possible to pick out what people on the other side of the room are saying, as long as they are speaking the language that one can understand. The different codes allocated to CDMA channels are equivalent to the languages used in the different conversations (i.e., channels) in the room.

Radio Access Technologies

Table 2.3 indicates the principal access methods and radio bands employed by various mobile radio systems.

As can be seen from Table 2.3, GSM, GPRS, and EDGE technologies use the same access methods and radio spectrum. The key difference between each technology is the way that each allocates bandwidth or resources (channels) to each user; the amount of data carried by each system per channel varies widely.

Most third-generation (3G) systems make use of CDMA technology because it is generally accepted that CDMA is more efficient in terms of spectrum usage than TDMA systems. Additionally, CDMA-based systems employ advanced receivers that have superior noise-rejection properties.

WiFi (or WLAN) can use various access schemes and frequency bands but primarily uses a combination of CDMA and FDMA.

The high price paid for many of the 3G licenses worldwide has demonstrated that radio spectrum is a scarce resource, and one that must be used as efficiently as possible to realize a good return on investment.

TABLE 2.3 Principal Access Methods and Radio Bands

Technology	Access Method	Radio Bands
GSM	FDMA/TDMA	400, 800, 900, 1800, and 1900 MHz
GPRS	FDMA/TDMA	400, 800, 900, 1800, and 1900 MHz
EDGE	FDMA/TDMA	400, 800, 900, 1800, and 1900 MHz
UMTS (W-CDMA)	CDMA	1.9–2.2 GHz; U.S. 1.85–1.99 GHz
cdma2000	CDMA	450, 800, 1700, 1900 MHz
WLAN (802.11b)	FDMA	2.4 GHz

Cell Planning

Cell Planning Basics

Cell planning differs significantly from technology to technology, but the primary aim is always to ensure that a specified geographic area is served by a continuous *coverage* plan, where the radio sites provide the required number of channels (of the required type), reliably and efficiently. A summary is that the cell plan must account for:

- Coverage
- Capacity
- Quality

For FDMA schemes, reusing the same frequency many times within the same network allows for very efficient use of the radio spectrum. Remember that the frequency cannot be reused within a distance that will cause too much interference with the current cell. Because power is generally set to provide sufficient coverage for the cell in question, the smaller the cells, the lower the reuse distance. Hence, with careful power control, cell planning becomes purely a geometrical problem.

Second-generation (2G) systems are more interference tolerant than the analog first-generation (1G) systems, allowing frequencies to be reused closer together within the reuse plan. This potentially means that more capacity is available within each cell as more frequencies can be used with less radio sites. Or to look at it another way, the same capacity can be provided by fewer cells.

In practice, software tools are used to plan networks, and the final plan is often unrecognizable as a *formal reuse pattern* due to factors such as terrain, varying density of customer use, radio sites not reflecting exactly (or sometimes even close to) the initially planned cell center, etc. (see Figure 2.68).

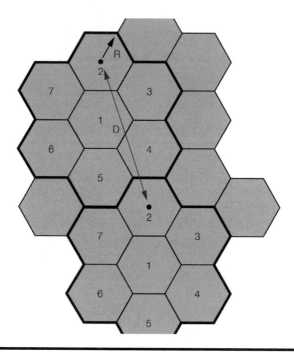

FIGURE 2.68 Traditional cell planning showing a seven-cell reuse pattern.

Sectorization

One simple way to reduce interference (for both FDMA/TDMA and CDMA systems) is to use sectored cells, wherein a single radio site can support up to six sectors (or cells) at a time. The directional antennae mean that transmitted energy can be directed, and received energy enhanced, allowing a single site to serve several sectors (or cells) at the same time, while interference from mobile devices outside the sector is minimized.

A sectored site is illustrated in Figure 2.69, showing the areas of intra-radio site overlap. As a user travels through these areas while making a telephone call, the user device would be handed over from one cell to the other, with the connection reconfigured at the same time within the radio site to keep the connection active.

CDMA Cellular Planning Principles

The CDMA planning concept initially seems simpler than for FDMA/TDMA systems. It is based generally on a single cell repeat pattern, where the same CDMA carrier frequency can be used in adjacent cells continually throughout the network. Users and control data are separated by the use of codes (Figure 2.70).

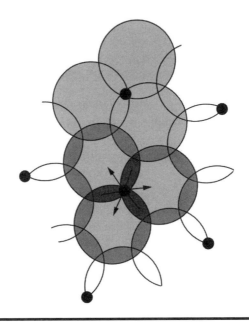

FIGURE 2.69 A sectored site.

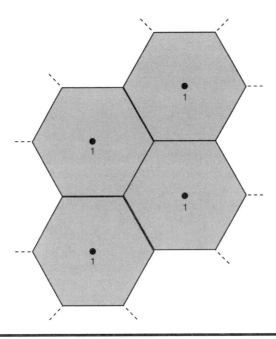

FIGURE 2.70 CDMA cellular planning. Codes are used to separate users and channels.

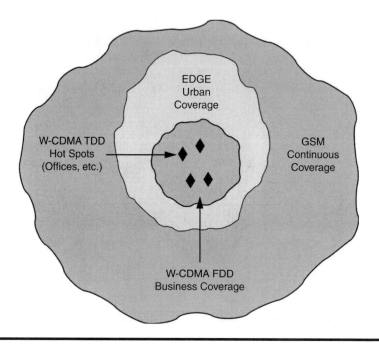

FIGURE 2.71 Examples of technology coverage areas.

In practice, planning for CDMA systems is more complex than for FDMA/TDMA systems, especially when considering the complex nature of the services and channels that need to be provided in advanced multimedia-capable CDMA systems.

Technology Coverage Areas

Although 2G cellular networks are generally planned to cover a high percentage of the country's population, enhancements (or technology upgrades) can be applied in a more selective way.

If an operator holds licenses for several radio access technologies, the network can be planned with the different technologies providing support in different geographical regions. The coverage is considered in terms of an overall business plan, with costs of providing the radio sites and potential revenue being the dominant factor (although other factors may be involved, such as requirements of the license).

Figure 2.71 provides an example where GSM and GPRS provide continuous coverage for voice and data (respectively) over a wide area. Any GSM or GPRS operator can upgrade to a technology called EDGE (Enhanced Data for Global Evolution) to provide higher data rates, and in this case it has been made available in the urban areas where a higher demand for data services is expected. If the operator also has a 3G W-CDMA (Wideband Code Division Multiple Access) license, this can be used to provide even higher speed data in the business area, where demand is greatest. In certain *hotspots* (offices, cafes, airports, etc.), a technology such as WiFi, or even a specialized version of W-CDMA (known as Time Division Duplex — TDD mode) can provide very high speed connections.

In practice, an operator may choose not to use EDGE to simplify the network, concentrating on GSM or GPRS and W-CDMA. In addition, FDD mode may be considered too expensive, with WiFi providing a suitable alternative. License agreements may require greater coverage for 3G. There are many different scenarios — varying country by country, network by network, and even region by region.

Review Questions

Q1. Cellular radio systems such as GSM benefit from _____.
   a. Forward error correction and encryption
   b. Analog forward error correction
   c. Encryption and simplex communication
   d. Error detection and analog coding

Q2. The process of using a frequency (or frequency pair) more than once in a network is known as _____.
   a. Frequency modulation
   b. Frequency reuse
   c. Frequency reallocation
   d. Frequency mobility

Q3. The process of allowing multiple users to access a radio system is called _____.
   a. Multiplexing access
   b. Access multiplexing
   c. Multiple access
   d. Modulation access

Q4. The area served by a frequency or set of frequencies is known as a _____.
   a. Footprint
   b. Radio site
   c. Location
   d. Cell

Q5. If a single radio site serves more than one cell, it is known as a _____.
   a. Double site
   b. Sectored site
   c. Directional site
   d. Focused site

Q6. Large cells that provide continuous coverage over a geographic area are usually known as _____.
   a. Pico cells
   b. Omnidirectional cells
   c. Micro cells
   d. Macro cells

2.3.3 GSM, GPRS, and EDGE

The GSM (Global System for Mobile communications) Specifications

The GSM system is defined by a set of published standards known as the GSM Specifications, first published in 1990 but significantly updated throughout the period since then. Phase 1 was published in 1990, with Phase 2 in 1995 bringing many enhancements. Since 1995 there have periodic updates (initially on a yearly basis until 1999, but now less frequently), allowing different features to be specified, standardized, published, and made available to the networks and customers. These later enhancements are known as Phase 2+ (Phase Two Plus) and differentiated by the year, or "release." They would be, respectively, Phase 2+ (Release 96), Phase 2+ (Release 97), etc.

GSM Phases and Releases:

- 1990 Phase 1
- 1995 Phase 2
- 1996 Phase 2+ (yearly releases)

Phase 2+ includes the details for GPRS, and hence forms the basis for the term 2.5G (generation 2.5).

- GSM is regarded as a 2G technology
- GPRS is regarded as a 2.5G technology

The Ongoing Success of GSM

The 1.5 billion subscriber milestone was passed early in 2005, and the number of subscribers continues to grow rapidly. GSM is the only 2G cellular system operational within Europe and enjoys high penetration levels in all Western European

TABLE 2.4 Current and Predicted GSM Subscribers

	2005	2006	2007	2008	2009	2010
Subscribers (billions)	1.57	1.78	1.91	1.92	1.83	1.65

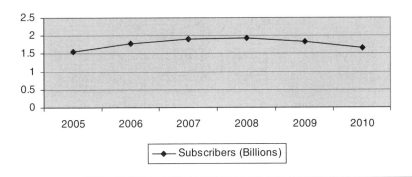

FIGURE 2.72   GSM predictions. (*Source:* **Informa Telecoms and Media.**)

countries. It is also the dominant technology in all areas of the world with the exception of the United States and Japan. Even so, both the United States and Japan have extensive coverage of GSM. In many countries, penetration is approaching (or at, or even exceeding) 100 percent (based on the number of contract and prepay subscriptions), although in developing countries, penetration may still be less than 5 or 10 percent.

In many forecasts, the number of GSM subscribers is predicted to start falling in 2008, but it should be noted that 3G W-CDMA subscribers are predicted to rise significantly at that time. As the two technologies are part of the same overall system or network, the overall number of subscribers for the "GSM family of technologies" is actually predicted to continue to rise significantly (Table 2.4 and Figure 2.72).

GSM Frequency Allocation

The spectrum assigned for the various GSM bands is shown below. In each case, there is an uplink (from mobile phone to base station) and a downlink (from base station to mobile phone). It is standard practice for the uplink to use the lower frequencies in cellular systems as the attenuation of this link will be (marginally) less than the corresponding downlink.

Transmitting and receiving on different frequency bands is known as Frequency Division Duplex (FDD). Each of the bands shown is divided into a number of radio

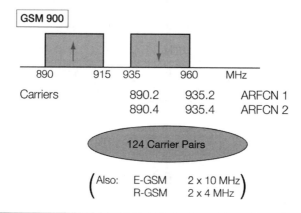

FIGURE 2.73 GSM 900 band spectrum. GSM 850 and 1900 bands may also be available, mainly in North America.

carriers. These carriers are separated in all cases by 200 kHz (0.2 MHz). The GSM 900 bands are each 25 MHz wide and therefore each contains 124 carriers. The "end" carriers are positioned at 200 kHz inside the band edges. That is, the lowest carrier in the 900 uplink band is 890.2 MHz and the highest is 914.8 MHz.

The carrier separation in the 1800 bands is the same as for the 900 bands. As the bands are each three times wider (at 75 MHz), this gives 374 carriers in each of the 1800 bands. The 1900 bands (United States) contain 299 carriers.

Carriers are assigned in pairs (uplink and downlink) and each pair is always separated by a fixed amount: 45 MHz at 900 and 95 MHz at 1800. A number called the Absolute Radio Frequency Channel Number (ARFCN) commonly refers to the pairs. In addition, GSM 850 and 1900 bands are also allocated on a required basis, predominantly in North America (Figure 2.73).

Time Division Multiple Access (TDMA)

Perhaps the most striking feature of the GSM air interface is its use of TDMA. TDMA is very commonly used in all areas of digital transmission and, of course, GSM is a digital technology.

Each carrier in GSM (uplink and downlink) is divided into eight timeslots and can be used by up to eight users at the same time (Figure 2.74). One advantage of this approach is that up to eight traffic channels are provided by a single radio transceiver. The timeslots are numbered from 0 to 7. A repeating timeslot is referred to as a physical channel. within standard GSM, one or two timeslots on the first Absolute Radio Frequency Channel Number (ARFCN) are utilized for signaling and control functions. This structure of timeslots described above is used in GSM, GPRS, and EDGE technologies, allowing all three to share the same radio resources.

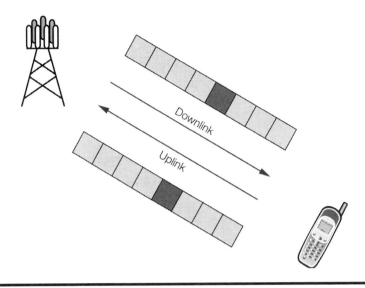

FIGURE 2.74   GSM timeslots.

For standard GSM, some of the data within the timeslot is used for user data and the remainder is used for error protection bits, allowing the data to be reconstructed should there be interference over the air (radio) interface. The overall data rate is 22.8 kbps, with speech requiring 12.2 kbps, and dial-up data connections (circuit-switched) requiring 9.6 kbps or 14.4 kbps (varies from network to network), with the rest of the capacity being used for error protection.

GPRS and Packet Data in GSM

GPRS adds a packet-based data capability to the GSM system, and brings IP networking into the GSM world. As well as modifications within the handsets and the radio interface, GPRS also involves the addition of new packet routers into the GSM core network, producing a core network with separate circuit- and packet-switched domains.

Using GPRS, each user only requires radio resources when a (usually Internet Protocol) packet of information is sent or arrives — it does not require the channel for the entire duration of the data transfer. In this way, the single timeslot can be made available for more than one user, allowing radio channels to be shared (by up to seven users in GPRS).

The overall data rates in GPRS are specified for each timeslot, varying from 9.05 kbps up to 21.4 kbps, depending on what is known as the *coding scheme*. There are four coding schemes (CS) in GPRS (known as CS1, CS2, CS3, and CS4). The data rates for each coding scheme are shown below. Note that each overall channel is actually designed to carry 22.8 kbps (as in standard GSM connections), and the

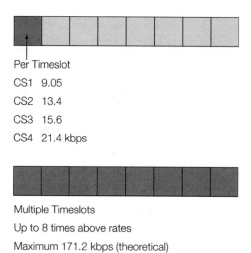

FIGURE 2.75 GPRS timeslot usage.

extra capacity is actually used for error protection (good error protection for CS1; not so good for CS4). In practice, only coding schemes 1 and 2 are widely used.

Multiple Timeslots

The data rates quoted in each coding scheme in GPRS are for a single timeslot (Figure 2.75). The data rate is usually increased by allocating multiple timeslots (on the same carrier frequency) to a single mobile station, or more often, to up to seven mobile stations. As an example, a mobile station using CS2 would have a user data rate of 13.4 kbps on a single timeslot. By operating on two timeslots, the mobile would therefore be able to operate at 26.8 kbps. For various reasons, no more than four timeslots in the downlink and one timeslot in the uplink would be allocated; and at CS2, this would result in 53.4 kbps in the downlink (network to mobile), and 13.4 kbps in the uplink.

Enhanced Data Rates for Global Evolution (EDGE)

EDGE is an enhancement to the GSM radio interface, which provides an *improved modulation scheme* and *improves spectrum efficiency*. EDGE uses 8 Phase Shift Keying rather than standard GSM Gaussian Minimum Shift Keying. It involves no core network changes, although its main use is in combination with the GPRS core network to make the most efficient use of the enhancement. An advantage of EDGE is that it does not require any further spectrum allocation for operators because it mainly involves just a change in modulation scheme. In theory (but not in practice), EDGE may support data rates as high as 473 kbps on a standard GSM carrier frequency.

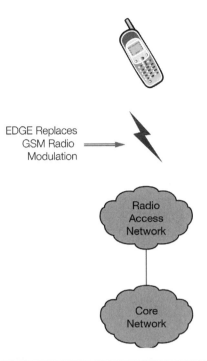

FIGURE 2.76 Enhanced Data for Global Evolution (EDGE).

A variation of EDGE was also defined for use with the North American TDMA system (Figure 2.76). In this context, EDGE is seen very much as a 3G radio access technology in its own right, because it is capable of delivering 3G data rates. As the North American network operators change over to GSM-based technologies, EDGE is still seen as a 3G technology due to problems with spectrum allocations for true 3G (W-CDMA). Elsewhere, EDGE is regarded as 2.5G, or sometimes 2.75G. For those operators with 3G licenses, EDGE may be bypassed altogether, with the network operator opting for GSM/GPRS, and W-CDMA infrastructure only.

The UMTS system makes no assumption that EDGE will be present in the 2G network from which UMTS evolves, but a 3G UMTS handset should be EDGE compatible, allowing *islands* of UMTS coverage with EDGE helping to deliver high data rates through the sea of GSM/GPRS that surrounds these UMTS islands.

Because an EDGE timeslot can support up to three times the data rate of standard GSM/GPRS timeslots, it can be used to increase the user's data rate, or to carry as much data as would otherwise be carried by three standard GSM/GPRS timeslots, therefore freeing up timeslots for voice. This is shown in Figure 2.77.

As is the case for standard GPRS, various coding schemes have been specified for EDGE/GPRS, as shown below. Note that the data rates are again specified per timeslot, and error protection is still extremely important. With no more than four timeslots provided in the downlink (and one in the uplink), the data rate would realistically be no more than 128 to 144 kbps.

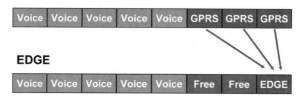

FIGURE 2.77 EDGE Usage.

The terminology used below is MCS — Mobile Coding Scheme — and incorporates the standard GPRS schemes (using Gaussian Minimum Shift Keying modulation — GMSK), as well as the EDGE coding schemes available with the new 8 Phase Shift Keying (8-PSK) modulation scheme. This time, the data rates are the rates available after much of the *extra* GPRS control data has been stripped off, and is closer to the actual user data rate than was previously quoted for the GPRS coding schemes.

Coding Scheme Modulation Type Date Rate (kbps):

- MCS-1   GMSK 8.8
- MCS-2   GMSK 11.2
- MCS-3   GMSK 14.8
- MCS-4   GMSK 17.6
- MCS-5   8-PSK 22.4
- MCS-6   8-PSK 29.4
- MCS-7   8-PSK 44.8
- MCS-8   8-PSK 54.4
- MCS-9   8-PSK 59.2

Review Questions

Q1. GPRS and EDGE use _____.
   a. FDMA
   b. CDMA and TDMA
   c. TDMA
   d. FDMA and TDMA

Q2. GSM can operate _____.
   a. In a number of radio frequency bands
   b. At one specified frequency only
   c. Only with separate frequencies in every cell in the network
   d. Only at the 900-MHz and 1800-MHz frequency bands

Q3. GPRS provides _____.
   a. Circuit-switched operation
   b. CDMA operation
   c. True multimedia
   d. Packet-switched operation

Q4. GPRS _____.
   a. Allows allocation of up to two timeslots to one or more users
   b. Allows allocation of a number of timeslots to one or more users
   c. Requires a dedicated timeslot for each user
   d. Uses a different modulation scheme than that used for GSM

Q5. EDGE _____.
   a. Uses a different TDMA multiplexing scheme from standard GSM/GPRS
   b. Uses a different modulation scheme from standard GSM/GPRS
   c. Would always be provided using a different radio site from GSM/GPRS
   d. Would never be provided in areas with GSM/GPRS coverage

Q6. EDGE can be used to _____.
   a. Reduce the requirement for timeslots for data applications
   b. Increase the overall number of timeslots available in the cell
   c. Provide 2 Mbps for each user
   d. Provide cdma operation

2.3.4 CDMA-Based Systems

Key Features Required of a 3G System

Despite the enhancements to 2G networks, it remains the case that many of the *future services* require higher data rates than available even with 2.5G (GPRS or EDGE). In addition, at any point in time, the future mix of services remains unknown; therefore, planning must be flexible, and both voice and data must be supported efficiently and effectively.

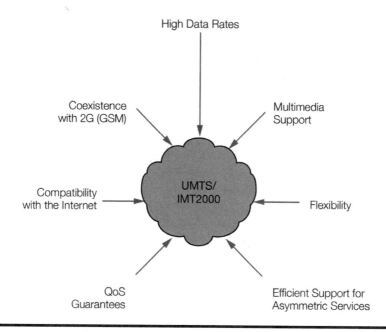

FIGURE 2.78 Requirements of a 3G network.

The International Telecommunications Union, through its IMT2000 (International Mobile Telecommunications 2000) initiative, began the process of trying to describe the required capabilities of a third-generation system in the mid-1990s.

Although requirements change, we can compile the key features required of a 3G system (Figure 2.78). Some examples are listed below. Some features are more difficult to achieve than others.

1. *High data rates:* initially from 384 kbps up to 2 Mbps, to enable applications such as large file transfers, mobile video and music, etc.
2. *Multimedia service support:* allows the system to multiplex voice, data, video, and other services on a single connection, and to be received simultaneously.
3. *Flexibility:* gives the ability to request *bandwidth on demand*, and variably set data rates to suit applications in progress.
4. *Efficient delivery of services:* includes asymmetric services such as web browsing. This could require the enabling of different bit rates on the uplink and downlink.
5. *Quality of Service control:* require guarantees over a range from real-time, low loss services such as speech or video telephony down to *best-effort* Internet access type services.
6. *IP support:* enables efficient interworking with the Internet and other IP-based applications.

7. *Coexistence with 2G:* allows coexistence and interworking with existing 2G networks (GSM) and services, in which operators have already invested a huge amount.
8. *Ease of global harmonization:* would allow users to gain access to their services wherever they are.

These requirements are common to most advanced cellular or wireless access systems, including W-CDMA, cdma2000, and also WiFi.

Resulting W-CDMA System for UMTS

The W-CDMA system that was born out of these 3G requirements brought the following key changes from the GSM radio interface to meet the service requirements for 3G. (W-CDMA is the radio access scheme adopted for the 3G Universal Mobile Telecommunication System — UMTS.) See Figure 2.79.

1. *Carrier spacing (bandwidth): 5 MHz in W-CDMA versus 200 kHz in GSM.* The wider bandwidth available to W-CDMA means that much more information can be sent, and that data rates for the user can be much higher. Indeed, in theory, the specification allows data rates of up to 384 kbps (see below for FDD mode), or even in later implementations up to 2 Mbps (for TDD mode). Further evolution of W-CDMA allows even greater data rates (especially with systems such as the High Speed Downlink Packet Access (HSDPA) implementation of W-CDMA). The wider bandwidth also allows a much more flexible system in terms of channel allocation.
2. *Negotiation of radio bearer properties to suit different QoS requirements.* In W-CDMA, quality (data rates and other channel properties) is controlled as part of "radio resource management," with radio resources applied to suit a particular application or user need. In GSM, quality was more a result of network and frequency planning, because all applications had access to the same sort of channels.
3. *Two coexisting modes: FDD & TDD.* These refer to the way uplink and downlink signals are separated. FDD (Frequency Division Duplex) is the most commonly applied mode in UMTS deployments, and is suited to wide area mobility; whereas TDD (Time Division Duplex) allows higher data rates to be more efficiently offered over limited areas, such as in small urban cells or hot spots.

Typical UMTS License Allocations

Figure 2.80 shows the spectrum allocations for the United Kingdom. Five operators are licensed, with Hutchison 3G being a new entrant to the market in the United

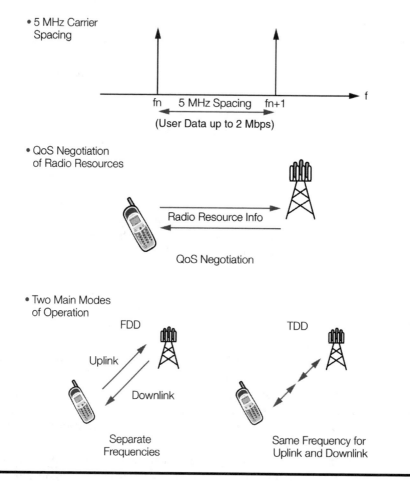

FIGURE 2.79 Key features of W-CDMA.

Kingdom. With the exception of Vodafone, all operators possess two or thee FDD channels (each requiring an uplink and downlink frequency) and a single TDD channel (requiring only a single frequency). In early UMTS networks, only FDD mode is available.

The IMT2000 Family

In parallel with W-CDMA, those in the cdmaOne community were developing their next-generation system (cdma2000), and those in the North American TDMA community were looking toward EDGE as a solution to their own 3G needs. A single global standard became both politically and technically unlikely; but after some compromises were reached, the ITU announced a "family concept" for IMT2000 terrestrial radio interfaces.

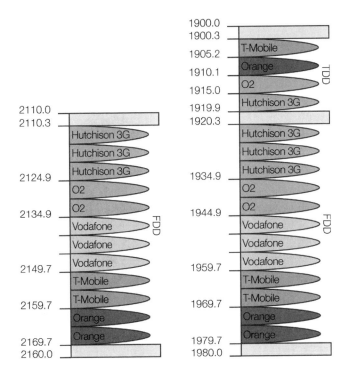

FIGURE 2.80 3G licenses in the United Kingdom. (All frequencies in MHz.)

In particular, the following IMT2000 (3G) systems were defined:

- TDMA-based systems:
    - TDMA with EDGE
    - DECT (actually a European Cordless Telephony Standard)
- CDMA-based systems:
    - W-CDMA FDD mode
    - W-CDMA TDD mode
    - cdma2000 (evolution of cdmaOne)

The TDMA-based systems have either never been considered as practicable 3G cellular technologies (DECT) or a less attractive business case than CDMA solutions. Therefore, W-CDMA and cdma2000 are the only (increasingly) widely available 3G systems. Even the requirement for TDD mode seems to be challenged by systems such as WiFi. This leaves W-CDMA and cdma2000 as the main 3G standards.

cdmaOne and cdma2000

Although GSM is by far the most successful technology family, there are other successful systems, which include:

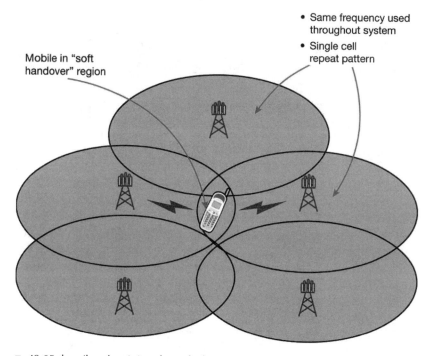

- IS-95 describes the air interface of cdmaOne.
- Radio channels are separated by codes; not frequency or time.
- Uses wideband frequency carriers (1.25 MHz), each with numerous channels separated by different codes.
- Mobiles can be served by more than one base site at the same time: soft handover.

FIGURE 2.81 The cdmaOne concept.

- *cdmaOne*. CDMA has been used as a technology for 2G cellular systems since the mid-1990s. The systems are based on a standard that originated in the United States, known as IS-95, and collectively around the world are referred to as cdmaOne systems. (IS-95 refers specifically to the Air Interface, and cdmaOne is the collective term for that and other standards that make up the entire system.)

cdmaOne systems now have a market presence virtually throughout the world, with the exception of Western Europe. For some time in the early 1990s, there was much discussion regarding the rivalry between GSM (TDMA) and CDMA technology. The primary reason for the subsequent success of GSM lay in the fact that it was first to market. Having been proven as a successful European standard in the early 1990s, it was rapidly adopted worldwide and thus became established as the world standard. The main features of cdmaOne are shown in Figure 2.81. The soft handover concept is common to the cdmaOne/cdma2000 family and to W-CDMA used in UMTS.

Technology Principles ■ 217

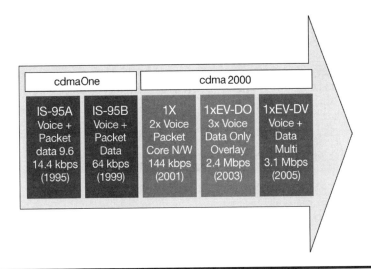

FIGURE 2.82 The cdmaOne/cdma2000 family of technologies.

CDMA was first launched in 1995. The technical standards were developed by Qualcomm. Using techniques such as soft handover, spread spectrum, and the use of advanced receivers makes CDMA an inherently reliable technology. Those who support CDMA point out its spectral efficiency, especially when compared to other 2G technologies. When launched under the IS-95A specifications, performance was similar to that of a GSM network. IS-95B saw the ability to provide 64-kbps data bearers. See Figure 2.82.

- cdma2000 is the name given to represent the family of CDMA technologies that meets the ITU's IMT-2000 requirements for 3G mobile networks.
- 1x EV-DO (Data-Only) is an overlay network that offers asymmetric connections using only one 1.25-MHz carrier.
- 1x EV-DV (Data-Voice) combines data and voice, with a download speed up to 3.1 Mbps.

Review Questions

Q1. CDMA systems _____.
    a. Employ hard handover techniques only
    b. Employ soft handover only
    c. Can employ soft handover, softer handover, or hard handover techniques
    d. Employ softer handover techniques only

218 ■ *Introduction to Mobile Communications*

Q2. UMTS uses _____.
   a. Wideband CDMA channels of around 5 MHz
   b. Wideband CDMA channels of around 1.25 MHz
   c. Wideband FDMA channels of around 1.25 MHz
   d. TDMA channels of 200 kHz that can support up to eight users

Q3. UMTS _____.
   a. Has a common core network with cdma2000
   b. Has a common radio access network with GSM/GPRS
   c. Must use entirely unique components
   d. Has a common core network with GSM/GPRS

Q4. cdma2000 _____.
   a. Is applied in its own part of the radio spectrum
   b. Can be applied in the same spectrum band as cdmaOne
   c. Uses the same radio band as W-CDMA
   d. Uses unlicensed spectrum

Q5. Which is *not* correct?
   a. W-CDMA can provide higher data-rate channels than GSM/GPRS.
   b. W-CDMA provides a more efficient radio scheme than GSM/GPRS.
   c. W-CDMA can operate over a greater range than GSM/GPRS.
   d. W-CDMA provides support for multimedia services.

Q6. The two modes of W-CDMA are _____.
   a. TDMA and FDMA
   b. FDD and TDD
   c. cdmaOne and cdma2000
   d. CDMA and FDMA

2.3.5 Review Questions

In an example system, each macro cell can support up to 30 simultaneous calls (+ two signaling channels) using four frequency pairs in each cell.

   a. In a seven-cell repeat pattern, how many frequency pairs would be needed in total?
   b. If the plan were changed to a four-cell repeat pattern with the same number of frequency pairs available, how many frequencies would be available per cell?
   c. If four channels were needed for signaling in each cell in the four-cell repeat pattern, how many calls could be supported in each cell?

d. Over an area that requires support for up to 300 users making calls simultaneously, and using simple calculations (ignoring probability theory and the standard traffic measure of Grade of Service), approximately how many cells would be required in the:
   i. Seven-cell repeat pattern?
   ii. Four-cell repeat pattern?
e. If three cells could be supported from a single radio site, how many sites would be required over the area for the:
   i. Seven-cell repeat pattern?
   ii. Four-cell repeat pattern?
f. Which configuration would:
   i. Be more cost-effective?
   ii. Provide less interference (and therefore a better quality radio signal)?
g. Additional frequency pairs may be needed to provide extra capacity in busy areas such as shopping malls or sports stadiums. These would generally be provided in the form of (delete as appropriate):
   i. Macro cells
   ii. Micro or pico cells
   iii. Umbrella cells

In practice, the repeat pattern used in a network may be difficult to identify due to the many extra factors that must be taken into consideration. In fact, commercially available planning tools often provide a best-fit solution, but would use much of the logic set out in the exercise above.

Practice Questions

Q1. Radio propagation _____.
   a. Is a way of describing how the energy in the radio signal travels
   b. Describes the medium in which the radio energy is found
   c. Gives the radio energy its direction
   d. Allows the radio signal to be dissipated

Q2. Cellular radio systems such as GSM use _____.
   a. Ground wave propagation
   b. Sky wave propagation
   c. Satellite communications
   d. Line-of-sight propagation

Q3. Cellular radio systems such as GSM benefit from _____.
   a. Forward error correction and encryption
   b. Analog forward error correction
   c. Encryption and simplex communication
   d. Error detection and analog coding

Q4. Errors in digital systems will _____.
   a. Always be noticed by the user
   b. Never be noticed by the user
   c. Be irrelevant
   d. Only be noticed by the user if the errors exceed the capabilities of the error correction process

Q5. Radio systems _____.
   a. Are more costly than installing copper cables
   b. Are too unreliable to replace copper links
   c. Take too much power to use in remote sites
   d. Can provide a cost-effective alternative to copper links

Q6. Radio can _____.
   a. Provide support for many varied scenarios and systems
   b. Support only point-to-point links
   c. Be used only over long distances
   d. Provide short-range connections only

Q7. A mobile network operator will _____.
   a. Use the most appropriate frequencies for the chosen mobile system
   b. Have a license to operate within a specific frequency band, or set of frequency bands, within a specific country
   c. Share frequencies within a frequency band, or set of frequency bands, with other network operators
   d. Always use a single radio channel to provide services in a defined geographic region

Q8. If a centrally located radio site serves a single cell, it is known as _____.
   a. A directional site
   b. A sectored site
   c. A single site
   d. An omnidirectional site

Q9. Cells that are used to provide additional capacity in areas such as shopping malls or sports stadiums are known as _____.
   a. Additional cells
   b. Macro cells
   c. Micro or pico cells
   d. Multi cells

Q10. For a system such as GSM, a pico cell located within the area of a macro cell would _____.
   a. Use the same frequencies as the macro cell
   b. Always share the macro cell frequencies
   c. Use different frequencies from those used in the macro cells
   d. Be known as an umbrella cell

Q11. Which of the following is not a recognized radio access scheme?
   a. Frequency Division Multiple Access (FDMA)
   b. Code Division Multiple Access (CDMA)
   c. Amplitude Division Multiple Access (ADMA)
   d. Time Division Multiple Access (TDMA)

Q12. GSM uses _____.
   a. FDMA
   b. CDMA and TDMA
   c. TDMA
   d. FDMA and TDMA

Q13. GSM and GPRS _____.
   a. Share the same core network elements
   b. Require different core network elements
   c. Are both provided independent of the core network requirements
   d. Must both use IP technology

Q14. EDGE _____.
   a. Uses a different TDMA multiplexing scheme from standard GSM/GPRS
   b. Uses a different modulation scheme from standard GSM/GPRS
   c. Would always be provided using a different radio site from GSM/GPRS
   d. Would never be provided in areas with GSM/GPRS coverage

Q15. EDGE can be used to _____.
   a. Reduce the requirement for timeslots for data applications
   b. Increase the overall number of timeslots available in the cell
   c. Provide 2 Mbps for each user
   d. Provide cdma operation

Q16. CDMA systems _____.
   a. Use several frequencies within the reuse pattern
   b. Use a repeating seven-cell repeat pattern
   c. Use a single cell reuse pattern
   d. Cannot employ frequency reuse

Q17. W-CDMA frequency allocation _____.
  a. Usually provides an operator with only a single FDD carrier
  b. Usually provides an operator with a number of FDD carriers
  c. Would usually be either an FDD or TDD carrier only
  d. Is in the existing GSM frequency band

Q18. CDMA systems _____.
  a. Employ hard handover techniques only
  b. Employ soft handover only
  c. Can employ soft handover, softer handover, or hard handover techniques
  d. Employ softer handover techniques only

Q19. UMTS uses _____.
  a. Wideband CDMA channels of around 5 MHz
  b. Wideband CDMA channels of around 1.25 MHz
  c. Wideband FDMA channels of around 1.25 MHz
  d. TDMA channels of 200 kHz that can support up to eight users

Q20. cdma2000 _____.
  a. Is applied in its own part of the radio spectrum
  b. Can be applied in the same spectrum band as cdmaOne
  c. Uses the same radio band as W-CDMA
  d. Uses unlicensed spectrum

Chapter 3

Mobile Network Infrastructure and Supporting Systems

Module 3 Learning Objectives

- Identify the architecture of a mobile core network and describe the functionality of the main network elements.
- Identify the elements that are added to the core network to support packet data and describe the functions of these elements.
- Explain typical mobile network procedures, such as Location Update and Call Setup
- List typical IN services and identify the architecture necessary to deliver these.
- Describe how messaging services are integrated into the core network.
- Understand the motivation for moving to all-IP networks.
- Describe how the Internet Protocol operates and identify the other protocols and applications that it supports.
- Make a list of service platforms and explain their role in the mobile network.
- Outline the need for OSS/BSS systems and describe the typical systems you expect to find as part of OSS/BSS system.
- Show what elements or information can be used to bill for services used in mobile networks

3.1 Mobile Network

3.1.1 The Overall System

The Requirements

For any cellular network, the procedures can be split into different requirements. The basic requirements include categories for:

- Establishing the radio connection to the network
- Managing the location of the individual users
- Establishing a full call connection

The categories above are in priority order, because a radio connection must be established before information on a user's location can be exchanged; and the location is required before a call can be established.

There are also extra procedures required for:

- Establishing a packet data session (e.g., for accessing the Internet)
- Sending text messages
- Handling supplementary services

The requirements do not change from system to system, but the way they are categorized and the specific procedures involved will differ markedly from one family of technologies to another.

To explain the general procedures used within cellular networks, we consider the procedures for that of the GSM family of technologies — including GSM and GPRS, but also applicable to EDGE and UMTS. Because the core network is common to the different radio access technologies in the family, the procedures for establishing the call, establishing a data session, sending a text message, and handling supplementary services will be essentially the same (with relevant modifications).

However, the very different radio access techniques mean that procedures required for establishing the radio connection will be different in each case.

Architecture and Procedures in the GSM Family of Technologies

Categorizing Procedures

In line with the requirements, the GSM family procedures can be split as shown below. This categorization is only used for procedures between the user device and core network but is useful to illustrate the formal procedure requirements. The core network itself tends to categorize procedures based on the signaling and control systems required.

First, for standard operation, the categories are:

- Radio Resources (RR)
- Mobility Management (MM)

- Connection Management (CM)
  - Call Control (CC)
  - SMS (SMS)
  - Supplementary Services (SS)

Adding packet-based operation also gives us:

- Radio Resources (RR)
- GPRS Mobility Management (GMM)
- Session Management (SM)

When explaining the procedures, it is useful to look at the major requirements and to explain what happens in various scenarios (e.g., turning the mobile device on, making a call, accessing the Internet, etc.). This means that explanations of how procedures fit together generally require a mix of procedures from each category — much more useful as an aid to understanding than listing each procedure in each category in alphabetical order.

To explain the main procedures, the different network elements used in the GSM family must be defined. Initially, basic GSM elements will be considered, followed by the additions required for GPRS. The equivalent elements required for W-CDMA (for UMTS) operation are also noted. Note that the network elements required for EDGE are as for GPRS, but with the radio system elements upgraded to provide the new modulation scheme and radio channel structure.

Overall GSM Architecture

As with any network, the components of a GSM network can be divided into access and core networks (Figure 3.1).

- *Radio access.* The radio access part of the network consists of a set of base transceiver stations (BTS) that provides the radio communication with the mobile handset. A group of BTS will be controlled by a base station controller (BSC), with this combination of BSC and BTS elements known as the base station subsystem (BSS).

  In UMTS, the Node B is equivalent to the BTS, and the radio network controller (RNC) is equivalent to the BSC.

- *Core network.* The radio access network connects to the core network at an element known as the mobile switching center (MSC). This acts like a local telephone exchange to all the mobiles currently registered in the MSC area. A Gateway MSC (GMSC) can also be defined, and describes an MSC that provides access between the core GSM network and another interconnected network, such as the public switched telephone network (PSTN).

  The MSC will communicate with a number of databases, known as location registers. These location registers hold relevant information regarding the subscriber's identity, the subscriber's current cell location, and the services

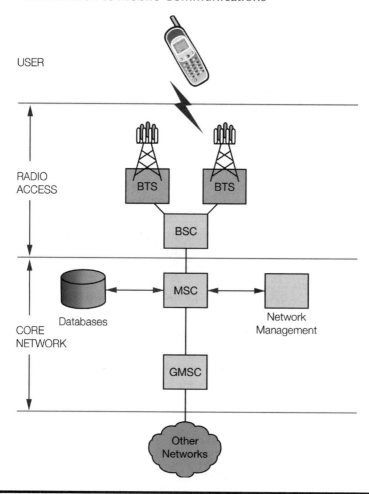

FIGURE 3.1 Basic GSM network elements.

subscribed to. They allow the MSC to connect and direct services and traffic (voice or data) appropriately. The location registers are known as the home location registers (HLRs) and the visitor location registers (VLRs).

The MSC, HLR, and VLR are also an integral part of UMTS (at least initially).

A network management function will also connect into the MSC to control overall network operation.

GPRS Network Elements

Additional elements are required to provide the General Packet Radio Services (GPRS), including core network *GPRS support nodes* (GSNs), comprising the Serving GSN (SGSN), and Gateway GSN (GGSN) (see Figure 3.2). These elements use IP (Internet Protocol) technology to route information to and from the mobile handset. They are effectively IP routers with modifications to allow for managing

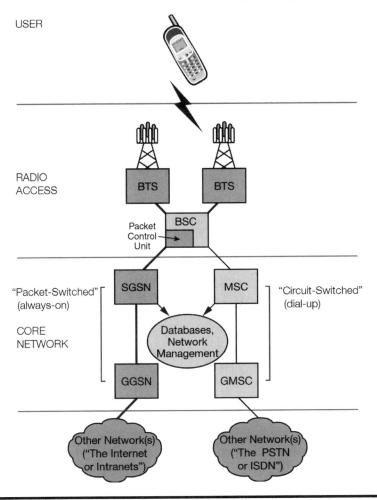

FIGURE 3.2 GPRS network elements.

subscriber access, mobility, and IP sessions. A GPRS data session would be handled by the GSN rather than the MSC, which is used for voice.

In addition, GPRS operation requires the BTS to be GPRS-capable, and a packet control unit must be available (usually at or within the BSC). These two base station subsystem (BSS) elements allow the data packets to transfer in the right format on the correct GPRS radio channels.

Identities

To ensure the various elements can be referenced individually within the relevant procedures, they are identified in a variety of ways.

Each core network element will have its own unique number that can be used within the signaling and control messages. This number uses a standard format that allows identification of first the country, then the network, and then the element itself. Whether the network element is an MSC, a GPRS support node (GSN), or a location register (LR), each one will have one of these unique numbers. The numbers are called E.164 numbers, from the International Telecommunications Union document number that specifies the format.

The cell itself must also be identified, and this takes the form of a unique cell-ID. This again uses a standard format, and in this case it allows identification of the cell as one of a number within a group of cells that form a specified location area. There is therefore also a unique location area ID. It is also important to be able to identify individual users. There are various identities assigned to a user:

- *IMSI (International Mobile Subscriber Identity)*. The IMSI is used in the network to identify the subscriber identity module (SIM), or UMTS SIM (USIM); that is, the SIM card within the mobile. The IMSI is held within SIM card, as well as in the location registers in the network. This is the number that the network uses to identify the subscriber (or subscription).

  Associated with each IMSI is a whole range of other information, including the services supported, current location (down to location area [LA] or routing area [RA] only), and status for that mobile. It is unique to the mobile in question because it reflects the mobile country code, mobile network code, and subscriber identity within that network.

  The TMSI (temporary mobile subscriber identity) replaces the IMSI over the radio interface. It is shorter than the IMSI and is only used locally within the visitor location register (VLR) or mobile switching center (MSC) area. It changes at regular intervals by the VLR to maintain security.

  The P-TMSI (Packet TMSI) is essentially similar to the TMSI but is used in the Serving GPRS Support Node (SGSN) rather than in the VLR. Again, it is used as a replacement for the IMSI and changed at regular intervals by the issuing SGSN.

- *MSISDN (Mobile Station ISDN)*. The number used by callers is the MSISDN. This is the number that a subscriber would give to friends and family. Multiple MSISDNs can be mapped to one IMSI. The structure of an MSISDN is in accordance with ITU-T E.164, as in the case for network elements.

- *IP address (Internet Protocol address)*. User equipment can also have IP addresses (version 4 or 6) for use in data transactions. They may be assigned dynamically by the network operator (effectively the Internet service provider — ISP), or may be kept permanently by the user (static assignment). For various reasons, dynamic assignment is much more common.

IMEI (International Mobile Equipment Identity). The IMEI identifies the actual handset (hardware), and allows a check to be made for stolen handsets (or other handsets that need to be traced) in a database called the equipment identity register (EIR).

Review Questions

Q1. Basic cellular procedures must provide for managing _____.
   a. Radio connections, user mobility, and end-to-end connections
   b. GPRS and GSM
   c. Only the end-to-end connection
   d. The overall user experience

Q2. The GSM element that provides functionality similar to a local telephone exchange is the _____.
   a. HLR
   b. BSC
   c. BTS
   d. MSC

Q3. The GSM element that provides a temporary database is the _____.
   a. VLR
   b. BSC
   c. GMSC
   d. MSC

Q4. The GPRS element that provides a connection to the MMSC, Internet, or WAP gateway is the _____.
   a. SGSN
   b. BSC
   c. GGSN
   d. PCU

Q5. The identity that a user would give to friends and family is the _____.
   a. IMSI
   b. IMEI
   c. MSISDN
   d. IP address

3.1.2 The User Equipment and Radio Access Network

Handset Types

The user gains access to telecommunications services and applications through the handset. Unfortunately, there is often confusion when referring to the handset because it is known by various terms, including:

- Mobile equipment
- Terminal equipment

- User equipment
- Handset
- Telephone
- User device
- The mobile
- Mobile device
- Mobile terminal

The mobile handset sector is the area of mobile networks that develops fastest, with constant innovation leading to many different device types and associated applications. Within the industry, a number of categories of mobile handsets have been defined, including:

- *Basic phone.* Despite its name, the basic phone is, in reality, a very complicated device. Its functionality will depend on the radio access technology being deployed. However, even a basic GPRS handset will be capable of multiple timeslot operation, offering high-speed, always-on connectivity.
- *Smart phone or VAS phone.* These devices offer more in the way of value-added services to the user than simply voice- or WAP-based services. They tend to have a larger screen and may be considered a mobile phone with built-in PDA functionality. These phones tend to include Bluetooth and SyncML technologies to allow for synchronization to a PC. If supplied with an Open Architecture OS, these devices will offer the user a wide range of features.
- *Connected computing devices.* This suggests almost any portable type computer, ranging from the simplest PDA to a laptop. Such a device can connect to a GPRS-capable mobile or alternatively a GPRS adaptor card that might slot into the PCMCIA slot, for example. Connecting a laptop to a GPRS-enabled device will get around the issue of the small screen size on the terminal itself.
- *PDA and Pocket PC.* This is really a subset of the above category and represents a major growth area in the marketplace. Such devices offer many of the features that users require from PCs, such as calendars, games, word processing, and an increasing number of applications generally associated with the home PC. Some offer the ability to double as music players, or even play video. Increasingly, these devices will be able to connect to a GPRS-capable device for communication or be combined into a single device.

Basic Requirements

The requirements of the end user will often focus on features and capabilities and the overall ease of use (usability), and of course the end user would like maximum functionality for minimum cost.

For a network operator, the handset represents the "front end" of the network and is the platform through which users gain access to the rich set of services

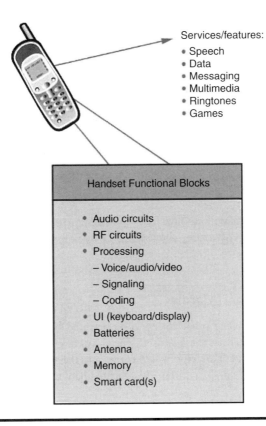

FIGURE 3.3 Handset functional blocks.

offered. The network operator is also very interested in the cost of the handset because in many markets the handset is subsidized by the operator — and therefore the operator bears some of the real handset costs.

A mobile handset, irrespective of radio technology, contains a number of generic functional blocks. These blocks are illustrated in Figure 3.3.

Audio circuits. To process audio signals from the microphone to the loudspeaker and to any connected devices (e.g., headsets).

RF circuits. This module deals with the modulation and demodulation processes and is also involved in all the frequency generation requirements.

Processing. A number of processors will typically be involved in undertaking all the voice coding, signal processing, channel coding, and signaling tasks. This is a very complex part of the overall handset.

User interface (UI). The UI consists of the display and the keyboard, and these devices typically interface to the main processor within the handset. More sophisticated handsets may include elements such as voice recognition and handwriting recognition functions as part of the overall UI.

Battery. The source of energy for all the other elements, the battery will be surrounded by circuitry related to recharging and to reporting the status of the battery so that it can be displayed on the UI.

Antenna. Often an integrated element on current handsets, the antenna is the interface between the handset and the radio channel connected to the system. Technology improvements will refocus attention on this critical element as techniques such as multiple input multiple output (MIMO) are introduced.

Memory. In addition to the various memory elements that accompany the processor, the phone may have other memory capacity provided by the SIM card and by memory cards or sticks.

Current 3G phones will contain many noncellular technologies, such as video codecs and picture compression algorithms, for which a license fee is payable to third-party innovators.

SIM Cards and Smart Cards

In GSM and UMTS, the subscription is represented by a smart card (the SIM or USIM) that plugs into the handset. The Subscriber Identity Module (SIM) effectively separates the subscription from the actual handset. This allows a customer to upgrade his handset without the need to change subscription or telephone number.

The SIM card provides a number of important functions with respect to connection procedures and security. It holds subscriber-specific information about services, identities, contacts, and (increasingly) applications. This card can also be used to support transaction services such as m-commerce.

SIM cards have increased in processing power and memory over the years, and can support a range of different applications and features. In 3G, the SIM is regarded more as a software module to be included on a smart card, maybe with other SIM modules and applications.

The GSM Radio Network Elements (BSS)

In GSM systems, the core network Serving MSC connects via the A interface to the Base Station Subsystem (BSS) (Figure 3.4). In the GPRS domain, connection to the BSS is from the SGSN via the Gb interface. The Gb interface is a GPRS interface located between the SGSN and the PCU.

The BSS consists of a Base Station Controller (BSC), which controls the activities of a number of Base Transceiver Stations (BTS), whose radio coverage is defined in terms of cells. In turn, a number of BSCs will be controlled by one MSC.

Mobile Network Infrastructure and Supporting Systems ■ 233

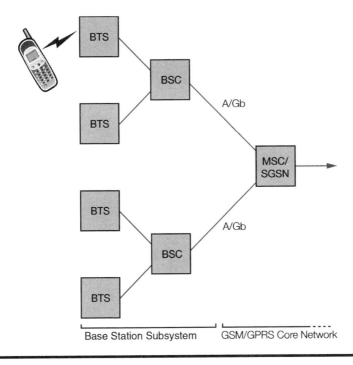

FIGURE 3.4 The GSM base station subsystem. The packet control unit (PCU) is a key element of BSS for GPRS and is located in the BSC. Note that the packet control unit (PCU) is a key element of BSS for GPRS and is located in the BSCC.

The Base Transceiver Station

The BTS is designed to provide the radio interface on the network side. It provides the necessary processing for the radio channels, as well as the modulation, transmitter/receiver, and antenna system. Each BTS can provide support for a single cell, or multiple cells (typically three), using sectored antennae. Figure 3.5 shows a typical three-sector GSM radio site (or BTS).

The Base Station Controller

The BSC manages radio resources for BTSs in its area, including radio-channel setup, frequency hopping (because in GSM, cells operate at different carrier frequencies), and handovers between cells under its control. The MSC, which controls these BSCs (or may involve more than one MSC, should this be required), handles handover between cells under the control of different BSCs. Note that in GSM, there is no interface specified between BSCs.

FIGURE 3.5 Typical GSM radio site.

Enhanced Data Rates for Global Evolution (EDGE)

EDGE is a technology at the air interface to GSM that provides an improved modulation scheme and improved spectrum efficiency. EDGE uses 8 Phase Shift Keying rather than standard GSM Gaussian Minimum Shift Keying. This has the effect of increasing data rates, with rates as high as 473 kbps in a 200-kHz GSM carrier.

The main architecture requirement for EDGE is usually just a "card" change or software upgrade for the BTS — as long as the BTS was purchased after the late 1990s.

Edge involves no core network changes, although many see its main use in combination with other changes that deliver high data rates using a GPRS core.

The UMTS system makes no assumption that EDGE will be present in the second-generation network from which UMTS evolves, but in practice many networks will include areas of coverage that are designated as standard GSM/GPRS (2G), EDGE (2.5G), or W-CDMA (3G). Any areas designated as W-CDMA or EDGE will usually include 2G coverage as well.

EDGE Evolution

EDGE Evolution introduces a higher-speed version of EDGE, effectively allowing a range of high-speed services to be accessed without 3G coverage being in place. EDGE Evolution standardization is part of 3GPP Release 7.

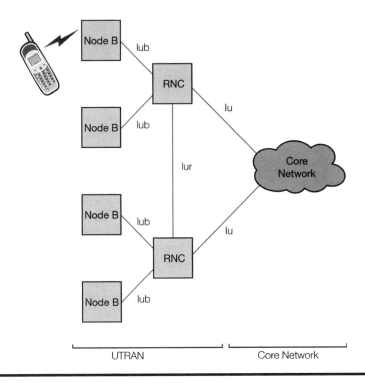

FIGURE 3.6 The UMTS Terrestrial Radio Access Network (UTRAN).

The Elements of the UTRAN

The UMTS Terrestrial Radio Access Network (UTRAN) does not reuse elements of the existing GSM radio network, although some comparisons can be usefully drawn.

In terms of architecture, the UTRAN is divided into individual radio network systems (RNSs), where each is controlled by a radio network controller (RNC), connected to several Node B elements that may in turn serve one or more cells (see Figure 3.6).

Comparison can be drawn with GSM, where Node B is broadly equivalent to the BTS, and the RNC similar in architectural hierarchy to the BSC. The interface between RNC and Node Bs is defined, and is assigned the label Iub. Another significant new addition is a brand new interface, the Iur, which is required to cope with soft handover, a new feature of CDMA systems that arises because adjacent cells are operating on the same frequency band.

The Node B

The term "Node B" refers to the base station equipment that communicates with the subscriber's handset via the radio link (and, of course, with the main network via a telecom link). It provides radio resources for a UMTS network and uses UMTS channel allocation to communicate with the handset. It provides all the RF processing, enabling transmission and reception information to and from the mobile terminal. This information is encoded using the W-CDMA scheme.

The Radio Network Controller

The RNC (radio network controller) controls the operation of multiple Node Bs, managing resources such as allocating capacity for data calls and providing critical signaling such as call setup, plus switching and traffic routing functionality. Compared to 2G systems, it is broadly equivalent to the BSC, but also includes some functionality of the MSC. In particular, it enables autonomous radio resource management by the UTRAN by allowing the RNC to communicate directly (via the Iur interface), thereby eliminating this burden from the core network. So, all handover processes, even where moving between cells controlled by different RNCs, are kept within the UTRAN. Compare this with the situation in GSM, where handover between different BSC areas required involvement of the MSC, and hence the core network.

HSDPA (High-Speed Downlink Packet Access)

HSDPA is a specific implementation of W-CDMA, where several users can share a high-speed channel (in the downlink network to user equipment) that has been optimized for maximum data throughput. The overall channel speed can vary by changing the modulation scheme (adaptive modulation). However, once the modulation scheme is set, users are allocated time within the channel.

Obviously, each user will require different data rates and be at varying distances from the Node B, and this is provided by allocating (scheduling) different amounts of time to each user, and changing the ratio of "user data" to "error correction data" within the overall high-speed channel. Those that are farther away will generally have more error correction data and less user data included. If they require high data rates, this can still be achieved by allocating (scheduling) increased time. The individual users effectively have their own channels allocated within the overall HSDPA channel. HSDPA mainly affects the radio interface; however, the handset, Node B, and RNC will need HSDPA capability.

Connecting the Radio Access Network Elements Together

The radio access networks should be connected together by transmission networks that provide a suitable number of channels to cater to both the user's traffic (voice or data) as well as the required signaling and control.

Mobile Network Infrastructure and Supporting Systems ▪ 237

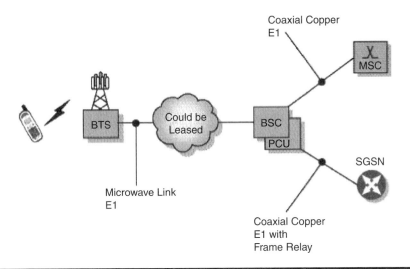

FIGURE 3.7 Typical GSM/GPRS radio network connections.

The GSM, GPRS, EDGE, cdmaOne, or 3G (W-CDMA or cdma2000) radio interfaces are considered separately as they deal with connecting the user equipment to the radio access network (known as the Base Station Subsystem in GSM). Here we consider the interfaces between the radio site (BTS or Node B for GSM and UMTS) and the radio network controller (BSC or RNC for GSM and UMTS), and between the controller and the core network (MSC and GGSN for both GSM and UMTS).

As always, the following requirements are important:

- Providing enough capacity (for current and forecast usage)
- Ease of management – including fault alarms
- Fault tolerance, incorporating alternative routes if possible
- Costs, both installation and ongoing

Decisions must be made on how the elements are connected together with respect to:

- Transmission media
- Transmission system
- Whether leased lines are used

Although different options can be used, Figure 3.7 shows how the radio elements are often connected together for GSM-, GPRS-, and EDGE-based systems. The basic way of organizing the channels in the radio network of GSM or GPRS is to use the PDH (Plesiochronous Digital Hierarchy) E1 link.

It is assumed here that the BSC, MSC, and SGSN are all colocated in the same building; hence, coaxial cables are used to connect those elements (due mainly to cost). If the BSC is remote, then a more appropriate medium might be optical fiber (with SDH).

As usual, the BTS are located appropriately to provide radio coverage for the users over a specified area, and are often connected via microwave links to the BSC. Alternatively, the connections can use copper or even fiber. Leased lines can be used.

For UMTS, the radio elements should be connected initially using ATM (Asynchronous Transfer Mode), with an appropriate underlying transmission system carrying the ATM *cells*. This system is much more flexible and allows for flexible allocation of channels and resources through the radio network. Eventually it is envisaged that most 3G networks will use IP (Internet Protocol) technology instead of ATM, thereby reducing costs appropriately.

WLAN and Mobile Networks

The existing GPRS/GSM and newer 3G networks have for a long time offered data services to their subscribers.

Original circuit-switching data at rates of 9.6 kbps do not allow a great deal of flexibility and severely restrict the service and content that can be offered. The advent of GPRS has improved this matter significantly. The packet-switching service can offer much higher rates to the user and at the same time offer resource efficiencies to the network operator. However, the data rates are still in the range 10 to 40 kbps. Even if EGDE is deployed in the network, the data rate will *only* be a maximum of hundreds of kilobits per second. The advent of 3G networks improves on this data rate, thus allowing a greater range of services to the users.

These data rates, no matter how impressive, are simply blown away by the rate available through WiFi hotspots — where typical data rates can be as much a 2 Mbps. The obvious difference here is that the cellular networks are providing wide area coverage for voice and data, whereas the WiFi service is limited to a very specific place and does not yet offer mobility services. There may be advantages to the cellular operators by allowing some form of interworking between the WiFi systems and their own networks. It is likely that this can be achieved in several different ways, depending on available technologies and potential agreements between WiFi and cellular operators.

Some cellular operators have formed alliances with the WiFi service providers, allowing the cellular subscriber access to the hotspots under the single mobile subscription. Mobile phone manufacturers are already producing equipment that supports the WiFi radio in multi-mode devices. This allows the phone itself to use the WiFi hotspot for data download or even in some cases to make phone calls using VoIP protocols.

In due course it may become common for the cellular operators themselves to build and maintain their own WiFi infrastructure, connecting directly to the mobile

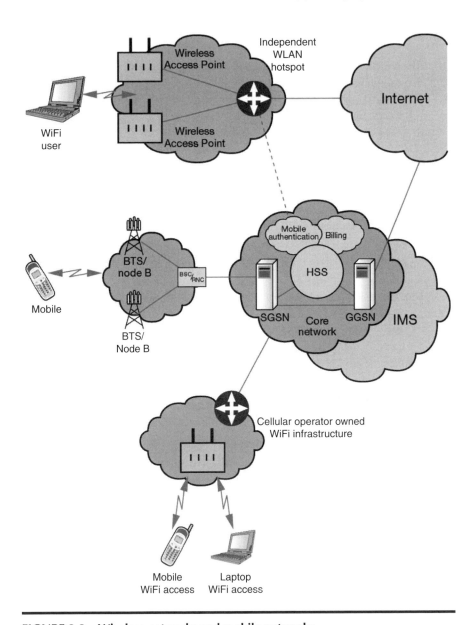

FIGURE 3.8 Wireless networks and mobile networks.

core network. Using mobile IP and a 3G core network, this offers users seamless mobility between mobile networks and WiFi systems under a single subscription.

In Figure 3.8, WiFi is illustrated as a stand-alone technology that connects directly to the Internet, or as an integral part of the cellular operator's network — effectively providing alternative access. Note that IMS refers to the IP Multimedia Subsystem that is designed to provide support for advanced multimedia services.

Review Questions

Q1. The handset is very often referred to as the _____.
   a. Mobile handheld
   b. Handset device
   c. User device
   d. User equipment

Q2. Which element controls the activities of the BTS?
   a. HLR
   b. BSC
   c. PCU
   d. MSC

Q3. For GPRS, the radio elements required are _____.
   a. The same as for GSM
   b. Modified GSM elements (and also PCU at the BSC)
   c. New Node Bs
   d. New SGSN and GGSN

Q4. Implementation of EDGE in the network would often require _____.
   a. A new BSC and BTS
   b. A completely new radio network
   c. Just a card change or software upgrade
   d. 3G infrastructure

Q5. W-CDMA requires new radio network elements as follows:
   a. PCU at the BSC
   b. Node Bs and RNC
   c. SGSNs and GGSNs
   d. Card change at the BTS

Q6. HSDPA is _____.
   a. A specific implementation of GPRS
   b. A specific implementation of EDGE
   c. An uplink high-speed W-CDMA connection
   d. A specific implementation of W-CDMA

3.1.3 Circuit-Switched and Packet-Switched Core Networks

Core Network Requirements

The requirements of the core network can be divided into what is required for circuit-switched services (mainly telephony), and for packet-switched services (access the Internet or corporate intranet, for example).

In addition, we need to consider the control and signaling in each of these two *domains*. For circuit-switched services we use Signaling System Number 7 (SS7). For the packet-switched network, a separate control system has been developed as an integral part of the packet-switched infrastructure. In GPRS, this control is known as GPRS Tunneling Protocol (GTP), although SS7 is also used on some interfaces, where the GPRS infrastructure interacts with non-GPRS-specific elements such as the Home Location Register (HLR).

The requirements of the core network are:

- Circuit-switched service requirements
- Packet-switched service requirements
- Signaling and control requirements

The Circuit-Switched (CS) Domain

The entities specific to the CS domain are MSC, GMSC, and VLR (Figure 3.9). The GSM core network elements also form the basis for the circuit-switched domain in UMTS, albeit with some enhancements to support the higher data rates and other requirements of UMTS services. There are equivalent elements in cdma-based systems (cdmaOne and cdma2000).

Mobile Switching Center (MSC)

The MSC provides the interface between the radio system and the fixed network, performing all necessary functions to handle CS services to and from mobile terminals. As such, an MSC will interface with several base stations. In effect, it is a telephone exchange that performs switching and signaling functions for mobiles within its designated area of control. It should take into account the allocation of radio resources and the mobile nature of users, which impact the location registration and handover between cells.

Gateway MSC

For incoming calls, the GMSC provides routing to the appropriate MSC where the destination mobile terminal is located, after having interfaced with the databases within the home environment.

Visitor Location Register (VLR)

The visitor location register (VLR) is used by an MSC to retrieve information for mobile terminals currently in its area. A mobile terminal registers with the VLR as it enters the area, at which point the VLR and HLR (home location register — see

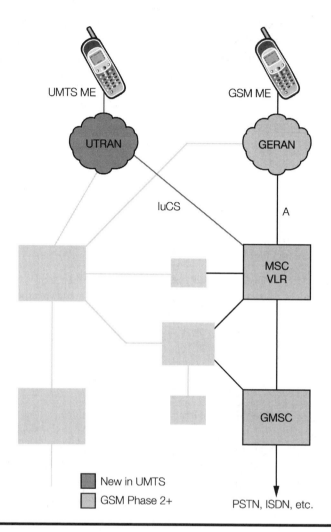

FIGURE 3.9 The circuit-switched domain.

below) exchange information on the subscriber and his service capabilities. It is the VLR that tracks the current location of the mobile terminal although the HLR will know on which VLR the subscriber is registered.

An additional element that is required in UMTS is the Interworking Function (IWF). In broad terms, an IWF provides the functionality to allow interworking of differing networks such as ISDN, PSTN, and PDNs — Packet Data Networks (i.e., protocol conversion). A new element required for the CS part of the core network in UMTS is an interworking to enable the core network to operate with both the existing 2G and new UMTS radio access.

Mobile Network Infrastructure and Supporting Systems ■ 243

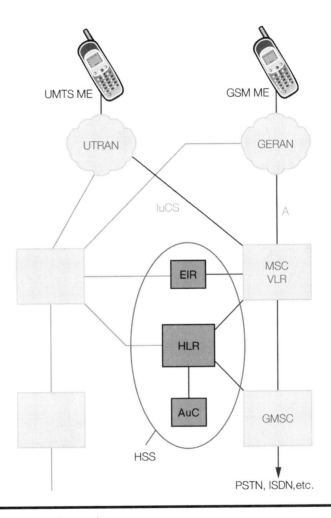

FIGURE 3.10 Location registers.

The GSM Location Registers

The core GSM elements also include some additional databases, which are also carried forward into UMTS (with appropriate modifications as required):

- *HLR.* The home location register (HLR) contains subscriber information, and is the register to which a subscriber is assigned (Figure 3.10). It will also contain information enabling charging and packet routing of messages to the area where the mobile is currently registered (for GPRS support), plus various location-service-related information if that is also supported. Subscriber information consists of:

- The IMSI (International Mobile Subscriber ID)
- Mobile Station ISDN numbers (telephone numbers)
- Packet Data Protocol addresses for GPRS
- Location indicators for location services
- Information on service access and restrictions

Authentication Center (AuC). The AuC stores data for each subscriber to allow the subscriber (IMSI) to be authenticated and to allow ciphering of communication over the radio path. In short, it allows the mobile to use the network. The data required for these two processes is transmitted via the HLR to the VLR, MSC, and SGSN as required.

- *Equipment ID Register (EIR)*. The EIR (Figure 3.10) is responsible for storing the International Mobile Equipment IDs (IMEIs) in the GSM system. These classify equipment as white-, gray-, or blacklisted, and thus enable service to be prevented to stolen or uncertified terminals.

This set of databases and registers can be grouped together for simplicity and defined as the *home subscriber server* (HSS).

The Packet-Switched Domain

The circuit-switched side of the network is limited to 64 kbps by its ISDN-based switching capability, whereas GPRS allows direct interconnect with data networks of much higher bit rates (see Figure 3.11). Therefore, GPRS is a prerequisite for the introduction of UMTS. The GPRS-specific entities are described below:

- *SGSN (serving GPRS support node) and GGSN (gateway GSN)*. The GGSN and SGSN have functions and architectural positions comparable to the GMSC and MSC/VLR in the circuit-switched domain (see Figure 3.11). They are IP routers, which allow direct transmission between mobile terminals and data networks such as the Internet and corporate intranets.
  - The *SGSN* includes a location register function that stores subscription information and location information for packet-switched services for each subscriber registered in the SGSN.
  - The *GGSN* stores subscription information and routing information for each subscriber for which the GGSN has at least one PDP context (effectively a data session) active. This information is used to send packet data destined for a GPRS terminal through to the SGSN where this terminal is registered. Once again, a new interworking function is required so that the SGSN can communicate both with the new UMTS terrestrial radio access network and the existing GSM base station subsystem.

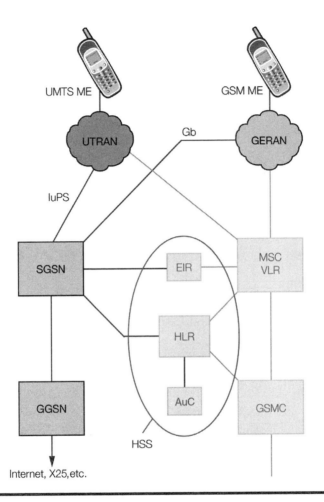

FIGURE 3.11 The packet-switched domain.

Signaling and Control in the Core Network

Different applications, features, and services within the modern telecommunications network are supported by various sets of signaling messages. The messages are exchanged between various network elements in a standard way, and via a common underlying signaling network called Signaling System Number 7 (SS7) (Figure 3.12).

The message sets, formatting rules, and operation of the signaling entities are all standardized and form part of the SS7 specification. SS7 provides a highly reliable and resilient packet-switched network over which the network control information

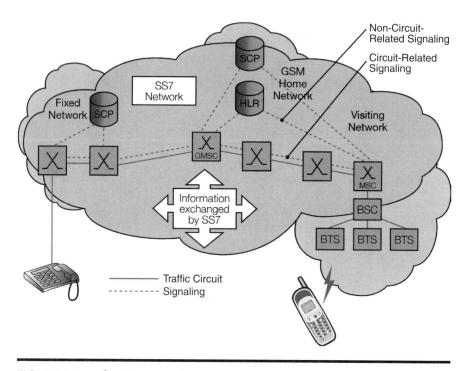

FIGURE 3.12 The SS7 network.

(in the form of SS7 messages) can be transferred. Hence, call control, supplementary services, and intelligent networks are all supported by SS7 in the fixed network, while CAMEL, Mobility Management, Short and Multi-Media Messaging, and GPRS (General Packet Radio Service) can be added to the list for mobile networks. SS7 is also an integral part of UMTS.

For security and ease of management, SS7 is only provided within the core of the network and not generally extended into the access network.

Connecting the Core Network Elements Together

As for the radio network, the core network elements must be connected together by transmission networks that provide a suitable number of channels to cater for both the user's traffic (voice or data), as well as the required signaling and control (Figure 3.13). The main difference here is that there is initially a distinct difference between the circuit-switched (predominantly voice) network requirements and the packet-switched (data) network requirements.

The requirements are essentially similar for 2G, 2.5G, and 3G networks, although as more flexibility and different channel types are introduced with the 3G systems, there needs to be better control and prioritizing of data packets through

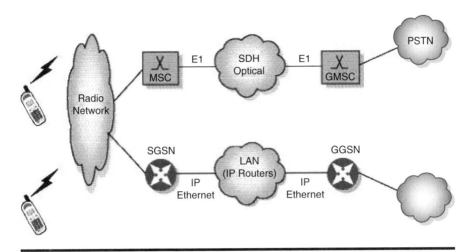

FIGURE 3.13 Typical GSM or GPRS core network connections.

the packet-switched network (i.e., Quality of Service [QoS] mechanisms). This would generally be provided by the transport network; and because this is based on IP technology, ATM (Asynchronous Transfer Mode), or more likely, MPLS (Multi-Protocol Label Switching) can be used.

The circuit-switched core network can be combined with the packet-switched core network at the transmission or transport level, leading to greater efficiency and lower costs. Finally, the circuit-switched core network can be discarded altogether as voice is carried (with appropriate priority or Quality of Service) through the packet-switched network. These changes reflect the Third Generation Partnership Project (3GPP) release 4 and release 5 standard core network specifications.

Now consider the interfaces between the MSC/GMSC and between the GSN (GPRS Support Nodes).

As always, the following requirements are important:

- Providing enough capacity (for current and forecast usage)
- Ease of management, including fault alarms
- Fault tolerance, incorporating alternative routes if possible
- Costs, both installation and ongoing

Decisions must be made on how the elements are connected together with respect to:

- Transmission media
- Transmission system
- Whether leased lines are used

Although different options can be used, the diagram below shows how the core network elements are often connected together in networks that provide "best-effort" Quality of Service.

The basic way of organizing the channels in the circuit-switched network (for voice mainly) is to use the PDH (Plesiochronous Digital Hierarchy) E1 link. This can be carried between MSC sites using SDH (often over fiber optic) to give better flexibility, management, and fault tolerance.

For the packet-switched network, the *best-effort* requirement would mean that the transport or transmission network can be simple as no QoS mechanisms are required. The connection may simply be via LAN (local area network) type Ethernet. Interconnections between two different operator's networks would use appropriate WAN (wide area network) connections.

Review Questions

Q1. The requirements of the core network can be summarized as requirements for _____.
   a. Packet switched, voice switched, and signaling/control
   b. Voice and data
   c. Control and services
   d. Packet switched, circuit switched, and signaling/control

Q2. In GSM-based networks, the telephone exchange is known as the _____.
   a. HLR
   b. LE
   c. MSC
   d. SGSN

Q3. For GPRS, the SGSN would always be located in the _____.
   a. Home network
   b. Radio network
   c. Circuit-switched network
   d. Serving network

Q4. The core network signaling and control is provided (mainly) by _____.
   a. SS7
   b. IP
   c. ATM
   d. PDH

Mobile Network Infrastructure and Supporting Systems ▪ 249

Q5. The circuit-switched and packet-switched core network elements could typically be connected together (respectively) by _____.
   a. IP and voice circuits
   b. Optical fibers and twisted copper wire
   c. IP routers
   d. PDH and Ethernet

3.1.4 Section Review Questions

Q1. On the simplified GSM and GPRS network diagram in Figure 3.14, label the following elements using their full titles:

   BTS   BSC
   PCU   MSC
   GMSC   VLR
   HLR   SGSN
   GGSN

Q2. Group together the elements that provide the following:

   The radio network
   The circuit-switched core network
   The packet-switched core network

Q3. Annotate (on the diagram in Figure 3.14) the methods of connection that would be suitable within each part of the network (radio, circuit-switched core, packet-switched core, signaling/control). Include, for example, PDH, SDH, Ethernet, fiber, copper, radio, ATM, and IP where relevant. Where different technologies or media are used on the same specific interface (connection), describe how they fit together.

Q4. The GSM element that provides a connection to the wider PSTN is the _____.
   a. VLR
   b. BSC
   c. GMSC
   d. MSC

Q5. GPRS _____.
   a. Requires a separate BSC but can use the existing GSM BTS
   b. Requires a separate PCU, BSC, and BTS
   c. Shares the existing GSM BTS and BSC but requires a PCU
   d. Shares the MSC but requires a PCU at the BTS

250 ■ *Introduction to Mobile Communications*

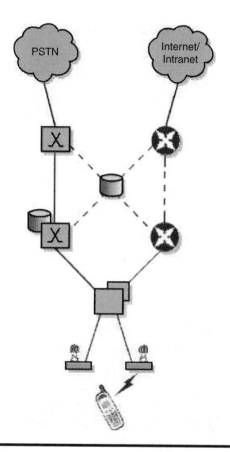

FIGURE 3.14 Simplified GSM and GPRS network diagram.

Q6. Basic cellular procedures must provide for managing _____.
   a. Radio connections, user mobility, and end-to-end connections
   b. GPRS and GSM
   c. Only the end-to-end connection
   d. The overall user experience

Q7. The GPRS element that provides a connection to the MMSC, Internet, or WAP Gateway is the _____.
   a. SGSN
   b. BSC
   c. GGSN
   d. PCU

Q8. The identity that a user would give to friends and family is the _____.
   a. IMSI
   b. IMEI
   c. MSISDN
   d. IP address

Q9. The handset is very often referred to as the _____.
   a. Mobile handheld
   b. Handset device
   c. User device
   d. User equipment

Q10. Which network element connects to the BTS?
   a. HLR
   b. BSC
   c. PCU
   d. MSC

Q11. For GPRS, the radio elements required are _____.
   a. The same as for GSM
   b. Modified GSM elements (and also PCUs at the BSC)
   c. New Node Bs
   d. New SGSNs and GGSNs

Q12. Implementation of EDGE in the network would often require _____.
   a. A new BSC and BTS
   b. A new radio network
   c. 3G or WiFi infrastructure
   d. A card change or software upgrade

Q13. W-CDMA radio elements are _____.
   a. PCUs at the BSC
   b. GSNs
   c. Node Bs and RNCs
   d. New cards at the BTS

Q14. HSDPA is _____.
   a. A specific implementation of GPRS
   b. A specific implementation of EDGE
   c. An uplink, high-speed W-CDMA connection
   d. A specific implementation of W-CDMA

Q15. The requirements of the core network include _____.
   a. Circuit-switched data, voice-switched, and signaling/control
   b. Voice, data, and radio
   c. Control and services
   d. Packet-switched, circuit-switched, and signaling/control

Q16. In GSM-based networks, the telephone exchange is known as the _____.
   a. VLR
   b. MSC
   c. SMSC
   d. SGSN

Q17. For GPRS, the SGSN would always be located in the _____.
   a. Home network
   b. Radio network
   c. Circuit-switched network
   d. Serving network

Q18. The core network Signaling and control is provided (primarily) by _____.
   a. Internet Protocol
   b. Signaling System Number 7
   c. Asynchronous Transfer Mode
   d. Plesiochronous Digital Hierarchy

Q19. The circuit-switched and packet-switched core network elements could typically be connected together (respectively) by _____.
   a. IP and voice circuits
   b. Optical fibers and twisted copper wire
   c. IP routers
   d. PDH and Ethernet

Q20. Radio elements are designed and planned to allow for _____.
   a. User mobility within a designated geographical area
   b. User mobility within a cell
   c. User mobility within a "hotspot"
   d. User mobility indoors

Q21. "Roaming" allows mobility within _____.
   a. The entire country
   b. A "Roaming partner's" network
   c. The GPRS network
   d. Any network worldwide

Q22. A SIM card _____.
   a. Is identified by the IMEI
   b. Identifies the network being used
   c. Identifies the GPRS network
   d. Identifies the subscription or subscriber

Q23. User equipment _____.
   a. Is always designed to be either a data-only or voice-only terminal
   b. Can be either 2G/2.5G (only) or 3G (only) compatible
   c. Can be multi-access technology and multi-function
   d. Can always connect to any network type

3.2 Mobile Network Procedures

3.2.1 Example GSM Procedures

GSM Procedures

Mobility Management States

For standard GSM operation, the mobile terminal can exist in one of three different states, described as (1) detached, (2) idle, or (3) dedicated (Figure 3.15).

First, the mobile can be *detached* — either switched off or out of range. However, when *connected* to the network, the mobile is said to be attached, and can be in either *idle mode* or *dedicated mode*. In idle mode, it will be monitoring system information and measuring cell transmission strengths (often the mobile would be in the user's pocket or on the desk in this state). In dedicated mode, the mobile may be performing procedures such as call setup or location update.

- *Attached/detached*. It is important that the network knows when a mobile is available to accept or make calls, and a formal sequence of *attached* and *detached* procedures are available. Of course, if the mobile *wanders* out of range, the procedures cannot be used, but other procedures ensure that the system is eventually updated (periodic location updates). An attach always includes a location update so the core network knows where the mobile is (i.e., which cell) at that particular time.
- *Idle and dedicated mode*. To ensure maximum efficiency, once attached, a mobile will be either in idle mode or dedicated mode. In idle mode, the mobile does not have its own radio channels but shares a number of common channels with many other devices that are also in idle mode. This provides a great saving in terms the number of channels required in the overall system, and hence cost of the network. The mobile constantly monitors the network for information

254 ■ *Introduction to Mobile Communications*

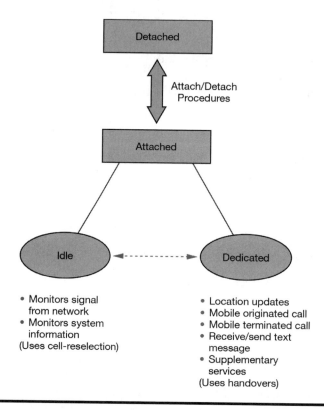

FIGURE 3.15 GSM states and modes.

such as a *paging* message (notifying the mobile of an incoming call), or *system* information, including the current cell identification and the radio parameters required by the mobile for effective connection when required.

When *dedicated* radio channels are required, the mobile automatically moves into dedicated mode, remaining there for as long as dedicated channels are needed (e.g., the duration of the voice call).

Idle mode:

- Monitors signal from network
- Monitors system information

Dedicated mode:

- Location updates
- Mobile originated call
- Mobile terminated call

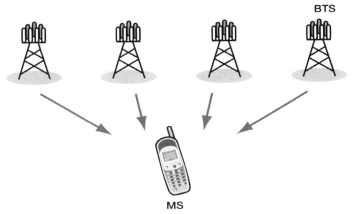

- MS not in call
- MS decides on cell selection and reselection
- Decision based largely on received downlink signal
- Network not informed of cell changes (unless the new cell is in a new location area)

FIGURE 3.16   GSM idle mode.

- Receive/ or send text message
- Supplementary services

In a case where a mobile moves from one geographical location to another, there are standard procedures defined for when the mobile transits from one cell area to another. These procedures differ, depending on whether the mobile is in idle mode or dedicated mode:

- Idle mode:
  - Cell selection and reselection
- Dedicated mode
  - Handover

The Mobile in Idle Mode

In the idle state (Figure 3.16), the mobile is switched on and has already attached to the network, but is not engaged in any active transaction (i.e., not in a call). In this state, the mobile will receive information broadcast by the cell it is currently monitoring, but will also listen for other cells (on different frequencies) in the immediate area. It is constantly assessing the current and *neighbor* cells to decide which is best to use. The procedures are known as cell selection (for selection when the mobile is first turned on) and reselection (for all subsequent decisions on changing a cell), as appropriate. Cell reselections occur on a regular basis for mobiles that are moving from area to area (e.g., in a car).

When reselecting to a new cell, the mobile works independently of the network. It makes decisions regarding cell reselection based on fairly complex calculations. By far the most significant part of these calculations is the (downlink) received signal strength as received by the mobile. Simply stated (and true in most cases), the mobile will reselect the cell that provides it with the best downlink signal level.

The network is not informed of the cell changes unless the new cell is in a *new location area*; that is, a group of several cells used by the core network to monitor the location of a user device.

The Mobile in Dedicated Mode

In dedicated mode, the mobile is usually engaged in a call (alternatively, it may be involved in other transactions such as text messaging (SMS) or performing a location update). Dedicated mode implies that there is a two-way dedicated channel assigned to the mobile.

Handovers

In this situation, moving to a new cell is referred to as a handover (as opposed to idle mode cell reselection) and it involves more complex processes. It is the network that makes the decision and will, at the appropriate time, command the mobile to change to a new cell to continue with the call. This decision is based on several factors associated with:

- Uplink radio signal conditions (assessed by BSC)
- Downlink radio signal conditions (assessed by mobile)
- How the mobile is receiving other neighboring cells (downlink)

Although the network makes the decision about handovers, the mobile plays an important role in the process, in that it must report the radio conditions that exist in the downlink to the network (BSC). When in dedicated mode, therefore, the mobile is constantly taking measurements of the serving and neighboring cells and reporting those measurements to the serving cell. This process is referred to as mobile-assisted handover (Figure 3.17).

Location and Routing Area Updates

For effective mobility management (managing the location of a user), regular updates of the user's location are needed by the network element that is assigned this task — the visitor location register (VLR). There is one VLR (at each MSC) responsible for knowing the whereabouts of every *attached* mobile in the area covered by the MSC. The process of updating this information is known as a *location update* (see Figure 3.18).

Mobile Network Infrastructure and Supporting Systems ■ 257

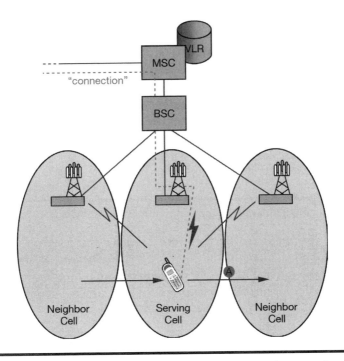

FIGURE 3.17 Dedicated mode handovers (1= mobile monitors serving and neighbor cell signals; 2 = measurement report sent to BSC every 1/2 second; 3 = on reaching A, neighbor cell is assessed as the *better* cell to use; 4 = BSC instigates handover; 5 = new connection made via the *neighbor* cell (new serving cell) and BSC; and 6 = the old connection is dropped).

Normal location update
- On transiting from one cell to another in a different location area
- Periodically (as required by the network requirements broadcast to all mobiles)
- On switch on (attach)

FIGURE 3.18 GSM location update types.

For efficiency and flexibility, the level of accuracy is to a *location area*, rather than a cell, although a location area can be as large or small as the network operator decides. In practice, they tend to be quite large and are made up of tens of cells.

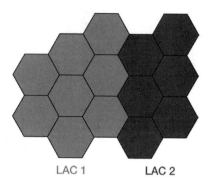

FIGURE 3.19 GSM location area.

A location update initially occurs when a mobile is switched on (often called a registration in this instance). The other two scenarios when a location update is performed are on a periodic basis (the period to be used is broadcast to all mobiles in the system information on the broadcast channel — available to them in idle and dedicated mode), and on transiting a location area (LA).

There is a very similar procedure required for GPRS. This uses the *routing area* rather than the *location area*, although physically, the two areas can be set to encompass the same cells.

Location Areas — Location areas are each identified by a location area code (LAC) and might cover between 20 and 100 cells (Figure 3.19). It is common practice for a location area to equate to a BSC area. For GPRS operation, a further set of areas is defined, and these are called *routing areas*. This allows the SGSN to track a GPRS-enabled mobile in the same way that the VLR tracks the GSM mobile in idle mode (Figure 3.20).

The rules for both location areas and routing areas are as follows:

- A GSM location area may not extend beyond a MSC/VLR serving area. There may be several location areas within a MSC/VLR area but any one location area must be fully enclosed within a MSC/VLR area.
- A GPRS routing area must be fully enclosed within a location area. It is generally the case that there will be multiple routing areas within a location area. Also, a routing area must be fully contained within an SGSN area.

Location Update Procedures

These procedures are initiated from idle mode and are performed to keep the core network (VLR) informed of the mobile's location. This is useful so that, for example, the network knows which cells require paging messages for incoming calls to a specific user.

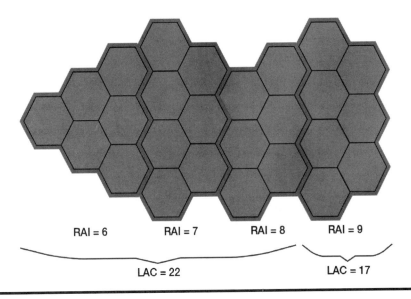

FIGURE 3.20 GPRS routing area.

Location updates are used either to update the location area (LA) within the visitor location register (VLR), or where a mobile has transited into a cell in a new MSC/VLR area, it can be used to inform the new VLR of its presence.

This has the advantage of allowing the new VLR to communicate with the mobile's home location register (HLR) to retrieve the mobile's service profile and other required data. It also informs the HLR that the mobile is now located within that VLR area. This is used to route any incoming calls from the home network (Figure 3.21).

Routing Area Update Procedures

Routing area update procedures are very similar to location updates, but are relevant to the Serving GPRS Support Node (SGSN). The RA is used as the update area. Under certain circumstances, a combined LA/RA update can be performed but these are not common.

Review Questions

Q1. If the mobile is involved in a call, it is in which mode?
    a. Detached
    b. Connected
    c. Idle
    d. Dedicated

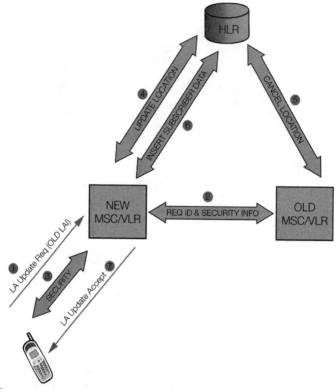

- Radio connection is established before 1.
- For location update within the same MSC/VLR area, delete 2, 4, 5, and 6.
- For RA update, the same basic procedure is followed, replacing MSC/VLR with SGSN and changing the messages/parameters.

FIGURE 3.21 Location update (new MSC/VLR area).

Q2. In idle mode, the mobile performs _____.
    a. Location updates
    b. Cell reselections
    c. Cell updates
    d. Location reselections

Q3. In idle mode, when changing cells within the same location area, a mobile will _____.
    a. Inform the VLR
    b. Do a location update
    c. Update the HLR
    d. Perform a cell reselection

Q4. A location area consists of _____.
    a. A single cell
    b. All the cells within an MSC area

c. A number of cells as determined by the network operator
d. A BSC area

Q5. If a mobile enters a new cell while in a call, it performs a _____.
a. Cell update
b. Location update
c. Handover
d. Cell reselection

3.2.2 Making Calls

Calls to Mobile Networks

Making a call to a mobile network involves keying in the called subscriber's MSISDN (mobile subscriber's ISDN) number. This number routes the call to the called subscriber's home network via a gateway mobile switching center (GMSC).

Here, signaling interactions with the home location register (HLR) occur. The HLR, in turn, interacts with the visitor location register at the mobile switching center at which the called subscriber is currently registered.

Eventually, although it takes very little time, new *routing* information is sent back to the GMSC to replace the subscriber's MSISDN number. The new information allows the call to be routed to the network in which the called subscriber is currently registered and, more specifically, to the serving MSC (either at home or abroad [roaming cases]) (Figure 3.22).

Calls from Mobile Networks

For outgoing calls from a mobile network, the serving MSC acts as a local exchange for the subscriber that is making the call — with the call being routed directly from the serving MSC (rather than having to route back through the subscriber's home network if abroad).

In all cases, however, the call is in accordance with any service information being held in the VLR (having been previously sent from the subscriber's HLR when first registering at this MSC). This may include such restrictions as *call barring* or advanced interactions required for prepaid subscribers (Figure 3.23).

3.2.3 Example: GPRS Procedures

GPRS Mobility Management

The most striking difference between GPRS and GSM in terms of *mobility management* (MM) is that there are no handovers in GPRS. This is because the GPRS mobile is never on a dedicated circuit-switched connection. All channels

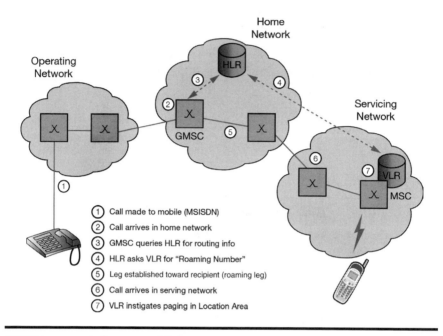

FIGURE 3.22 Calls to a mobile.

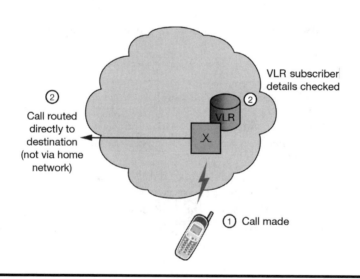

FIGURE 3.23 Calls from a mobile.

are packet-switched and shared in GPRS. Consequently, the GPRS mobile performs cell reselection regardless of whether or not it is engaged in data transfer. The three MM states for GPRS and the state diagram that relates them are shown the Figure 3.24.

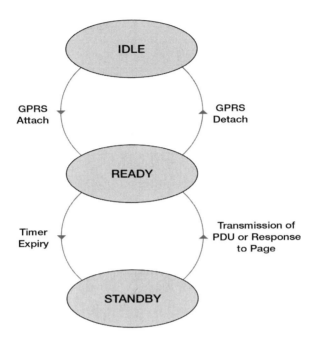

- Cell reselection performed by mobile
- Informs network for each cell change when in the ready state
- Informs network for each routing area change when in the standby state

FIGURE 3.24 GPRS mobility management (MM) states.

- *Idle.* In the idle state, the GPRS mobile is detached from the network. The MS can be switched on and performing cell reselection, but the network does not hold any location information about the mobile. No point-to-point data can be exchanged.
- *Ready.* In the ready state, the mobile is able to exchange data with the serving base station. The mobile will move from the *idle* to the *ready* state when it performs a GPRS attach. This will normally be done prior to the mobile activating a "PDP context" (see later) for the exchange of user information.

When in the ready state, the mobile will perform *cell reselection.*

This is essentially the same cell reselection process that is carried out in GSM. That is, the mobile makes the decision to reselect based (largely) on signal levels received from the serving and neighboring cells. The difference between this case and the GSM idle state is that the mobile must inform the network of each reselection to a new cell. This is necessary so that exchange of user data can continue. Effectively, the mobile will detach from one cell and reattach onto the new cell. In this state, there is no need to page the mobile for incoming data.

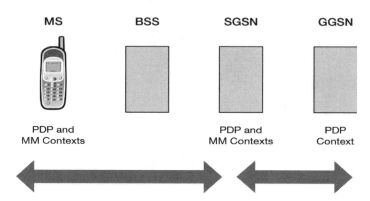

FIGURE 3.25 PDP context association between MS, SGSN, and GGSN.

- *Standby.* For purposes of efficiency and to save battery power, a mobile is able to fall back into standby state once there has been no data exchanged for a predefined time, and a preset timer has expired. The mobile would reenter the ready state once data starts to flow again. It is in ready state that *routing area updates* would be performed.

Packet Data Protocol (PDP) Contexts

The PDP context is central to the entire operation of GPRS. There are no dedicated circuit-switched connections as in GSM because GPRS is a packet-based network. The only thing that identifies a particular packet-based *transaction* is the PDP context. Strictly speaking, it is not a communications session but is best described as an *association* between the MS (mobile station), SGSN, and GGSN. In practical terms, it represents a temporary database of information pertaining to the transfer of packet-based data.

The PDP context exists at the MS, the SGSN, and at the GGSN (Figure 3.25). The information stored at these locations is largely common among those three locations. Alongside the PDP context is mobility management information that enables routing of packets from the MS to and from SGSN and GGSN.

It is instructive to consider the user equipment as consisting of two parts. First there is the GPRS-capable mobile station (with SIM) that is capable of establishing and maintaining GPRS *bearers*. That is, it is capable of transferring PDP data between itself and the SGSN and from there to external networks.

Sitting *on top of* the mobile equipment are the applications that make use of the GPRS bearer service. These applications may reside in a separate piece of hardware (e.g., a laptop connected to a GPRS terminal) or alternatively may be incorporated within the same physical device (e.g., a GPRS-capable PDA). It is a function of the GPRS terminal to attach to the network, but it is the function of the application to initiate the PDP context.

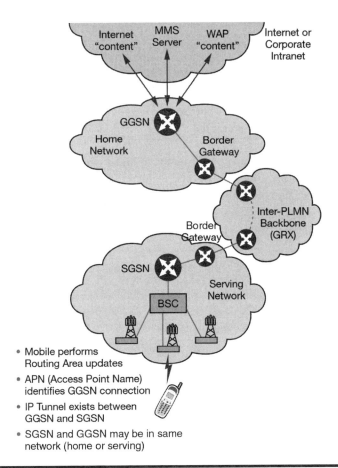

FIGURE 3.26 GPRS connections.

GPRS Connections

To move information from content and application servers to the handset and vice versa, a path through the network must be defined. This is achieved using the PDP context procedure, where the handset, SGSN, and GGSN all store data relating to this virtual connection (Figure 3.26).

The GGSN defines the gateway to the content — maybe the Internet, MMS server, or a WAP gateway. The GGSN is identified by a special identity known as the Access Point Name — effectively a URI (Universal Resource Indicator, as used for Web pages, etc.).

Because billing and other administrative functions may be based on this GGSN, it does not make sense in most instances to change this gateway dynamically. Hence, even when a user is roaming abroad, the GGSN is usually located in the subscriber's home network, and a path is defined through the serving network and an inter-PLMN backbone network (incorporating GPRS Roaming Exchanges (GRX)), to the home network. The border gateways define the edge of the administratively separate networks with associated firewalls.

An alternative arrangement may be that the SGSN and GGSN are located in the same network. This would certainly be the case where the subscriber is attached via the user's home network, but may also be true, for example, where connection settings in the handset for Internet access use a generic APN (access point name), shared by different operators, allowing access to the nearest GGSN (in the serving network).

Using different procedures, the *connection* can be maintained. If necessary, the PDP context information can be moved from SGSN to SGSN without having to reestablish either the connection or the PDP context (Figure 3.26).

Review Questions

Q1. For incoming calls, the call is routed _____.
   a. Directly to the recipient GSM subscriber
   b. Via the GSM subscriber's home network
   c. Via the HLR and VLR
   d. Via the GMSC but not the serving MSC

Q2. For outgoing calls, the call is routed _____.
   a. Always via the home network
   b. From the MSC directly toward the recipient
   c. Via the HLR
   d. From cell to cell

Q3. A call between two GSM subscribers _____.
   a. Would involve one or two GSM networks only
   b. Could involve a number of transit networks as well as one home and two serving GSM networks
   c. Would always be routed directly between serving GSM networks but include transit networks
   d. Would always be routed via the two subscribers' home networks

Q4. The three GPRS states are _____.
   a. Idle, ready, and standby
   b. Idle, attached, and detached
   c. Idle, dedicated, and detached
   d. Dedicated, standby, and detached

Q5. The *access point name* identifies the _____.
   a. Application
   b. SGSN connection to the GGSN

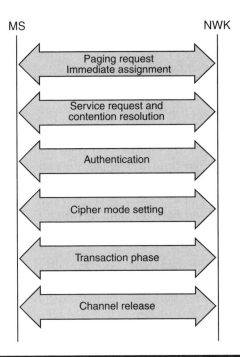

FIGURE 3.27 General sequence of procedures for the GSM family of technologies.

    c. Radio connection
    d. GGSN connection to the external network or server

Q6. The GGSN would usually be in the _____.
    a. Home network
    b. Serving network
    c. Radio access network
    d. Circuit-switched network

3.2.4 Procedure Sequences

For all cellular systems, each transaction (incoming or outgoing call, data session, text message, etc.) carried out by the user requires a number of separate procedures to occur in a set sequence. Each transaction will have its own unique requirements, but the general sequence will usually be common to all transactions.

The general sequence used for the GSM family of technologies is shown in Figure 3.27. At this stage, this sequence ignores the complex requirement for setting up the radio resources. In reality, the radio resource procedures would occur within the sequence shown at any point where there is a requirement to allocate, change, or release radio connections.

Note that the standard sequence illustrates the required procedures in six identifiable steps. This sequence is referred to as the elementary procedure. Before any communication with the core network, there must be a radio connection in place. The mobile must also identify the user to the network prior to any service request being granted.

The first step in the sequence is the assignment of an appropriate channel, which may be after a paging message has been received from the network in case of an incoming call. This is followed by the mobile handset requesting the service that is required, as well as the security procedures (authentication of the user, and ciphering of the radio channel). Both security procedures are optional; and although available in most networks, there are exceptions.

These mobility management procedures are used for establishing a *mobility management* (MM) connection with the core network, including the authentication of the user. Once complete, *connection management* (CM) procedures are used to set up and manage the transaction itself, including establishing outgoing or incoming calls, setting up and managing data sessions, or sending or receiving a text message. Once complete, all relevant channels will be released.

The elementary procedures will be initiated while a handset is in idle mode, requiring it to move into dedicated mode for the duration of the sequence, before falling back to idle mode once the sequence is complete. If another transaction is required while one is in progress, the mobile uses the existing MM procedures already in place as the basis for the new transaction. For example, no further security procedures, or even a paging message, will be required for a mobile that receives a text message while in the middle of a voice call. For GPRS and UMTS packet-data operation, similar considerations apply.

3.2.5 Radio Resources

Radio Resource Procedures for GSM, GPRS, and EDGE

The general purpose of radio resource procedures is to establish, maintain, and release radio connections. To be effective, the radio resource procedures are complex and include the cell selection/reselection and handover procedures. Other example procedures are described in Figure 3.28.

Radio Resource in Idle Mode

The mobile handset continuously monitors the system information broadcasts on the broadcast channel, interpreting the information appropriately. This information contains parameters such as the identity of the serving cell and information on how to access the network.

Once the cell is selected, the mobile leaves the idle state to register its location with the network. In the absence of any other activity, it then falls back to the idle mode, to

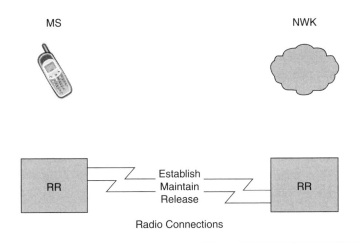

FIGURE 3.28 Radio resource management.

monitor the paging channel (to identify any incoming calls), updating its location as required (periodically, or as location area or routing area boundaries are crossed).

Radio resource in idle mode:

- Monitor system information broadcast (contains required information)
  - Cell selection
- Cell reselection
- Location registration and updating
- Monitor paging channel

Radio Resource Connection Establishment

This is instigated from idle mode. The RRC connection request is sent from the mobile to the network. This may be to make a call, transfer data, or update a location or routing area. Once the connection is established, a confirmation is sent to the mobile.

The mobile now moves into the connected state, sending a Radio Resource *Connection Setup Complete* indication to the network on the dedicated channel that has just been assigned.

Paging

For idle mode mobiles, paging is initiated over the required location area, or routing area for incoming calls, or for session setup in the case of packet services. The core network node (MSC / VLR or SGSN) initiates the paging procedure.

The paging message is broadcast using Radio Resource procedures on the paging channel that each idle mode mobile monitors within the cell(s).

Other Radio Resource Procedures

Radio resource procedures are extremely comprehensive and cover a wide range of requirements. Some of the procedures not dealt with here include:

- Power control (uplink and downlink)
- Handovers (can be extremely complex in UMTS/W-CDMA)
- Connection release
- Packet mode procedures

Radio Resource Procedures for UMTS (including HSDPA)

UMTS (Universal Mobile Telecommunication System) procedures follow the general requirements for GSM and GPRS, except that the radio interface and channel structure are more complex and therefore require more (and more complex) procedures. This is also true for HSDPA (High-Speed Downlink Packet Access), which is a specific implementation of W-CDMA used on the UMTS radio interface.

The mobility management (MM) and connection management (CM) procedures are essentially very similar to those discussed above, except that more options are generally available with UMTS because of the multimedia-capable nature of the connections. In fact, the procedures for the GSM family of technologies are contained within a common set of documents (available through the Third Generation Partnership Project [3GPP] Web site at www.3GPP.org).

The documents are arranged such that all procedures relating to MM and CM will be in a single document, irrespective of access technology.

Radio resource (RR) management is where the main differences occur between the technologies and, in this case, the procedures for UMTS (W-CDMA) are separated from those for GSM, GPRS, and EDGE by the use of different documents detailing the RR procedures.

First, the radio elements are known as the Node B, which is equivalent to the BTS in GSM, GPRS, and EDGE, and the radio network controller (RNC), which is equivalent to the BSC. The term "radio network subsystem" (RNS) is used in place of the base station subsystem (BSS) used in GSM, GPRS, and EDGE.

To illustrate just two of the major differences between UMTS (W-CDMA) and GSM, GPRS, and EDGE, two procedures relating to radio resources (RR) are described below.

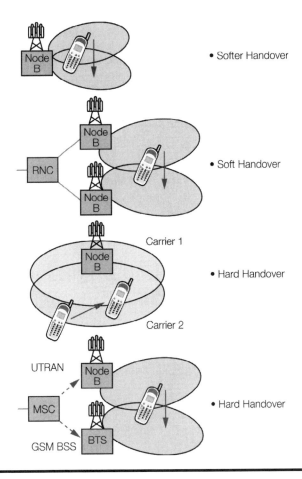

FIGURE 3.29 Handover types in UMTS.

Handover Types

There are three different handover types in UMTS (Figure 3.29):

1. The softer handover is where the mobile receives the signal via two radio interfaces at the same time, with both interfaces provided by the same Node B.
2. The soft handover is where the mobile receives the signal via two radio interfaces at the same time but the interfaces are provided by different Node Bs.
3. The hard handover is the same sort of handover found in GSM, where the new radio resources are established, and the mobile makes contact via the new resources before the old resources are released. Communication only ever occurs via a single interface. This can be used for handover between W-CDMA frequencies, or between different systems.

Power Control

Fast power control is required in UMTS to minimize the interference in the system. It is performed 1500 times per second (much, much faster than for GSM, GPRS, and EDGE), and is designed to adjust the power to meet a defined signal-to-interference ratio (SIR). The SIR is set to give the required quality while minimizing the contribution to overall system interference.

Review Questions

Q1. The basic procedure sequence is referred to as _____.
    a. Standard procedures
    b. Elementary procedures
    c. Mobility management procedures
    d. Fundamental procedures

Q2. Which of the following is *not* a radio resource procedure?
    a. Radio resource connection establishment
    b. Paging
    c. Power control
    d. Call establishment

Q3. Which is incorrect? UMTS handover types include _____.
    a. Soft handovers
    b. Softer handovers
    c. Static handovers
    d. Hard handovers

Q4. Radio resource procedures occur between the _____.
    a. Mobile and core network
    b. Mobile and radio network
    c. Radio network and core network
    d. Mobile and application

Q5. UMTS power control occurs _____.
    a. 1500 times a minute
    b. Every half a second
    c. More slowly than GSM
    d. Much faster than GSM to minimize interference in the system

Q6. Handover between UMTS and GSM would be a _____.
    a. Soft handover
    b. Softer handover
    c. Hard handover
    d. Not allowed

3.2.6 Section 2 Review Questions

Q1. The cells within the cellular system are grouped in various ways; match the area with the correct description, and put the areas in order of their size from largest to smallest:
   a. The area covered by a frequency or set of frequencies from a single BTS
   b. A number of cells that are grouped together and used to determine when the mobile needs to update the SGSN with its current position
   c. All the cells within the area covered by one or more BSCs connected to the same MSC
   d. A number of cells that are grouped together and used to determine when the mobile needs to update the VLR with its current position

Area	Description	Size
Location area		
Cell		
Routing area		
MSC area		

Q2. Insert the correct *mobility* procedure:
   a. In idle mode, the GSM mobile would perform _____.
   b. In dedicated mode, the GSM mobile would perform _____.

Q3. For an incoming (mobile terminated) call, while the receipt subscriber is roaming abroad:
   a. Draw two clouds identifying the home and serving networks.
   b. Add the originating party, GMSC, HLR, MSC, VLR, BSC, BTS, and terminating party in the correct positions on your diagram.
   c. Draw the path (or interactions) for the incoming call, identifying the path or interactions that are for the voice circuit and those that are for signaling (SS7).
   d. Identify which party pays for which leg of the call.

Q4. For a GPRS connection to the Internet in the roaming scenario:
   a. Draw two clouds identifying the home and serving networks.
   b. Add the user terminal, Internet, GGSN, HLR, GRX, SGSN, BSC, PCU, and BTS in the correct positions on your diagram.
   c. Draw the path (or interactions) for the connection, identifying the paths (or interactions) that are for the data and those that are for signaling (SS7).

Q5. Cellular network procedures are designed to allow services to be provided while _____.
   a. The user is stationary
   b. The user is mobile
   c. The user is connected to any cellular network
   d. The user remains within a specific location area

Q6. GPRS procedures allow for _____.
   a. Mobile voice connections
   b. General services within the radio network only
   c. Mobile data services
   d. Efficient text messaging services only

Q7. In general, cellular procedures _____.
   a. Are essentially the same for the roaming and non-roaming cases
   b. Are very different for the roaming and non-roaming cases
   c. Only work while roaming
   d. Are the same as in the fixed network

Q8. Procedures would be carried out in the following order of priority:
   a. Services, radio, mobility
   b. Services, mobility, radio
   c. Radio, services, mobility
   d. Radio, mobility, services

Q9. UMTS procedures differ from those used for GSM, GPRS, and EDGE mostly in terms of _____.
   a. Mobility
   b. Radio
   c. Services
   d. Text messaging

Q10. The mobile can be switched off, in which case the network regards it as _____.
   a. Detached
   b. Attached
   c. Idle
   d. Dedicated

Q11. If the mobile is switched on and registered in the VLR as "switched on," then it must be _____.
   a. Detached
   b. Attached
   c. Idle
   d. Dedicated

Q12. A location update would *not* occur _____.
   a. When selecting a cell in a new location area
   b. Periodically
   c. On switch-on
   d. On cell reselection within the location area

Q13. A routing area _____.
   a. Is used for GPRS and must be enclosed within a single location area
   b. Is used for GSM as a more accurate way of specifying area
   c. Requires a location update if a mobile enters from another routing area
   d. Is not only required for W-CDMA operation

Q14. The handover is performed only in the following mode:
   a. Detached
   b. Attached
   c. Idle
   d. Dedicated

Q15. For incoming calls, the call is routed _____.
   a. Directly to the recipient GSM subscriber
   b. Via the GSM subscriber's home network
   c. Via the HLR and VLR
   d. Via the GMSC but not the serving MSC

Q16. For outgoing calls, the call is routed _____.
   a. Always via the home network
   b. From the MSC directly toward the recipient
   c. Via the HLR
   d. From cell to cell

Q17. The connection from the mobile to the SGSN is described in terms of the mobile being _____.
   a. Connected, disconnected
   b. Attached, disconnected
   c. Attached, detached
   d. Idle, ready

Q18. The PDP context denotes _____.
   a. The connection to the SGSN
   b. A "tunnel" between the SGSN and GGSN
   c. An association between the peer applications
   d. An association between the mobile, SGSN, and GGSN

Q19. The SGSN would be in the _____.
   a. Home network
   b. Serving network
   c. Radio access network
   d. Circuit-switched network

Q20. The basic procedure sequence is referred to as _____.
   a. Standard procedures
   b. Elementary procedures
   c. Mobility management procedures
   d. Fundamental procedures

Q21. The procedure sequence is broken into how many elementary steps?
   a. 4
   b. 5
   c. 6
   d. 7

Q22. Which of the following is *not* a radio resource procedure?
   a. Radio resource connection establishment
   b. Paging
   c. Power control
   d. Call establishment

Q23. Which is incorrect? UMTS handover types include _____.
   a. Soft handovers
   b. Softer handovers
   c. Static handovers
   d. Hard handovers

Q24. Radio resource in idle mode would not include _____.
   a. Monitoring system information broadcast
   b. Cell reselection
   c. Handover procedures
   d. Monitoring the paging channel

3.3 Supporting Systems

3.3.1 Service Platforms — General

Service Platforms

Software and Service Platform Requirements

There is a whole range of service platforms required in modern networks. They include platforms for value-added services (VAS), such as voicemail, Short Message Service SMS), and Multimedia Message Service (MMS). They also include mobility databases, including home and visitor location registers (both of which play an important role in service provision for GSM, GPRS, and UMTS); generic service platforms, such as service control points (used in Intelligent Networks) and service nodes. Finally, they include content servers, which provide access to basic and advanced information, graphics, pictures, music, and video.

Service platforms include:

- Short Message Service center
- Multimedia Message Service center
- Voicemail platform
- Content servers
- Service nodes
- Intelligent Network platforms (and CAMEL)
- Home location registers
- Visitor location registers

With the reduction in the cost of a voice call, the service platforms are assuming greater significance as a way of providing value-added services in an attempt to maintain and increase the average revenue per user (ARPU). In networks that are optimized for data, such as ADSL, GPRS, and UMTS/3G, they provide the main mechanism to grow the number of subscribers and to increase revenue.

The user would either be using these advanced data networks as a bearer to access content and services from an independent provider, or be directed by network resources (network-provided service platforms) to a point in the network where the service can be accessed (network-provided service platforms or third-party service provider platforms).

Network operators would like their customers to access the operator-provided content and services so that they can capture the ensuing revenue, but increasingly, third-party service providers are seen as an integral part of the overall system. In fact, work has been done within industry and standards bodies to provide standard ways of connecting third-party providers to the network so they can be accessed by customers. Examples of third-party content and services include information such as movie schedules, restaurant opening times and locations, credit card payment for calls, navigation system updates (traffic updates), etc.

Third-party service providers are:

- Increasingly common
- Connected via standard interfaces
- Accessed via network service platform

Vendors

Provision of *service platforms* of whatever sort is considered big business. For some vendors, it is seen as a secondary function, helping to make the core product (often telephone exchanges and routers) more attractive in that an integrated solution can be provided, with the subsequent reduction in administration, testing, and required employee knowledge base. Some vendors are seen as specialists in the area, relying on reputation, and above all on an excellent product (or product range). Example equipment vendors include, but are not limited to:

- Alcatel
- Ericsson
- LogicaCMG
- Lucent
- Nokia
- Nortel
- Telcordia

Capacity

Capacity issues for service platforms can be quite complex, with huge variations in usage throughout the day (which can be expected and predicted), and in addition, variations occurring as a result of specific applications. These variations in demand could be triggered by advertising campaigns for certain content, televoting (by standard circuit-switched call, or messaging, for example), or perhaps by world events dictating mass network usage (New Year's celebrations, as an example).

In all cases, the service must be effectively managed. This includes the service and application handling itself, as well as any congestion condition. Any request for service

FIGURE 3.30 Logical view of the service platform.

that is not satisfied directly results in lost revenue. In addition, if congestion is encountered, and appropriate announcements or responses are not given to the customer, they will be dissatisfied and reluctant to try later, again resulting in lost revenue.

One of the main requirements therefore is to balance the cost of equipment against potential usage, taking into account lost revenue through congestion.

This is easier to achieve for some platforms than for others. For example, platforms that primarily provide database applications (such as a home location register in GSM) can be planned, with forecasts of subscriber growth allowing a steady scale-up of network resources. A known history of usage patterns for the HLR on a daily and seasonal basis can be catered for.

Other platforms are not so easy to plan. These are generally the service platforms that support a range of services (or a single service) that are influenced by less predictable variables. For example, a *service control point* (explained later) handling a televoting service will hit congestion levels early (and remain there) if a vote that had forecast to receive two million votes actually receives three million. Although mechanisms do exist within the technology to handle the congestion, the lost revenue is still a huge problem. In any case, scalability is a big factor in service platform provision (Figure 3.30).

Upgrading the network must also be transparent to the customer, leading to the requirement for a flexible architecture. This would allow one to view a single functional entity in terms of access to the service, but with the functionality actually provided by a number of physical entities (servers), as shown in Figure 3.31. This gives the required flexibility and scalability, allowing the network to grow at a planned rate, while ensuring that spare capacity is calculated to cater for peaks, but not be so great that it is wasted investment.

Redundancy

Some services and applications may be more critical than others. The high revenue services should be protected from congestion when a mass service scenario is expected. In addition, services and applications can be very different, with widely varying requirements.

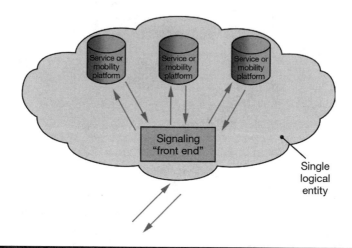

FIGURE 3.31 Single logical database provided by distributed architecture.

For both of these reasons, it is usual for a network operator to implement a number of different service and content platforms. Even when the platform is generic (such as an Intelligent Network service control point), it would generally be the case that different platforms would support different services. There may be a platform for prepaid billing, one for dialed network services, one for fraud control, one for mass calling type services, etc.

In reality, all of these chosen examples could be provided by a single platform but common sense dictates otherwise. Having a single platform implemented for each service would not be ideal either. First, capacity requirements may require more than one platform, but the other compelling reason is that of resilience and redundancy.

Continued availability of any service, even under platform failure, is extremely important; hence, it makes sense for a service to be available for use by the required customers in at least two platforms. The platforms would ideally be sited in different geographical locations (Figure 3.32).

The Open Service Architecture (OSA) Concept

OSA defines an architecture to enable operators and third-party developers (e.g., value-added service providers (VASPs)) to make use of network functionality through an open standardized interface (called an application programming interface [API]). It provides applications with access to *service capability servers*, and thus provides the "glue" between the applications and the service capabilities of the network (Figure 3.33).

In this way, applications become independent of the network. But within the network, features such as Intelligent Networks and CAMEL provide support for the required services. The actual applications, however, can be executed in application

Mobile Network Infrastructure and Supporting Systems ■ 281

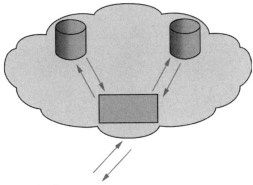

- Physically separate databases
- Information and data duplicated, but could be handling different services, serving different regions, or load sharing

FIGURE 3.32 Redundancy and resilience.

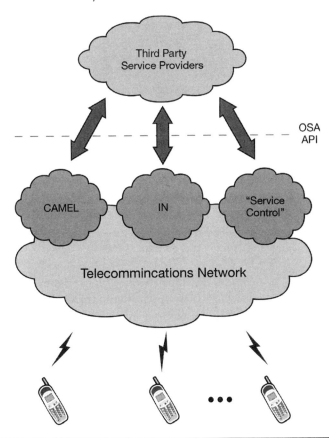

FIGURE 3.33 The Open Services Architecture.

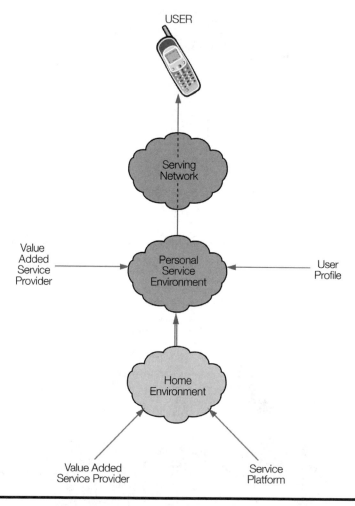

FIGURE 3.34 The virtual home environment (VHE).

servers physically separated from the core network entities. They may be part of the operator *domain*, or may be third-party applications. The OSA API is secure and independent of vendor-specific solutions and programming languages, operating systems, etc.

The Virtual Home Environment (VHE) Concept

In the VHE (Figure 3.34), users are consistently presented with the same personalized features, user interface, customization, and services — in whatever network they are located, or terminal they may be using (assuming that capabilities in the network and terminal exist).

In defining the VHE, it is useful to introduce the concept of the home environment. This can be synonymous with the user's home network and subscribed services, but can also include other value-added service providers (VASPs), which are accessed through this home network service provider. The home environment provides (and controls) the personal service environment in association with the user's own personal profile.

The serving network describes the network to which the user is attached at the time, and thus may be a network in which they are roaming (when traveling abroad). In the VHE concept, this network should be invisible to the user, with services transported seamlessly. It may be another mobile network, but could equally also be applied to a fixed network, the Internet, etc., depending how the users choose to access their services at any one time.

VHE also takes into account the possibility of *value-added service providers*, who may be part of neither the home nor the serving environment. For example, a banking service may be provided directly from a bank VASP. Users should still be able to transparently access these services whether or not they are in their home network.

Location Platforms

3G systems (including UMTS) are designed from the outset to provide for accurate location of user equipment (mobile handsets), and this allows for providing advanced *location-based* services. The location information is collated and managed by location platforms (primarily the serving mobile location center and the gateway mobile location center). This information can then be accessed and used by the service platform in a variety of different services.

In UMTS, the location determination can be achieved in three main ways (discussed shortly), and may also be provided as part GSM or GPRS.

The location information can be used by the PLMN operator, emergency services, value-added service providers, and for lawful interception by authorized agencies. The PLMN could use location information for a variety of purposes, including handover optimization. Emergency services can radically improve the overall response time if automatic mobile terminal location is provided. Value-added services can be significantly enriched and in some cases enabled — for example, downloading a map showing where the subscriber is and how to reach a local address.

Location clients

- PLMN operator
- Emergency services
- Value-added services
- Lawful interception location

The quality of location information is defined in terms of horizontal accuracy (10 to 100 meters, according to the application), vertical accuracy (up to 10 meters), and response time (no delay, low delay, or delay-tolerant criteria). It can also be defined by priority (e.g., emergency services have highest priority), time stamping (vital for some applications such as lawful interception), security measures (to ensure controlled access to user location information), and privacy.

Quality of Location Information

- Horizontal accuracy
- Vertical accuracy
- Response time
- Priority
- Timestamp
- Security
- Privacy

In general, the user can restrict access to location information except for emergency or lawful interception purposes. For lawful interception, the user will not be aware of location tracking.

Location Techniques

The three primary ways of obtaining the location of a handset are:

1. Based on the cell (cell identity, cell ID) where the handset is located
2. Based on GPS (Global Positioning System) — chips within the handset
3. Observed time difference of arrival (OTDOA) method — unique to CDMA-based systems

The cell ID method i.e., the network using knowledge of the cell currently in use by the mobile) may provide sufficient accuracy if either the mobile is using a micro cell or pico cell, or if the accuracy requirement is low (i.e., if the mobile is using a macro cell).

Network-assisted GPS is based on inclusion of a GPS receiver within the mobile. This can provide location coordinates that can be translated by the network into a format suitable for the client application.

The OTDOA method involves measurements at the mobile of the time of arrival of transmissions from a number of Node Bs. These measurements, together with measurement carried out by serving Node Bs, are processed to give a position for the mobile. This method is used in UMTS where the mobile can access several Node B interfaces simultaneously.

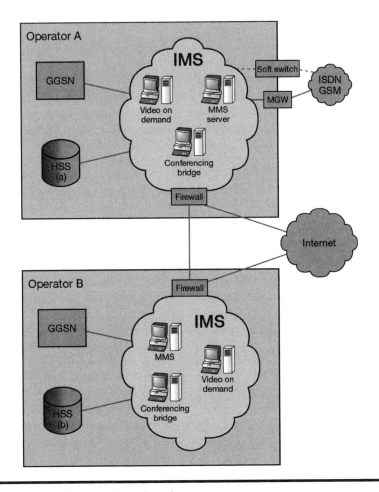

FIGURE 3.35 The IP multimedia sub-system.

IP Multimedia Sub-system (IMS)

This is an IP domain designed to provide appropriate support for real-time multimedia services. The user equipment communicates with the IMS using GPRS. The IMS is directly connected to the GPRS network via the GGSN (Figure 3.35).

The IMS provides service to the mobile user such as:

- Real-time communications using voice, video, or multimedia messaging
- Audio conferencing and videoconferencing
- Content delivery services (e.g., video audio or multimedia download)
- Content streaming services (e.g., video, audio, or multimedia streaming)
- Multimedia messaging (MMS)

The IMS of one operator can be connected to the IMS of another, allowing multimedia services to be accessed from visited networks and allowing users on different networks to exchange content.

Connections to the Internet also allow the exchange of multimedia content with other non-mobile users (e.g., VoIP service or video telephony).

Connections to existing networks like the ISDN via media gateways allow the VoIP service to extend to conventional users. Interconnection between the IMS and other IP networks is via the firewall to protect the system from external attacks.

Review Questions

Q1. Service platforms do not include _____.
   a. Messaging platforms
   b. Telephone exchanges / MSCs
   c. Voicemail platforms
   d. Content servers

Q2. A particular service _____.
   a. Would always be provisioned on a single service platform
   b. Could be provisioned on distributed service platforms
   c. Would be provisioned on a single server, together with all the other services provided by the network
   d. Could never be provided by the telephone exchange/MSC

Q3. The OSA API _____.
   a. Is secure, but vendor specific
   b. Provides completely open access for service provision
   c. Provides an interface between the network and customers
   d. Provides a secure, standard interface between the network and third-party service providers

Q4. In 3G systems, there is (are) _____ primary way(s) to determine location.
   a. One
   b. Two
   c. Three
   d. Four

Q5. IMS is an acronym for _____.
   a. IP multimedia services
   b. IP multimedia subsystem
   c. IP message switching
   d. IP multiplexing switch

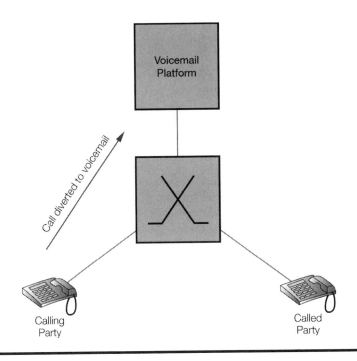

FIGURE 3.36 The voicemail platform.

3.3.2 Messaging Platforms

Voicemail Platforms

While voicemail appears to be a very simple service to provide to customers (Figure 3.36), the fact is that it is an extremely important platform for operators — especially in terms of revenue generation.

First, any call that is answered by a voicemail announcement is technically terminated. All of the networks involved in the delivery of the call will receive revenue. Second, on receipt of a voicemail message, many users will return the call, generating yet further revenue for the operators.

In many cases, basic voicemail services are provided free of charge or included in the monthly tariff. However, operators can charge a premium for premium voicemail services, where the user is able to keep messages stored for a longer period. Combining voicemail with calling line ID, operators can provide a service that allows the user to call the person who left the message by pressing a single key.

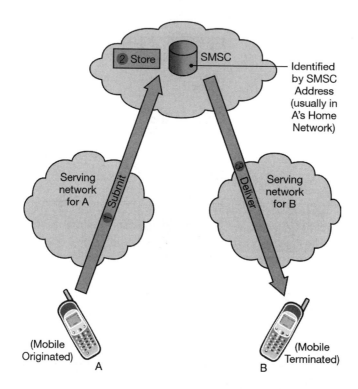

- Maximum of 160 characters per message (depending on language)
- Alternative e-mail delivering/submission can be used
- Enhanced Messaging Service (EMS) allows for more advanced features

FIGURE 3.37 Short message service center (SMSC; point-to-point).

Short Message Service SMS

SMS allows for the exchange of short alphanumeric messages between a mobile station and an SMS service center (SMSC). The messages can be either mobile terminated (from SMSC to mobile) or mobile originated (from mobile to SMSC). Of course, in most SMS interactions, the mobile originated service is followed almost immediately by the mobile terminated service (of the same message, but different users). Messages are limited to 160 characters (depending on the language used).

Mobile Terminated Point-to-Point

These messages are sent from an SM service center (Figure 3.37) to a mobile station. Upon receipt of the message, the mobile station will return a confirmation to the SM service center. The message may be received when the mobile is not being used, or even while it is being used for a voice call.

Mobile Originated Point-to-Point

Here the message is from the mobile to the SM service center, and a confirmation of receipt (not necessarily delivery) is given. The generally accepted *problem* of mobile originated SMS is the inputting of the text to the phone, which can be a slow and laborious process. One increasingly popular short-cut is to edit the message on a computer, laptop, or PDA (personal digital assistant) and then transfer the message to the phone. On mobile phones themselves, intelligent text recognition software is becoming increasingly sophisticated.

Enhanced Messaging Service (EMS)

An extension of SMS, EMS allows ringtones, operator logos, and other simple visual images and icons to be sent to compatible devices. Pictures, sounds, animation, and text can be sent to a device in an integrated package. No modification of the SMSC is necessary because EMS uses the User Data Header (UDH) that is already present in SMS messages. However, EMS-compatible handsets are required. The EMS standard is included as part of the standard 3GPP feature set.

SMS Cell Broadcast

The SMS cell broadcast service transmits the same message to all mobiles within a particular cell (or group of cells). The message limit in this case is 93 characters, and the mobile must be in idle mode to receive the message. No acknowledgment of receipt is given. Cell broadcast does not generate revenue because broadcast messages, which offer no confirmation of receipt, cannot be charged. They tend, therefore, to be seen as a value-added feature to attract customers. Uses might include advertising (e.g., other network features) or the broadcasting of PSTN local area codes such that a mobile user is able to distinguish between local and long-distance calls (tariffs may vary between the two).

The Multimedia Messaging Service (MMS)

MMS is a non-real-time service, often seen as a natural progression from the GSM Short Message Service. Like SMS, messages can be stored before being forwarded to the recipient whenever they are available or they request to see the message. It combines different networks and integrates messaging systems that already exist in these networks, for example, SMS in GSM and so-called "Instant Messaging" via the Internet.

MMS is designed to support either standard e-mail addresses or standard ISDN telephone numbers; and WAP (Wireless Application Protocol) development also provides significant support for MMS.

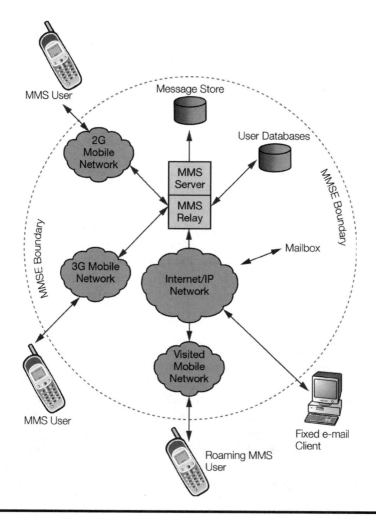

FIGURE 3.38 Multimedia Messaging Service (MMS).

The user terminal operates in the Multimedia Messaging Service Environment (MMSE). MMSE provides the service elements such as delivery, storage, and notification, which may be located in one network or distributed across different networks.

The basis of connectivity between the networks is provided by IP (Internet Protocol) and its associated set of messaging protocols, enabling compatibility between 2G and 3G wireless messaging and Internet messaging.

The architectural elements of MMS are shown in Figure 3.38 and include:

- The MMSE (Multimedia Messaging Service Environment) describes all the elements that provide the complete service to a user. In the case of roaming, the visited network is included within this environment.
- The MMSC (Multimedia Messaging Service Center) consists of the MMS relay and the MMS server. The MMS relay facilitates transfer between different messaging systems and can generate charging data, thus enabling the service to be billed. The MMS server is responsible for storing and handling incoming and outgoing messages.
- The MMS user databases contain subscription information, etc.
- The MMS user agent resides on the user equipment or on a device attached to it (such as a PC). It is an application layer function providing users with the ability to view, compose, and handle messages.

Review Questions

Q1. The voicemail platform is connected to _____.
  a. A switch/MSC
  b. An HLR
  c. A VLR
  d. A BSC

Q2. The SMSC is a service platform that supports _____.
  a. Voicemail
  b. Multimedia services
  c. Text messaging
  d. Multimedia services and control

Q3. SMS point-to-point services are specifically termed _____.
  a. Mobile originated and mobile terminated
  b. Packet switched and circuit switched
  c. Text messaging and multimedia messaging
  d. Single and multimedia services and control

Q4. MMS is a service or application that _____
  a. Allows a single media type to be sent from mobile to mobile
  b. Allows pictures to be sent via SMS
  c. Requires 3G connections to exchange multimedia information
  d. Allows multimedia information to be sent from mobile to mobile

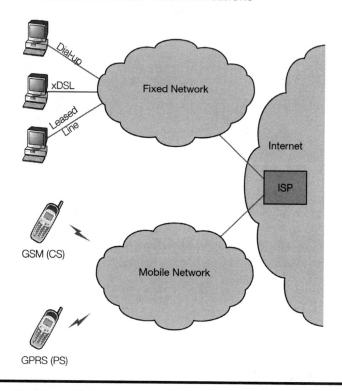

FIGURE 3.39 Accessing the Internet.

3.3.3 Accessing the Internet

Methods for Accessing the Internet

Figure 3.39 illustrates how fixed and mobile devices access the Internet. Dial-up connections using circuit-switched connections in fixed or mobile networks provide relatively low bandwidth access to Internet service providers (ISPs). The introduction of ISDN services in the fixed network offered the user the ability to use two or more 64-kbps connections to a compatible ISP. One of the main drawbacks to these types of connections is that the user is often paying for the duration of the call — even when their browser is idle.

The introduction of packet-switched, always on, services through xDSL technology in the fixed network and GPRS in the mobile network environments, allows users access to much higher speed data sessions. The method of charging for these services varies enormously, depending on the network. Digital subscriber line customers are most often offered unmetered packages for a fixed monthly fee. Mobile operators tend to charge, based on the amount of data downloaded.

The primary difficulty regarding displaying Internet-type information on a device such as a mobile phone is the complexity of Web pages, which tend to be very animated and multi-colored, and contain much more than the raw information often sought by the user, which is often a limited amount of text.

The problem lies in the fact that the *excessive* amount of data takes too long to download over the limited data rates of 2G systems such as GSM and also that this information cannot be meaningfully displayed on a screen of limited size. Other considerations involve the limited amount of processing power and memory that is available on mobile phones and similar devices.

Various developments have made the mobile Internet much more accessible. One is the Wireless Application Protocol (WAP), developed mainly by the GSM/UMTS community, and seen very much as a standard solution for that technology. Another is called i-mode — developed by the Japanese company NTT Docomo but now available in other networks worldwide. Finally, BREW provides support for content access mainly for the cdmaOne and cdma2000 networks.

Wireless Application Protocol (WAP)

WAP was developed initially to support access to information using a compact coding scheme, and it has evolved to allow more advanced media types (including video) to be supported as handsets themselves evolve, and as access speeds increase. In addition, extra features mean that WAP can be used to support such applications as MMS (Figure 3.40).

WAP Servers

WAP is seen as an enabling technology, with the WAP server at the heart of content provision. To provide WAP content, the service or content providers (often the network operator) need to hold the information within a WAP server, or WAP servers (Figure 3.41). These servers need to be accessed through the network itself; hence the correct (standard) interfaces need to be provided, with correct addressing schemes, access rights, billing, coding schemes, etc.

WAP content can be provided by the network operator, or by third-party content providers. Many operators try to ensure that their customers primarily access the network operator's content rather than a third-party provider's content for reasons of revenue and brand. The Vodafone Live! Portal is an example of this kind of strategy. Technically, operators can control access using the information in the subscription data (held in the HLR and SGSN in UMTS- and GPRS-based networks).

i-Mode

The i-mode system is seen more as an *ecosystem*, rather than just an enabling technology. It describes the brand, concept, access, and content. It was designed from the outset to allow third-party content providers to connect via standard interfaces, and to deliver content and information in a relatively straightforward manner.

A compact coding scheme allows easier content delivery despite the limitations of the mobile network and mobile handset.

294 ■ Introduction to Mobile Communications

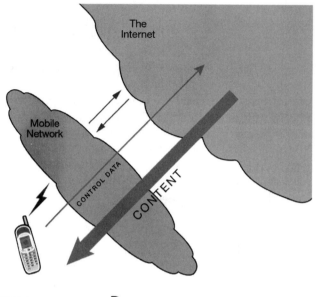

FIGURE 3.40 Wireless Application Protocol (WAP).

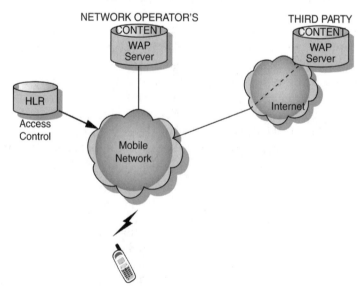

FIGURE 3.41 WAP servers.

i-mode originated in Japan but has been adopted as a way of delivering content by many GSM/GPRS and UMTS network operators. If a network has implemented the i-mode system, it can be marketed as i-mode in the same way that systems such as Vodafone Live! have been marketed.

BREW (Binary Runtime Environment for Wireless)

BREW is also seen as an ecosystem. It has been associated mainly with cdma-based systems; however, technically, it is available for use with GSM/GPRS and UMTS networks — that is, it is mobile technology agnostic.

BREW provides a whole range of features, applications, and services that can be accessed by the user via the BREW software on the handset. BREW application developers are provided with the development and testing tools they need to make the applications available to the users. Operators benefit from a flexible system that provides a platform for a variety of advanced services and features.

The BREW distribution system (BDS) is the network-based system that distributes the content and applications.

Review Questions

Q1. Internet access is more efficient by _____.
   a. Circuit-switched connections
   b. Packet-switched connections
   c. Accessing via the OSA API
   d. Service platform connections

Q2. WAP content _____.
   a. Is optimized for small screens and low download speeds
   b. Allows full HTML information to be displayed on mobile terminals
   c. Requires high-speed connections to be rendered on the mobile device
   d. Can only be provided by the network operator

Q3. i-mode is seen as _____.
   a. An enabling technology
   b. A service platform
   c. A content format
   d. An "ecosystem"

Q4. BREW is _____.
   a. Associated mainly with cdmaOne and cdma2000 systems
   b. A service type
   c. A specific messaging format
   d. Provided over broadband connections only

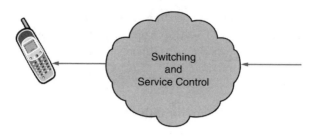

FIGURE 3.42 Pre-Intelligent Network service provision.

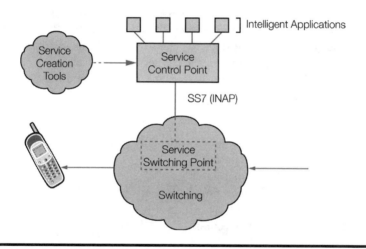

FIGURE 3.43 Intelligent Network (IN) service provision. SCP controls user interaction when required.

3.3.4 Intelligent Networks and CAMEL

Introduction to Intelligent Networks (INs)

Traditionally, switching equipment (telephone exchanges) would need upgrading each time a new service was added to the network (Figure 3.42). In some networks, the presence of hundreds or thousands of telephone exchanges meant that this was a very long and labor-intensive task.

The concept of Intelligent Networks (INs) allows the service to be provided within a *central* computer or service control point (Figures 3.43). Once the telephone exchanges have been upgraded to include the IN features, no further upgrades are required except for the updating of the tables that identify the IN triggers, such as the 1-800- or 0800-digit string to identify the toll-free or freephone service.

Intelligent Networks originally provided advanced features such as:

- Toll-free and freephone
- Premium rate
- Televoting
- Calling card services

These services are generally provided through service control points, which translate the service-specific dialed numbers (an 800 number perhaps) into standard numbers for routing to the actual destination (telephone) in the network. The special number is the initial trigger for the telephone exchange to contact the service control point, to translate the number, and to carry out special billing, such as reverse charging or collect calls, or premium rate.

These early services were soon followed by further advanced services based on the intelligence or control features within the service control point. Extra service features such as interaction with the user allow further customization of services.

IN separates service intelligence and switching. This means that new services can be quicker and cheaper to install, and that service creation and switching is split into two markets, thereby increasing vendor competition.

Most IN, including GSM Phase 2+ networks, use SS7 (Signaling System Number 7) protocols to enable the switches (known as service switching points, or SSPs) to communicate with databases known as service control points (SCPs). A standardized set of SS7 messages known as INAP (Intelligent Network Application Part) is used for interaction between the SSP and SCP.

The intelligent applications that control IN services are defined by the operator, and are not themselves standardized. This means that IN offers a route to operator differentiation, but also that in many cases the same services cannot be offered outside the network of that operator.

Service Control Platforms

Service control platform (SCP) is the name given to a platform that is, in some way, controlling services on a network, or in some cases, across network boundaries. One of the main advantages in using an SCP within a network is that it significantly reduces the upgrade costs within a network whenever a new service is introduced. Features of an SCP, shown in Figure 3.44, include:

- *Control of supplementary services.* These may include services such as conference calls, advice on billing, and provision of call-back services such as ring back when free. Supplementary services are seen as key revenue generators for network operators.
- *Number translation services.* In most modern networks it is possible to dial toll-free, local or premium rate services. These are, in effect, virtual numbers that have special tariffs. Number translation services (NTS) minimize the amount of configuration required on the network when a new service is provided.

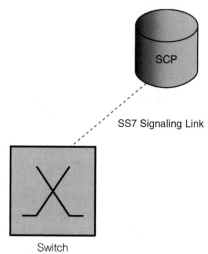

- Control of supplementary services
- Number address translation
- Intelligent routing
- Access control
- Control of announcements and user interaction
- Prepaid (and other) billing

FIGURE 3.44 Service control point features.

- *Intelligent routing.* With intelligent routing, calls are routed according to a set of either predefined or dynamic rules. The rules may concern using the cheapest route or the route that takes the most direct route.
- *Access control.* With the introduction of services such as prepaid, it is important for operators to control access to their networks and services by looking up the amount of credit that a user has before allowing the use of a service.
- *Control of announcements and user interaction.* For many services, such as the use of calling cards, it is necessary to provide voice- and keypad-based interfaces to users. Here, an SCP will control the playback of prerecorded messages and collect the user's input for authentication of PIN numbers, etc.

CAMEL (Customized Applications for Mobile Networks Enhanced Logic)

INs and Mobile Networks: The Problems

In fixed networks, the IN concept can be implemented in various ways (Figure 3.45). In each case, however, the interfaces are confined within a single network, allowing the defined procedures and information flows to be applied on an interface with known and defined endpoints, and within hardware belonging to the single network operator.

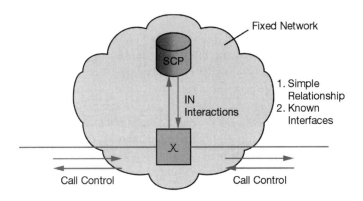

FIGURE 3.45 Fixed-network IN.

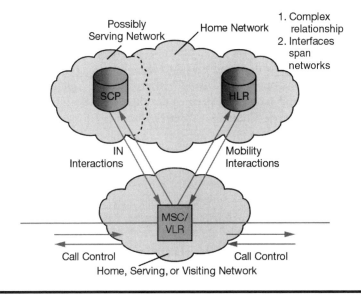

FIGURE 3.46 Mobile network IN considerations.

Testing is therefore more straightforward as each interface can be tested independently and, in addition, non-standard procedures and information can be used to satisfy particular problems (even if this is just a certain way of interpreting an ambiguous part of the standards). The IN procedures have been specified to interwork (mainly) with call control, hence a simple relationship exists between the two.

For mobile networks, the situation is more complex because roaming scenarios require that equipment in more than one network is involved (Figure 3.46). Considering all the roaming agreements each network will have in place, and the rapid change of each of these networks as new hardware (switches, etc.) is added, it is impossible to test each individual interface that will likely be involved in an IN call, or to interpret specifications in anything other than a single standard way.

The next big complication is that IN procedures need to interwork with call control, mobility management functions, and the additional features of mobile networks such as the General Packet Radio Service and the Short Message Service. Finally, any announcements should be supported by special announcement machines, the positioning of which needs careful thought. Location in the home network provides ease of management, but an international call leg would be needed to play announcements to roaming subscribers. Locating the announcement machines in serving networks leads to a higher cost of hardware and increased management costs, but reduced cost per call.

The Role of CAMEL

CAMEL has been specified to allow operator-specific services (IN services) to be provided for subscribers even while roaming abroad. The service control point (SCP) is generally located in the home network, allowing serving network switches to interact with it to provide the advanced services.

CAMEL is an acronym for Customized Applications for Mobile networks Enhanced Logic, and has been specified to allow for the introduction of IN services in mobile networks alongside both call control functions and mobility management.

In the later phases of CAMEL, it also allows for the provision of advanced services in support of GPRS, SMS, and supplementary services. The interfaces required for CAMEL are standardized to a high degree, allowing network elements from different networks to work together using the CAMEL Application Part (CAP) message set of Signaling System Number 7 (SS7). An example CAMEL procedure is shown in Figure 3.47 for clarity.

The aim is to allow Intelligent Network interactions with a service control point located in the subscriber's home network. CAMEL takes account of mobile originated, mobile terminated, and mobile forwarding cases, and also allows for both roaming and non-roaming cases.

Modifications at the MSC essentially give the switch the ability to recognize the requirement for a CAMEL-based service and build the required messages for the specified SCF. In addition, both the HLR and VLR must be modified to hold the relevant CAMEL subscriber details.

CAMEL has been defined in different phases to allow a staged approach to implementation. Each phase is a superset of the previous one. Not all networks will be at the latest phase of CAMEL, or even support CAMEL at all — but in many networks, CAMEL is used to support prepaid services.

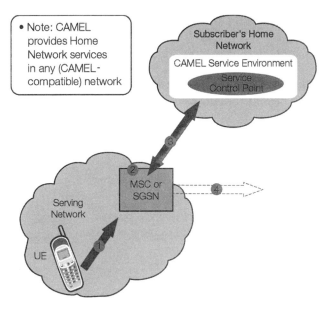

FIGURE 3.47   CAMEL interactions.

Review Questions

Q1. Intelligent Networks is a way of _____.
   a. Controlling services within a telephone exchange
   b. Controlling services within a service control point
   c. Providing effective operational support
   d. Controlling messaging services

Q2. A service control point is connected to a telephone exchange via _____.
   a. Voice circuits
   b. An IP network
   c. The OSA API
   d. Signaling System Number 7 (SS7)

Q3. CAMEL is the standard way of providing _____.
   a. Intelligent Networks in the fixed network
   b. Intelligent Networks in IP networks
   c. Intelligent Networks in 3G networks
   d. Intelligent Networks in GSM-based mobile networks

Q4. CAMEL procedures allow for _____.
   a. Roaming and non-roaming scenarios
   b. Non-roaming scenarios
   c. Roaming scenarios
   d. Home network calls only

Q5. CAMEL would typically be used to provide support for _____.
   a. Text messaging
   b. Calling line identity services
   c. Fixed-line voicemail
   d. Prepaid billing

3.3.5 Operational and Business Support Systems (OSS and BSS) and Billing

Operational Support System (OSS)

The Requirements

Support for the operational network must be comprehensive in terms of structure and procedures, and also in terms of systems and software tools. Broadly speaking, the requirement can be split into three main areas:

1. *Network management* deals with managing the operational network, including infrastructure, maintenance, and fault handling. The requirement is wide ranging, allowing the complete network to be viewed in overview, or detail, with network elements being handled remotely to keep the network running smoothly.
2. The *customer service system* allows for effective customer relations, covering areas such as new orders, billing queries, and technical issues. The requirement is easy to identify, and systems could be dedicated to this role, or be fully integrated within a wider *operational support system*. In either case, information must be available to the customer service representatives, relying on good data interfaces between related functions (for example, customer billing records need to be available to the customer service representative).
3. Other requirements can be grouped under the heading of *business management* and include sales and marketing functions, finance, personnel, and logistics, and a wide range of others.

These support systems can employ a range of different tools, or can be supported by an integrated system. However, it is unlikely that any one system would be able to provide complete support for all possible requirements.

In summary, support systems are required for:

1. Network management:
    - Operations and maintenance
    - Network fault handling
    - Infrastructure management
2. Customer service:
    - Customer handling
    - Orders handling
    - Billing
3. Business management:
    - Sales and marketing
    - Finance
    - Personnel
    - Logistics

Scale of the Problem

To illustrate the scale of the problem, the figures below are typical of a large network operator. The numbers speak for themselves. Tens or even hundreds of systems can be used to support the different functional requirements of the operator, leading to duplication, inefficiency, and inaccuracies in the data held. A large telco could require support for:

- 10 million customers
- 300,000 bills per day (monthly bills)
- 10 to 20,000 new orders per day
- 10,000+ faults fixed per day
- Hundreds or thousands of terminals for OSS functions
- Two million transactions per day (OSS)

It is not atypical for network operators to maintain tens or even hundreds of separate systems.

Operations System

The operations system must be effective across the whole range of procedures and requirements, the main ones being:

Operations requirements:
- Effective management of:
  - Resources
  - Staff
- Planning to meet:
  - New services
  - Increased capacity
  - Scheduled maintenance
- Handling of:
  - New orders (including response to marketing campaigns)
  - Problems and quality issues
  - Billing

Resources and staff must be effectively managed to maximize efficiency, while ensuring the right resources are made available for the required activities.

At the same time, one must pay attention to planning and preparation for new services, increasing requirements for capacity, and for scheduled maintenance.

The ability to handle new orders, including those in response to marketing campaigns, deal with problems and quality issues, and provide an effective billing system form the core of operations in most networks. These items are the subject of various Operational Support System (OSS) software tools, designed to make the whole process easier and more manageable.

Work scheduling, project management, purchasing, and resource allocation are central features in maintaining a workable and effective network.

Business Support

Good administration and support systems are required to ensure the continued effectiveness of operations, and to allow the organization to grow and evolve to meet new challenges.

A comprehensive customer care system is one of many administrative requirements, allowing for the required quality of customer relationship management (CRM). The sales force depends largely on the right systems and processes to be in place to maximize their effectiveness, by achieving maximum return on their efforts. Effective administration in this area allows for a reduction in the duplication of effort, maintenance of contacts, and following leads with maximum confidence.

Costs are, of course, a major part of the equation, with the inevitable requirement to keep costs as low as possible within the overall requirements. This implies sourcing administration systems that maximize efficiency through integration, allowing for the support of a number of related processes and requirements by a common software tool.

Mobile Network Infrastructure and Supporting Systems ■ 305

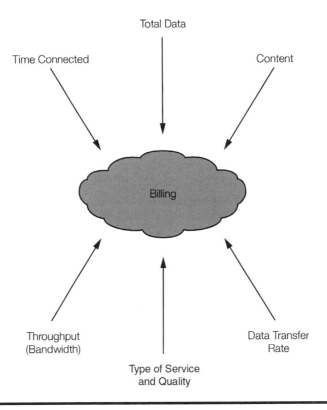

FIGURE 3.48 Services billed in fixed and mobile networks.

In addition, the *administration* and *backup systems* must provide the support required for bringing new products and services to market in good time.

An operator is always looking to improve business processes and management, using integrated support systems and software tools to provide the required level of administrative support. Administrative and backup requirements include:

- Comprehensive customer care
- Maximum effectiveness of sales force
- Simplified and improved business processes
- Reduced costs
- Reduced time to service for new products

Billing

What Can Be Billed?

As Figure 3.48 illustrates, there are a variety of ways in which billing can be achieved in both fixed and mobile networks.

- *Time connected.* From the moment that the call is connected, the customer is charged at the relevant rate set by the service provider.
- *Type of service/Quality of Service.* Depending on the type of service requested, whether it be data or voice, the quality of the service can be increased for voice or decreased for data as required. High quality is required for voice, to enable it to be audible at the receiving end. This higher quality is then reflected in the cost to the subscriber.
- *Total data.* A charge is made for the total amount of data sent or received, regardless of the time connected.
- *Content.* The content will dictate the charge levied. The relevant content supplier will possibly pay the overheads to the service provider, therefore encouraging more custom to their service. This is particularly applicable to media delivery.
- *Throughput (bandwidth).* If a line is not busy, then the redundant channels can be utilized to enable transmission of the data at a far greater rate. The billing would reflect the fact that additional bandwidth was used to transmit this data.
- *Data transfer rate.* For all services, the charge levied to the subscriber can be fixed to the speed of data delivery. This is particularly used in leased-line applications.

Billing in the Mobile Network

There are currently two main methods of billing within the mobile system: prepaid and post-paid.

1. *Post-paid.* The phone is used and a bill is produced at the end of the allotted period (usually monthly). The bill usually details the service (often this will be voice), the destination number, time of the call, and duration of the call. In addition, any advanced or premium rate services, messaging, voicemail access, and free calls will also be detailed. The user often pays by direct debit from a bank account.
2. *Prepaid.* An amount is prepaid to the network operator via the use of top-up vouchers, credit cards, or ATM bank machines. The Internet can often be used to make the transactions easier for the customer. This will then allow the customer a certain amount of minutes of calls, etc., whether peak or off-peak.

In summary:

- Post-paid: monthly bill (rental and calls)
- Prepaid: top-ups by voucher, ATM bank machine, credit card, etc.

The mobile network must cater for billing of circuit-switched services (voice primarily), packet-switched services (often data volume), as well as billing of messaging interactions (SMS and MMS) and other service platform interactions (e.g., content).

Mobile Network Infrastructure and Supporting Systems ■ 307

An integrated platform that can cope with all types of billing, both prepaid and post-paid, is the ideal solution, but it is rare to find such a platform implemented within the network. Different platforms are generally used for post-paid and prepaid. In addition, a technique called *mediation* is used to integrate different inputs (IP data volume, content charges, etc.) onto a common billing platform. This is especially useful for packet-switched services.

Review Questions

Q1. Which of the following is *not* a network management requirement?
 a. Operations and maintenance
 b. Network fault handling
 c. Customer care
 d. Infrastructure management

Q2. OSS is an acronym for _____.
 a. Operational Support System
 b. Operational System Supplements
 c. Outsourced Support Systems
 d. Outsourced Supplementary Services

Q3. A voice call would usually be billed _____.
 a. As content
 b. By throughput
 c. As a one-off event
 d. By time connected

Q4. The main billing methods in a mobile network are _____.
 a. Prepaid and monthly
 b. Flat fee and post-paid
 c. Flat fee and Prepaid
 d. Prepaid and Post-paid

Q5. In developing countries, the dominant billing method is _____.
 a. Prepaid
 b. Flat fee
 c. Monthly
 d. Post-paid

3.3.6 Section 3 Review Questions

Q1. Choose four types of service platform, and insert your choices in the table below. Then:
   a. Identify and describe the services provided by each platform.
   b. State how the connection between the user and the platform is achieved. Be as detailed as you can.
   c. Subjectively assess the potential revenue that can be generated by the relative. You can differentiate between present and predicted revenues for each type of service. We are not looking for exact figures, just the relative importance of each platform in terms of revenue generation. Use the industry outlook and other research sources to help you.
   d. Additionally, write down any factors that you think might be inhibiting the take-up of these services.
   e. Finally, apart from revenue generation, are there any other compelling reasons for a network operator to provide these services?

Q2. Identify four services often provided by third-party service providers.

Q3. List three advantages of using the OSA/Parlay API for the third-party service provision.

Q4. Accessing the Internet via a mobile device has several potential drawbacks. Identify three of the drawbacks in the table below, and then briefly state how technologies such as WAP and i-mode are designed to minimize the impact.

Drawback	WAP/i-Mode Solution Includes:

Q5. What does CAMEL stand for?

Q6. Why is CAMEL different from Intelligent Networks?

Q7. Identify three services or features that might be provided using CAMEL.

Q8. Which feature or service would you describe as the principle reason CAMEL has been adopted by mobile network operators?

Q9. Does CAMEL provide support for (tick the appropriate boxes)? Add appropriate comments, including the phase of CAMEL required (if you know it).

	Yes	No	Comments
Mobile Originated Calls			
Mobile Terminated Calls			
Billing for Voice Calls			
GPRS (including billing)			
Text Messaging			
Roaming Cases			
Non-Roaming Cases			
Content Billing			

Q10. Service platforms do not include _____.
   a. Messaging platforms
   b. WAP servers
   c. Voicemail platforms
   d. Telephone exchanges/MSCs

Q11. A particular service _____.
   a. Would always be provisioned on a single service platform
   b. Could be provisioned on distributed service platforms
   c. Would be provisioned on a single server together with all the other services provided by the network
   d. Could never be provided by the telephone exchange/MSC

Q12. The OSA API _____.
   a. Is a secure, vendor-specific interface between the network and third-party service providers
   b. Provides a secure, standard interface between the network and third-party service providers
   c. Provides an interface between the network and customers
   d. Provides completely open access for service provision

Q13. In 3G systems, there is (are) _____ principal way(s) to determine location.
   a. One
   b. Two
   c. Three
   d. Four

Q14. IMS is an acronym for _____.
   a. IP Multimedia Services
   b. IP Multiplexing Switch
   c. IP Message Switching
   d. IP Multimedia Subsystem

Q15. The voicemail platform is connected to _____.
   a. An SCP
   b. An HLR
   c. A switch/MSC
   d. A BSC

Q16. The SMSC is a service platform that supports _____.
   a. Voicemail
   b. Multimedia services
   c. Text messaging
   d. Multimedia services and control

Q17. In developing countries, the dominant billing method is _____.
   a. Prepaid
   b. Flat fee
   c. Monthly
   d. Post-paid

Q18. A voice call would usually be billed _____.
   a. By amount of data (volume)
   b. By time connected
   c. As a one-off event
   d. By Quality of Service

Q19. MMS is a service or application that _____.
   a. Allows sending a single media type from mobile to mobile
   b. Allows sending of pictures via SMS
   c. Requires 3G connections to exchange multimedia information
   d. Allows sending multimedia information from mobile to mobile

Q20. Internet access is more efficient by _____.
   a. Circuit-switched connections
   b. Packet-switched connections
   c. Service platform control
   d. SS7

Q21. WAP content _____.
   a. Can only be provided by the network operator
   b. Allows display of full HTML information
   c. Requires high-speed connections
   d. Is optimized for small screens and low download speeds

Q22. i-mode is seen as _____.
   a. A service technology
   b. An "ecosystem"
   c. WAP format
   d. A service platform

Q23. BREW is _____.
   a. Associated mainly with cdmaOne and cdma2000 systems
   b. A service type
   c. A specific messaging format
   d. Provided over broadband connections only

Q24. Intelligent Networks is a way of _____.
   a. Controlling services within an MSC
   b. Providing effective operational support
   c. Controlling services within a service control point
   d. Controlling messaging services

Q25. A service control point is connected to a telephone exchange via _____.
   a. Voice circuits
   b. Signaling System Number 7
   c. The OSA API
   d. An IP network

Q26. CAMEL is the standard way of providing _____.
   a. Intelligent Networks in the fixed network
   b. Intelligent Networks in IP networks
   c. Intelligent Networks in 3G networks
   d. Intelligent Networks in GSM-based mobile networks

Q27. CAMEL procedures allow for _____.
   a. Roaming and non-roaming scenarios
   b. Voice procedures only
   c. Roaming in any network
   d. Roaming network calls only

Q28. CAMEL would typically be used to provide support for _____.
   a. Prepaid billing
   b. Post-paid billing
   c. Fixed-line voicemail
   d. Text messaging

Q29. BSS is an acronym for _____.
   a. Business Support System
   b. Billing Support System
   c. Bridge Support System
   d. Binary Supplementary Service

Chapter 4

Handset, Services, Media, and Content Distribution

Module 4 Learning Objectives
- List the basic requirements for mobile handsets.
- Explain the role of standards bodies in the field of mobile handsets.
- Draw a diagram of the typical systems found inside a mobile phone.
- Explain the differences between the various types of memory found on mobile devices.
- List the technologies associated with displays and audio systems.
- Identify image and video codecs used in modern mobile equipment.
- Show types of serial interfaces found on mobile devices and describe typical applications.
- List the various OS used on mobile phones and outline the differences between them.
- Explain how DRM and other security systems are related to the mobile device.

4.1 Introduction to Handset Technologies

4.1.1 Handset Basics

The Basic Requirements

The mobile handset, in terms of its design and functionality, must satisfy three distinct sets of requirements (Figure 4.1). These relate to the network operators, the endusers, and the handset vendors themselves. The requirements of the end user will often focus on features and capabilities, and the overall ease of use (usability), and of course, the

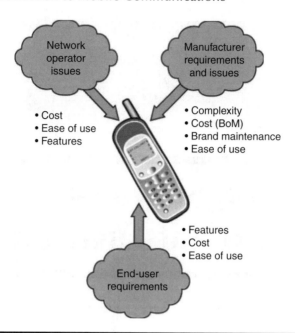

FIGURE 4.1 Handset requirements.

end user would like maximum functionality for minimum cost. Increasingly, the aesthetic appeal of the handset is important as it evolves from the functional devices of early handsets to *fashion items* that a user may wish to change every 12 to 18 months. Although users often look for complex feature sets in handsets, research indicates that a typical user frequently uses only 15 to 20 percent of the phone's capabilities.

For a network operator the handset represents the *front end* of the network and is the platform through which users gain access to the rich set of services offered. This again means feature sets and capabilities, and the ease of use is critical. The network operator is also very interested in handset cost because in many markets the handset is still subsidized by the operators, and therefore they bear some of the real handset costs.

Handset vendors are increasingly asked to provide evermore complex devices that are smaller and lighter than their predecessors yet cost no more or even less. The overall cost of the actual handset components is further complicated by the issue of technology licensing. A current 3G phone will contain many noncellular technologies, such as video codecs and picture compression algorithms, for which a license fee is payable to third-party innovators. Another major issue for handset vendors is brand maintenance; in the mind of the enduser, the phone may become associated with the network and not the original handset manufacturer, and the manufacturers expend a lot of effort in keeping their brand profile high.

The requirement to satisfy the sometimes-conflicting requirements of all parties means that the market for handsets has inevitably fragmented in terms of device capability and form factor. The devices produced for the prepay segment are very different from those aimed at the high end of the market.

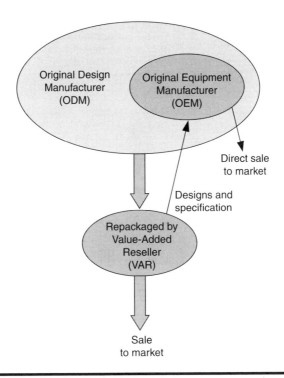

FIGURE 4.2 Equipment manufacturers.

Original Equipment and Device Manufacturers

In the area of manufacturing — and mobile handsets are no exception — the topic of original equipment manufacturers (OEMs) will normally surface. An OEM (Figure 4.2) is a manufacturing company that produces components or sometimes complete pieces of equipment to the specifications and design of another company, the value-added reseller (VAR). The VAR will either take components from one or more OEMs and integrate these into a system, or take an entire OEM assembly and repackage it. The level of repackaging might be as basic as relabeling products to adding greater value (e.g., software and other elements).

In some cases, OEM products are available direct to the market and in this case the products may be cheaper than the equivalent available from the VAR. This is particularly true for PC components. However, the commercial agreement between the VAR and their OEMs may prohibit this direct supply to market. Occasionally, an OEM takes a more extensive role in the design of components and systems to match the specifications of a VAR. In this case, they are referred to as *original design manufacturers*. Given the complexity of mobile handsets and the variety of components that they contain, it is not surprising that the role of OEMs in this marketplace is significant; no handset vendor would expect to self-manufacture the complete inventory of components that make up one of their handset products, as this would include diverse elements such as:

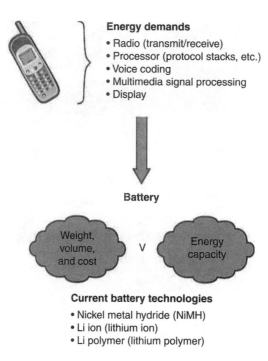

FIGURE 4.3 Battery technologies.

- Batteries
- Displays
- Microprocessors and memory chips
- Antennas
- RF circuits

Batteries

The handset battery is an essential component and is one of the largest contributors to the total weight and volume of the final product. It represents the balance between the competing demands of supplying energy and being as light and small as possible.

There are many processes within the mobile handset that are relatively energy hungry (Figure 4.3). These include the radio circuits, the processing for voice and multimedia signals, and the display. Various techniques are deployed both within the handset and in the specifications of the radio technology that aim at reducing energy consumption and therefore lengthening the life of the battery.

Although battery technology has come a long way since the early days of mobile networks, it can take up to ten years for a battery technology to proceed from concept to being fully commercially available.

The technologies encountered in today's handsets include:

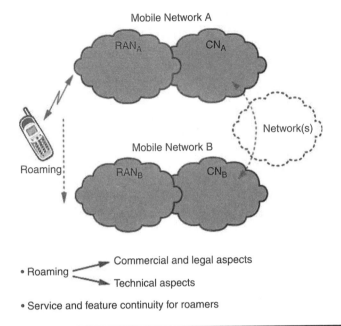

FIGURE 4.4 Network roaming.

- Nickel metal hydride (NiMH)
- Lithium ion (Li ion)
- Lithium polymer (Li polymer)

These represent a progression of more efficient battery technologies, measured in terms of energy storage per gram of battery weight, but also represent increasingly costly sources of energy.

The operational life provided by the battery is expressed as a combination of the standby time and the talk time, with handset products presently offering standby times of several hundred hours and talk times of ten hours or more. One issue that has surrounded the topic of batteries is the so-called memory effect, where a battery that is not fully discharged in one cycle *remembers* the point of discharge on the next cycle. This is not a problem with today's battery technologies.

Inter-Network Roaming

A critical feature of most mobile network standards has been the support of inter-network roaming, where a subscriber to one network is able to use his phone on another network (Figure 4.4). Roaming is represented by both a commercial agreement between two network operators to accept roamers and to provide billing data in an agreed format; it is also a technical issue where the handset must be capable of operating on both networks.

The GSM and UMTS standards have always included roaming as a key aspect, and phones sold to subscribers on one network should work on other networks using the same technology. However, the roaming user may not always have access to the same services when roaming and indeed the *look and feel* of the visited network may be somewhat different from the home network. Recently, work has been undertaken in the standards to minimize the difference seen by roamers across networks. This gave rise to the concept of the virtual home environment (VHE), which is based on Intelligent Network (IN) techniques and appropriate inter-network signaling.

The situation for some roamers is eased in some cases by the existence of agreements between networks that allow a roamer to dial his normal shortcode for, say, mailbox access, and for the networks to translate this number behind the scenes. While this may not be based on the VHE standards, it does start to solve the issues of look and feel.

Another factor aiding some roamers is the coalescence of the market around a number of large global operators. Once a mobile network operator owns multiple networks, it is clearly able to provide a consistent user experience across those networks.

Handset User Interfaces

Manufacturers often have entire teams or departments that are responsible for the design and implementation of user interfaces (UI), as this aspect of handset design is so critical. The complexity of handsets and the multitude of functions they contain make the UI critical in influencing the user experience. The UI supported by a device will depend largely on device capabilities and will range from a very simple UI for voice-centric handsets to very complex icon-oriented UIs, maybe incorporating touch-screen technology, for data-centric devices. The best UIs are of course those that are largely intuitive and that make redundant the complex (and costly to produce) user handbooks or operating manuals (Figure 4.5).

The dynamics of the market often come into conflict at the UI. The handset manufacturer will have spent a lot of time and money designing the UI, only for the network operator to insist in some cases that it be replaced, in whole or in part, with an operator-specific UI. This area of UI customization is another area in which the manufacturer must fight to preserve brand identity.

Human Interfaces

The issue of user-personalization is also influencing the UI; many end users want in some way to personalize the UI so it gives more ready access to the features and capabilities that they use most often. The human or user interface on handsets consists of a number of elements, including the display, the keypad or keyboard, the navigation keys, and any soft keys or hot keys. The precise configuration of the UI is determined by a number of factors, such as the device type and manufacturer preferences (Figure 4.6).

Handset, Services, Media, and Content Distribution ■ 319

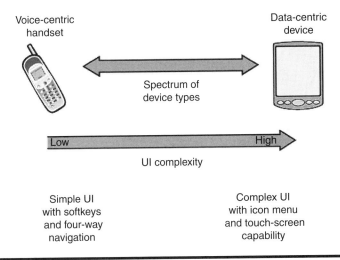

FIGURE 4.5 User interfaces. There is an increasing trend for operators to support operator-specific UIs; and there is an increasing requirement by users for UI personalization.

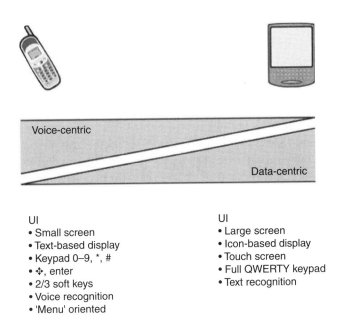

FIGURE 4.6 User interface (UI) requirements

At one end of the spectrum, a voice-centric handset would typically have a basic display that is relatively small and text based, with a standard 12-button keypad. Any additional entry functions would be by means of a navigation button, or roller,

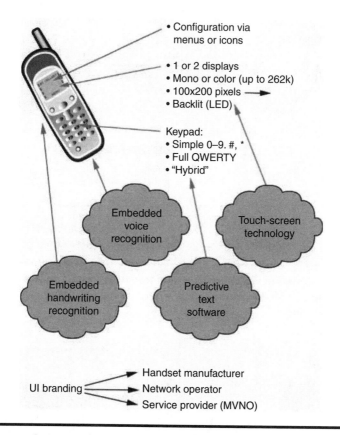

FIGURE 4.7 Elements of the handset user interface.

an *enter* key, and maybe two or three soft keys. The handset might also include some voice recognition software, although this is usually based on *pattern matching*. Users will be able to set up their phones using the manufacturer-supplied menu options. For text input, the same basic keypad is used, either in a *multi-tap* mode or combined with predictive text software.

Data-centric devices, on the other hand, will have larger displays, may well be icon-oriented, and will support touch-screen input. User input can either be from the touch-screen keypad, which is usually a full QWERTY keyboard, or in some cases by writing on the screen using a stylus. To support this option, the device requires text recognition software.

UI Components — The components of a user interface for a handset (Figure 4.7) include:

- Display(s) (two is common, particularly on *clam shell* designs)
- Keypad or keyboard

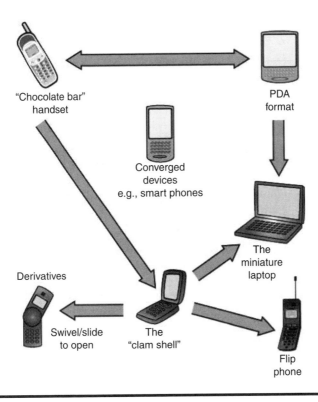

FIGURE 4.8 Handset form factors.

- Voice recognition software
- Handwriting-to-text conversion software
- Predictive text software
- Touch screen
- Configurations and settings menus

One of the most significant issues surrounding the UI, particularly the graphical user interface (GUI) or the welcome screen, is how to brand it. Of course, the handset manufacturer would like to display its brand here, but so would the network operator, the service provider, or the MVNO. Also, although the menu options are often designed by the manufacturer, the network operators increasingly would like a say in their design and functionality.

Handset Form Factors

The form factor or physical style of mobile phones has seen much in the way of innovation (Figure 4.8). The basic handset designs of early phones showed little differentiation between one manufacturer and another. The main design aim for many years was to create ever-smaller phones, with ever-increasing feature sets and with increasing standby and talk times.

However, the form factor is now seen as a critical means of product differentiation and a whole range of form factors are evident. These still include the basic chocolate bar, but also include the clam-shell design and its derivatives phones, which slide to open or rotate to open, or the flip phone, where a thin cover flips up to reveal the phone keypad.

At the other end of the market are PDA devices that have converged with phones to create handsets that include phone and PDA functions, and the basic PDA form has also morphed into devices that are, for all intents and purposes, the same form as a laptop computer.

Review Questions

Q1. A manufacturing company that produces components or sometimes complete pieces of equipment to the specifications and design of another company is known as a _____.
   a. Value-added reseller (VAR)
   b. Original equipment manufacturer (OEM)
   c. Equipment supplier
   d. Original design manufacturer (ODM)

Q2. Which of the following is a battery technology found currently in use with mobile telephones?
   a. Lead acid
   b. Hydrogen fuel cell
   c. Lithium ion
   d. Nickel cadmium

Q3. Which of the following cannot be attributed to the success of roaming between networks?
   a. Roaming agreements
   b. Mobile telecom standardization
   c. Market coalescence and global operators
   d. Prepaid billing

Q4. Which of the following is *not* part of the user interface?
   a. Screen or display
   b. Keypad
   c. Lithium ion battery
   d. Audio circuits

Handset, Services, Media, and Content Distribution ■ 323

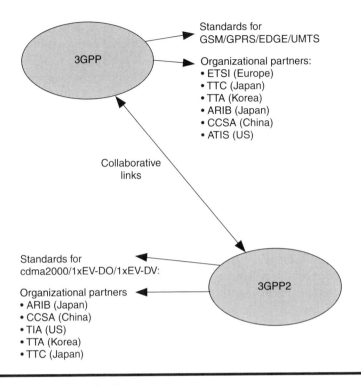

FIGURE 4.9 Standards bodies.

Q5. Which of the following handset form factors is most popular?
   a. Chocolate bar
   b. Flip phone
   c. Clam shell
   d. PDA

4.1.2 Handset and Standards Bodies

The large number of technologies incorporated in a handset is reflected in the fact that the standards on which this technology is based emanate from a diverse range of standards organizations and industry forums (Figure 4.9). From the perspective of the core cellular functionality, the most significant standards bodies are the two 3G bodies, the Third Generation Partnership Project (3GPP) and the Third Generation Partnership Project 2 (3GPP2).

3GPP is responsible for the UMTS standard and also assumes responsibility from ETSI for the GSM standard on which UMTS is based. 3GPP2 is responsible for the cdma2000 family of technologies, which includes the Evolution systems, 1xEV-DO (1x Evolution Data Optimized) and 1 x EV-DV (1x Evolution Data and Voice).

Both 3GPP and 3GPP2 have very similar structures, and indeed members. The two organizations are composed of organizational partners (OPs), which are national or regional standards organizations such as ETSI or the Telecommunications Industry Association (TIA). It is important to note that there are cooperative links between the two bodies as there is a lot of crossover technology that does not rely on the underlying network standards and can therefore be transferred.

In addition to the OPs, both the 3GPP and 3GPP2 have market representation partners (MRPs) and observers who contribute to the standards processes. The technologies covered by the two 3G bodies represent approximately 98 percent of the global users of mobile or cellular networks. In addition to the 3GPP and 3GPP2 standards, many other organizations have a bearing on handset design and functionality; these include:

- The Internet Engineering Task Force (IETF)
- The World Wide Web Consortium (W3C)
- The Open Mobile Alliance (OMA)
- The International Standards Organization (ISO)

The range of applicable standards in a handset is indicative of the convergence between telecommunications, Internet technology, and broadcasting and entertainment.

Testing and Type Approval

Before releasing a handset model into the marketplace and connecting it to a network, it must undergo a series of tests. These tests cover aspects such as regulatory requirements relating to electromagnetic compatibility (EMC) and spectrum tests, plus tests of functionality and interaction with other network components. The tests themselves will be performed by a combination of the handset manufacturer, network operator, third-party test houses, and in some cases industry bodies (Figure 4.10).

A handset must be type approved to operate on networks. *Full type approval* involves passing a sequence of tests normally specified by a technology body such as the 3GPP. The type approval may be done by an independent test house. For newer technologies on the market, an *interim type approval* (ITA) is sometimes agreed to; it is based on a limited sequence of tests and allows equipment for bringing the handset quickly to market.

During the design phase, it is common to apply interoperability tests (IOTs) to a handset. These are performed on a test network in a laboratory, often by the manufacturer. The IOT is testing a wide range of interactions between the phone and other network elements. Network operators will also have a regime of acceptance testing in which the phone functionality is checked against the network to ensure that the service interactions are correct. There are market initiatives aimed at reducing the requirement to repeat tests with several different networks. An example of

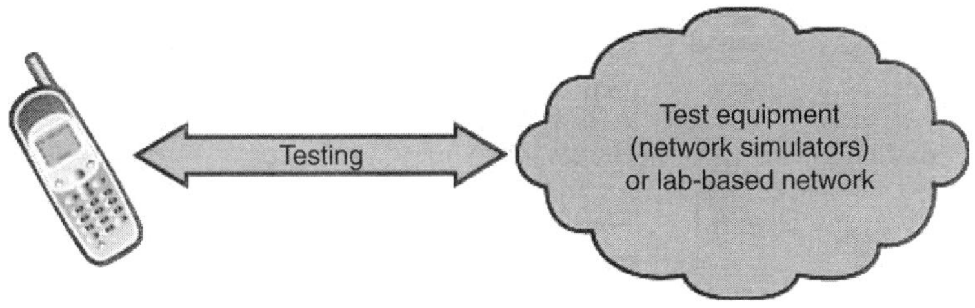

- Standardized test cases defined by standards body (e.g., 3GPP, 3GPP2) leading to full type approval (FTA) or, occasionally, interim type approval (ITA).
- Full interoperability testing (IOT) used for software development, usually lab-based.
- Network operator acceptance testing. In Europe, the RTTE Directive has an influence on approvals. The Global Certification Form (GCF) or PCS Type Certification Review Board is involved in standardizing type approval.

FIGURE 4.10 Handset testing.

this is the Global Certification Forum (GCF) for GSM and UMTS phones. The GCF is a partnership between network operators and equipment manufacturers that provides independent programs for interoperability testing.

Until recently there was also a requirement that equipment had to meet certain regulatory requirements for particular countries in which it may operate. In Europe this has been simplified through the introduction of the Radio and Telecommunications Terminal Equipment (R&TTE) Directive. The R&TTE Directive is a self-certification program for manufacturers to certify that their products meet statutory requirements after having been tested in a recognized facility.

Outside the GSM/UMTS arena there is a much greater emphasis on testing and approval by the network operators, particularly for cdma products. This is a result of the variances in network implementation for cdma, and can result in quite intensive and lengthy testing programs.

The Open Mobile Alliance (OMA)

The OMA, formed in 2002, includes network operators, handset manufacturers, infrastructure manufacturers, and content and service providers (Figure 4.11). The focus of the OMA is on services and service interoperability, irrespective of the underlying network technology. To achieve its aim, the OMA produces a series of standards known as enablers. Since its inception, the OMA has integrated a number of industry bodies and forums and, with it, the work items or technologies on which they focus. These bodies include the:

- WAP Forum
- Wireless Village
- SyncML Initiative

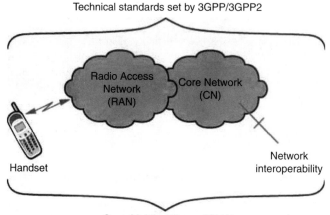

- Browsing
- Digital Rights Management (DRM)
- Location services
- Games
- Synchronization
- Push-to-talk over Cellular (PoC)
- Presence services

FIGURE 4.11   The Open Mobile Alliance.

- Location Interoperability Forum (LIF)
- Mobile Wireless Internet Forum (MWIF)
- Mobile Gaming Interoperability Forum (MGIF)

The output can already been seen to shape mobile services and, as a result, the design of mobiles. Examples are in the areas of digital rights management (DRM), where many handset vendors support the OMA DRM framework and browsing services. In a world divided by two dominant 3G standards — UMTS and cdma2000 — it will be the efforts of the OMA that help provide service interoperability across diverse networks and provide drivers for service and revenue growth. The OMA has a series of cooperation agreements or frameworks with the many standards bodies in the market, including:

- 3GPP/3GPP2
- ITU-T
- W3C
- CDMA Development Group (CDG)

- IETF
- Web Services Interoperability (WSI) organization

Open Service Access

The specification, design, and implementation of services in a telecommunications network were at one time limited to a very small pool of programmers, as services were normally written as additional software for network elements (e.g., switches). Even the advent of Intelligent Networks (IN) did little to improve this situation, as a service designer still needed an intimate knowledge of telecommunications networks and protocols.

Open service access (OSA) is an attempt to provide a much more open, yet secure, environment for service development by abstracting the network functions into a high-level view and providing access to these functions by means of standard, well-defined application programming interfaces (APIs) (Figure 4.12). The work on what is now known by some as OSA was started by the Parlay Group, which focused on the U.K. network of British Telecom (BT), which the regulator wanted to open up to third-party service providers. The Parlay Group works on the APIs and the model for service providers to access the network functions.

The OSA project of the 3GPP is now aligned with Parlay, and the names are now interchangeable. The OSA consists of an API that exposes, in an abstract manner, the service capabilities of the network. A service designer can manipulate these service capabilities to build innovative services (e.g., a service based on the location of a mobile device).

A major concern for network operators is security and the security of information; therefore, the OSA gateway that supports service providers also includes a framework element that is responsible for authenticating service providers and authorizing what functions (and information) that provider is able to see. Using standard network signaling protocols such as SS7-based protocols, the OSA gateway is able to manipulate the network elements to achieve the functions of a particular service, although the service designer is isolated from the detail of these protocols. The service provider in an OSA environment could be either within the network (i.e., the network operators themselves) or a trusted third party.

Live OSA-based mobile services are available in some networks, although the end user will not be aware that this is how the service is provided. These services include prepay, location services, virtual private networks (VPNs), and unified messaging (UM).

Web Services

Although the intention of the OSA was to abstract network functionality and isolate a service designer from the details of the networks and its protocols, it remains the thinking of many that the OSA API still relies on a relatively detailed knowledge of

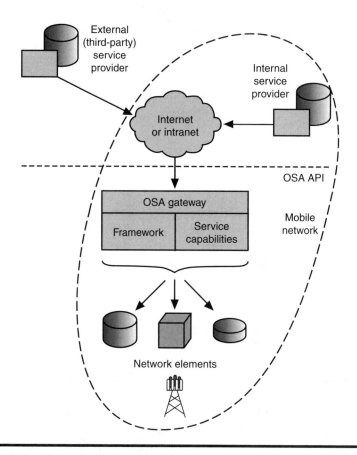

FIGURE 4.12 Open Service Access.

telecommunications networks. At the same time, the concept of delivering modular services using standard Web technology has gained a lot of interest, and these facts have given rise to the concept of *Web services* being deployed in mobile networks.

Web services (Figure 4.13) are based on standard Internet technologies such as the eXtensible Markup Language (XML) and related protocols such as the Simple Object Access Protocol (SOAP) and Web Services Description Language (WSDL). The concept behind Web services is that a *service provider* can describe and advertise the presence of a service on a service registry; other entities can then search these registries to find the service they need. Simple examples of Web Services would be stock quotes, currency conversions, or language dictionaries, and the model supports the distributed concept where the many services required by one user could be found across an entire series of service providers around the Web.

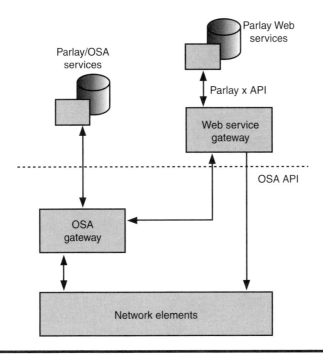

FIGURE 4.13 Web services. OSA/Parlay Web services defined in 3GPP TS20.100 part 1 to 14; and OMA enabler (OWSER) released in 2004.

The concept of Web services was introduced to telecommunications networks through the standards referred to as Parlay X. Parlay X represents a further abstraction and simplification of network functionality when compared to Parlay/OSA. A Web service gateway in the network will give service developers access to network functionality via the Parlay X API, which is based on Web protocols.

Review Questions

Q1. Which of the following technologies is *not* the responsibility of the 3GPP?
   a. cdma2000
   b. EDGE
   c. UMTS
   d. GPRS

Q2. Organizations such as the IETF and the W3C have a bearing on handset design.
   a. True
   b. False

Q3. Many tests are carried out on the mobile handset. Which of the following is *not* involved?
 a. Handset manufacturer
 b. Third-party testing house
 c. Network operator
 d. The 3GPP

Q4. A self-certification program for manufacturers to certify that their products meet statutory requirements is known as _____.
 a. R&TTE Directive
 b. EMC
 c. GCF
 d. Interim type approval

Q5. The focus of which of the following groups is on services and service interoperability, irrespective of the underlying network technology?
 a. 3GPP
 b. OMA
 c. OSA
 d. MWIL

4.1.3 Types of Handset and the Market

Handset Segmentation

Segmentation of mobile handsets is usually based on technology, with a correlation between technology and price; that is, the more complex or feature-rich the device, the more expensive it tends to be. However, more recently, and for the foreseeable future, alternative segmentation strategies have become increasingly important for both vendors and operators. They both seek to segment handsets and their markets in a variety of ways to cope with market trends and aid in the differentiation of their products to attract and retain subscribers:

- *Technology.* Technology is the traditional handset segment. Handsets are divided on a technical basis, with a direct correlation to pricing so that a handset incorporating less of the latest technology is invariably cheaper in the marketplace than one with greater technical prowess. This can now also be applied to features on the phone, such as the number of megapixels on a camera or the amount of storage capacity.

- *Lifestyle.* This is an increasingly ubiquitous segmentation strategy, reflecting the growing importance of the end user in the value web. This involves matching the functionality of a handset with the specific needs of the end–user (e.g., youth or style).
- *Price.* This strategy is beginning to be used more, owing to the move into developing markets where customer income is low. It is primarily based on the cost to manufacture the handset and the customer market segment at which it is aimed. It also includes customer purchase profiling.
- *Application specific.* This segments the market by optimizing components of the handset to focus on a specific application, such as games or music. It is not a technology-hungry mode, as the handset design is quite flexible and is a comparatively cheaper process because it eliminates many nonessential components.
- *Low-end phones.* The term "low-end handset" is used normally to describe those mobile phones that support only basic voice and data services and perhaps do not support many features such as color displays or cameras. These phones are usually targeted at the new user and prepaid user markets. While these phones are normally at the cheaper end of the market and have traditionally not supported much in the way of advanced features, the effect of technology trickle-down is beginning to be felt even in this market sector.

It is not unusual to find color screens, cameras, and even WAP-based Internet access available on some handsets. Polyphonic ringtones are very popular across all market sectors and a good revenue generator for the network operators; therefore, it is in the interest of the operator to enable the downloading and playing of various types of ringtones, even in the so-called low-end device market.

Also in this segment is the low-cost handset or ultra-low-cost handset. The GSMA is encouraging handset vendors to address the developing market by producing handsets at a very low price point, U.S.$40 to U.S.$50, for example. These devices have very few features in an effort to keep manufacturing costs to the bare minimum. Often, these devices will support only voice and text messaging and perhaps a limited selection of preprogrammed, non-changeable ringtones. While these handsets are designed to address the developing market to make mobile communication more affordable to those people that have little disposable income, they may also find a place in the senior (or "gray") market.

Low-Feature Phones

Low-feature handsets are generally geared toward more mature users and the replacement market. The main attraction of the mid-range handset is the more advanced feature set, in comparison to a low-end phone. Designs are more ergonomic, with shapes and keypads that are easy to hold and handle. Displays are generally bigger than those on low-end phones and are more likely to be color. The graphics quality and user interface are also expected to be superior, and this segment is most likely to see the first significant showing of tri-band handsets.

Music and Entertainment

Music entertainment handsets are generally equipped with a mobile music player for MP3 and AAC files, stereo FM radio, digital recorder, and Flash memory. This is an extremely popular market segment, largely due to the popularity of stand-alone music players such as the Apple iPod. Several handset vendors have launched mobile phones that address the music market directly. These players are often shipped to the subscriber with some form of access to a music downloading site, (e.g., iTunes). One challenge that faces the industry in this market sector is the issue of copyright and digital rights management (DRM). For the service to be popular, musical content must be easy to distribute and exchange between authorized music sites and vendors, as well as between the consumers of the music. This, however, is at odds with the management of the copyright holder's rights to the content. Industry bodies such as the Open Mobile Alliance (OMA) are working on methods to resolve these issues.

Feature Phones

This segment encompasses devices with high-technology capabilities and a variety of features but without harboring an advanced OS, which would put them into the smart phone device category. This segment has many of the attributes assigned to lesser segments and also incorporates advanced features such as video capture and playback, music, expandable memory slot, high-resolution screen, and megapixel camera.

Smart Phones

The smart phone is generally defined as a converged device whose primary function is that of a phone with added advanced computing capabilities or PDA functionality with an advanced operating system (OS). The category was born out of the amalgamation of mobile phones that offer PDA functionality with an advanced OS and a PDA with a WWAN connection. Smart phone devices invariably have a large color display, are inclined to have a larger form factor, and are feature-packed.

There are two primary types of smart phone. There are those that are mainly rich media devices with advanced OS and computing functionality that are basically mobile phones embedding a number of additional features outside the usual ones found on PDAs, such as MP3, camera, MMS, games, Java, or e-mail. These devices might be used in the business and corporate markets. At the higher end of the smart phone market, the devices tend to be PDA computers with phone capabilities, typically manufactured by handheld device vendors such as HP, PalmOne, and Toshiba; these devices are aimed at the business professional.

Review Questions

Q1. Which of the following *cannot* be considered a handset market segment?
   a. Lifestyle
   b. Price
   c. Form factor
   d. Application specific

Q2. Mobile devices that support basic voice and data service are often known as _____.
   a. Feature phones
   b. Low-end phones
   c. Smart phones
   d. Prepaid phones

Q3. Which of the following services is most likely found in a low-feature phone?
   a. Video streaming
   b. Color display
   c. Stylus entry
   d. E-mail client software

Q4. "A converged device whose primary function is that of a phone with added advanced computing capabilities or PDA-functionality with an advanced operating system" describes which of the following phone segments?
   a. Feature phone
   b. Smart phone
   c. Music phone
   d. PDA

4.1.4   Section Review Questions

Q1. State in order of importance the basic requirements of handset design and features for each of the following parties and, where possible, explain your answers.
   a. Network operator:
   b. Handset manufacturer:
   c. End user:

Q2. Choose a selection of mobile phones available today, from your home market or elsewhere, and put each of them into the following classifications. Explain your reasons for each classification.
   a. Low-feature phone:
   b. Feature phone:
   c. Smart phone:

Q3. A manufacturing company that produces components, or sometimes complete pieces of equipment to the specifications and design of another company is known as a _____.
   a. Value-added reseller (VAR)
   b. Original equipment manufacturer (OEM)
   c. Equipment supplier
   d. Original design manufacturer (ODM)

Q4. Which of the following is a battery technology found currently in use with mobile telephones?
   a. Lead acid
   b. Hydrogen fuel cell
   c. Lithium ion
   d. Nickel cadmium

Q5. Which of the following *cannot* be attributed to the success of roaming between networks?
   a. Roaming agreements
   b. Mobile telecom standardization
   c. Market coalescence and global operators
   d. Prepaid billing

Q6. Which of the following is *not* part of the user interface?
   a. Screen or display
   b. Keypad
   c. Lithium ion battery
   d. Audio circuits

Q7. Which of the following handset form factors is the most popular?
   a. Chocolate bar
   b. Flip phone
   c. Clam shell
   d. PDA

Q8. A self-certification program for manufacturers to certify that their products meet statutory requirements is known as _____.
   a. An R&TTE directive
   b. An EMC
   c. A GCF
   d. Interim type approval

Q9. The focus of which of the following groups is on services and service interoperability, irrespective of the underlying network technology?
   a. 3GPP (Third Generation Partnership Project)
   b. OMA (Open Mobile Alliance)
   c. OSA (Open Service Access)
   d. MWIL (Mobile Wireless Internet Forum)

Q10. Which of the following *cannot* be considered a handset market segment?
   a. Lifestyle
   b. Price
   c. Form factor
   d. Application specific

Q11. Mobile devices that support basic voice and data service are often known as _____.
   a. Feature phones
   b. Low-end phones
   c. Smart phones
   d. Prepaid phones

Q12. Which of the following services is most likely found in a low-feature phone?
   a. Video streaming
   b. Color display
   c. Stylus entry
   d. E-mail client software

Q13. "A converged device whose primary function is that of a phone with added advanced computing capabilities or PDA-functionality with an advanced operating system" describes which of the following phone segments?
   a. Feature phone
   b. Smart phone
   c. Music phone
   d. PDA

4.2 Handset Components and Architecture

4.2.1 Basic Handset Functions and Processing

Functional Blocks

A mobile handset, irrespective of radio technology, contains a number of generic functional blocks (Figure 4.14). These blocks include:

- *Audio circuits.* These are used to process audio signals from the microphone, to the loudspeaker, and to any connected devices (e.g., headsets).
- *RF circuits.* This module deals with the modulation and demodulation processes and will also be involved in all the frequency generation requirements.
- *Processing.* A number of processors are typically involved in undertaking all the voice coding, signal processing, channel coding, and signaling tasks.
- *User interface (UI).* The UI consists of the display and the keyboard, and these devices typically interface to the main processor within the handset. More sophisticated handsets might include elements such as voice recognition and handwriting recognition functions as part of the overall UI.

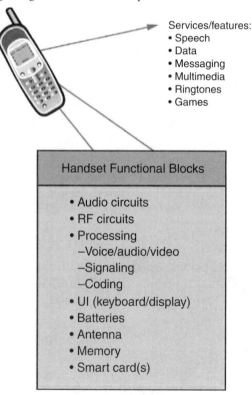

FIGURE 4.14 Functional elements of a handset.

- *Battery.* The source of energy for all the other elements, the battery will be surrounded by circuitry related to recharging and to reporting the status of the battery so it can be displayed on the UI.
- *Antenna.* Often an integrated element on current handsets, the antenna is the interface between the handset and the radio channel connected to the system. Technology improvements will refocus attention on this critical element as techniques such as Multiple-Input Multiple-Output (MIMO) are introduced.
- *Memory.* in addition to the various memory elements that accompany the processor the phone may have other memory capacity provided by the SIM card and by memory cards or sticks.
- *Smart cards.* In GSM and UMTS, the subscription is represented by a smart card (the SIM or USIM) that plugs into the handset. This card can also be used to support transaction services such as m-commerce.

Handset Architectures

A generic architecture for a basic 2G digital handset (Figure 4.15) can be divided into a series of functional blocks, to include:

- Analog baseband
- Digital baseband

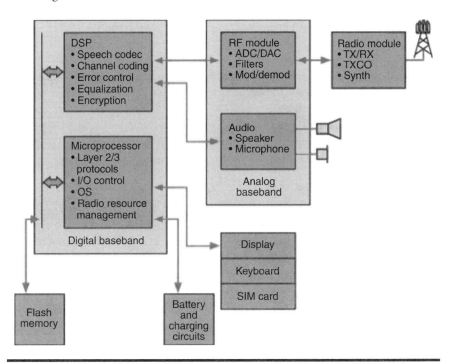

FIGURE 4.15 Handset architecture.

- Radio module
- I/O functions
- Battery

The digital baseband will typically contain both a main processor and a digital signal processor (DSP). The main processor is responsible for I/O tasks, supporting the layer 2 and 3 protocols (e.g., error control and signaling messages) and the management of radio resources. The main processor would interface with the keyboard, display, and SIM elements within the handset. The DSP is responsible for the layer 1 processes such as channel coding, error control, equalization, and encryption, and would typically also support the voice codec processes.

Associated with these processors would be the necessary memory elements such as Random Access Memory (RAM), Read-Only Memory (ROM), and any Electrically Erasable Programmable ROM (E2PROM) or Flash memory.

The analog baseband has two distinct analog functions: (1) those associated with audio signal processing and (2) those associated with RF signal processing. The RF signals would be connected to the radio module, which is responsible for generating the transmit signal from the handset, at the correct frequency and power level, and for receiving and amplifying the downlink signals at the antenna. The handset will also contain a battery with its associated charging and management circuitry.

Handset Processing

As handsets have evolved from simple voice-only analog phones through to complex 3G multimedia platforms, an increasing processing load has been assumed of the devices. Processing loads and capabilities can be compared using the measurement of "millions of instructions per second" (MIPS), and device manufacturers will usually quote the MIPS value for processor chips.

The processing requirement for a 2G GSM phone is in the order of 10 MIPS, with much of this processing requirement resulting from the voice coding function (Figure 4.16). The addition of a 2.5G technology such as GPRS raises this figure to somewhere on the order of 12 MIPS, although the processing complexity of a 2.5G phone will differ significantly across the range of device types encountered, and may be as high as 40 MIPS.

A UMTS (W-CDMA) handset operating in a 3G network requires a total processing capability in the region of 500 MIPS, with 40 percent of this requirement resulting from the relative complexity of the air interface. Adding features, particularly video processing, will increase the MIPS requirement and there is already discussion of phones with 1000 MIPS (1 GigaMIP) requirements.

An issue for handset manufacturers and chip designers has always been the processing limits of DSP chips, which is why handsets typically consist of multiple processors and hardware accelerators (which remove some of the repetitive tasks from the DSP and implement these in hardware).

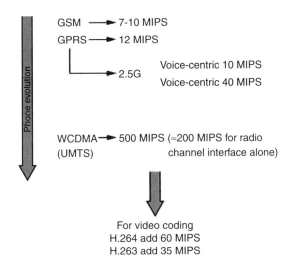

FIGURE 4.16 Processor requirements.

Increasing the processing load also has a tendency to increase the power consumption of the phone's electronics, which will shorten talk time and standby time unless battery capacities can also be improved.

Processor Architectures

The division of tasks between multiple processors within a handset is very common, and a typical architecture would include a microprocessor (or microcontroller), a DSP, and hardware accelerators. The role of the hardware accelerator is to remove from the DSP the more routine repetitive tasks, such as radio channel processing, leaving the DSP free to focus on other layer 1 tasks and vocoding (implementing a compression algorithm particular to voice) (Figure 4.17).

In this architecture, the hardware accelerator can be labeled as a coprocessor, although this is a generic term that may have other meanings in the context of handsets. Silicon manufacturers have in some cases produced single-chip solutions that contain the three processing elements; this is an attempt to reduce the area, volume, and cost of these vital handset components.

The typical task distribution in today's handsets places the emphasis on the accelerators rather than the DSP. However, as DSP technology improves, more and more of the total processing load could be assumed by the DSP, although of course by that time the evolution of services may be placing even more demands on the handset processors.

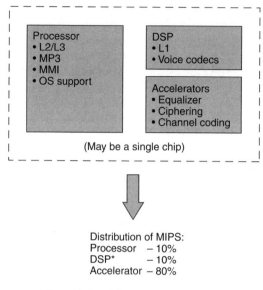

FIGURE 4.17 Processor architectures.

Coprocessors

The spread of processing load has led to a number of different strategies for the use of coprocessors (Figure 4.18). For example, in a dual-mode 2G/3G phone, a chip manufacturer might offer a main processor that is responsible for the 2G and 2.5G baseband functions and a companion chip that adds 3G baseband functions. These two processors can then be coupled to their corresponding radio modules. This solution is perhaps suited to a manufacturer that is looking to evolve a range of 2G/2.5G to support 3G capability. The basic core design of the handset can be reused and the 3G functions are added in parallel. Another possible solution is to use one processor for all the digital baseband processing and to use a separate co-processor to handle specific tasks such as multimedia services.

Review Questions

Q1. Which element within a mobile terminal is responsible for signal processing, voice coding and message processing?
   a. RF circuitry
   b. Audio circuits
   c. Processors
   d. Memory

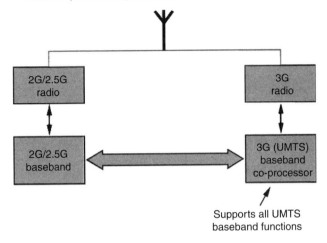

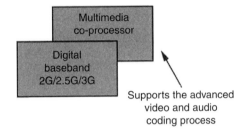

FIGURE 4.18 Coprocessors.

Q2. Which of the following elements do *not* belong to a typical 2G handset?
   a. Analog baseband
   b. Digital baseband
   c. 64-bit memory access
   d. I/O functions

Q3. A typical 3G mobile device requires how many instructions per second to operate?
   a. 7 to 10 MIPS
   b. 40 MIPS
   c. 500 MIPS
   d. 1000 MIPS

Q4. The processing element that assists the main processor and performs routine tasks for the DSP may be known as the _____.
   a. Microprocessor
   b. Process controller
   c. Hardware accelerator
   d. CPU

Q5. Handset manufacturers may include additional processors to perform functions such as multimedia support.
   a. True
   b. False

4.2.2 Memory Storage and Requirements

Types of Mobile Terminal Memory

Mobile handsets have always required elements of memory and storage as somewhere for the operating system code to reside and execute (Figure 4.19). In early phones, the user may have had access to a very limited area of memory to store contacts, as an alternative to storing these on the subscriber identity module (SIM).

The increase in volume of applications such as messaging and the ability for modern phones to support multimedia has increased the demands for storage capacity within the handset. A typical handset contains a range of memory or storage elements, which typically include the SIM card, hardware memory elements, removable cards or memory sticks, and, most recently, hard disk drives.

With users now storing pictures, videos, music, and games on their handsets, the total storage requirement has risen very rapidly; combined with this is the increased complexity of multimedia signal processing, placing additional demands on memory requirements.

FIGURE 4.19 Memory for mobile terminals.

Handset Hardware Memory

In common with any processing platform, a mobile handset requires a number of different memory types to support its processing sequences. The two broad categories of memory are Random Access Memory (RAM), which is as a working area and as memory for user and application data storage; the other type of memory is used to store the main program code and other applications. RAM is volatile and will lose its contents unless power is maintained, whereas the E2PROM or Flash memory used for the code storage is nonvolatile. (Figure 4.20).

The RAM types found in mobile handsets include Static RAM (SRAM) and Dynamic RAM (DRAM). SRAM will hold its contents as long as power is maintained and, unlike DRAM, does not require refreshment on a regular basis. DRAM memory stores bits of data as a charge on a capacitor; this charge must be regularly topped up to avoid it leaking away.

Electronically Erasable Programmable Read-Only Memory (EEPROM or E2PROM) is a nonvolatile memory type that can be programmed and reprogrammed many times by means of electrical signals. E2PROMs have life cycles of between 100,000 and 10 million write operations, although they may be read any number of times. A limitation of E2PROM devices is that only one memory location can be written to at any one time, this can be overcome by using Flash memory devices, which can have a number of locations written in one operation.

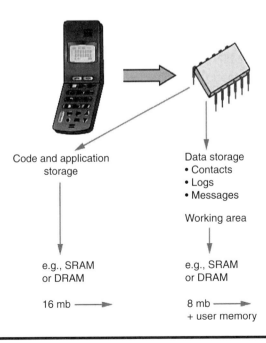

FIGURE 4.20 Hardware memory.

The amount of hardware memory in handsets has increased considerably in recent years as phones have included evermore complex operating systems and applications, and also as users have required more memory for storing personal data and multimedia files.

At the low end, the amount of E2PROM or Flash memory required might be 16 MB with an additional 8 MB of SRAM or DRAM. In addition, even low-end phones might offer users memory of between 5 and 20 MB.

The manufacturing requirement to get more and more memory in a handset, while at the same time decreasing the size of the device, has led to new techniques such as multi-chip packaging (MCP) where one physical chip package houses both SRAM and Flash.

Memory Growth

Driven by multimedia applications, the amount of hardware memory in handsets has increased dramatically from the relatively small amount in voice-centric handsets to a total of between 500 MB and 1 GB in 3G devices. The memory is typically a mixture of RAM and Flash (Figure 4.21).

Until recently there were two main types of Flash memory in production, and both of these are encountered in handsets; they differ in regard to the type of logic gate deployed in the memory matrix — either NOR or NAND. However, hybrid Flash memory devices are now appearing in the marketplace that combine the best features of the two memory types, and these may soon find their way into mobile products.

NOR Flash was developed first; and while it is a true random access memory (any location can be addressed when necessary), it suffers from relatively slow write and erase times and a more limited lifespan than NAND Flash, which is both faster and cheaper.

Increasing the memory capacity further in handsets is unlikely to be achieved simply by the deployment of higher density memory matrices and will be supplemented by other forms of memory, or storage, specifically:

- SIM-based storage
- Internal hard disk drives
- Removable cards or sticks

The total storage available in the latest 3G devices is already several gigabytes, achieved through a combination of all of these elements.

(U)SIM Cards

GSM and UMTS handsets are fitted with smart card devices, the subscriber identity module (SIM), and Universal SIM (USIM), respectively (Figure 4.22). These

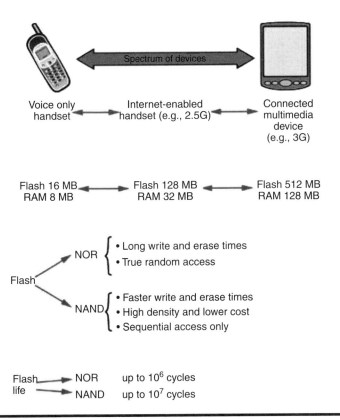

FIGURE 4.21   Growth in memory requirements.

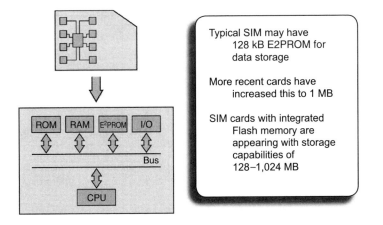

FIGURE 4.22   The Universal Subscriber Identity Module (USIM).

cards are used to store some of the network-specific and user-specific data that is used to influence the interaction of the phone with the network and for security procedures. The smart cards are based around a processor and its associated memory chips. In addition to supporting basic handset functionality, the SIM or USIM has always provided some storage for user data such as contact lists.

While early SIM cards had only a few kilobytes of storage capacity, a typical SIM now has between 128 kB and 1 MB of storage, implemented using E2PROM. Some manufacturers have released SIM products with integrated Flash memory, giving additional storage capacity of up to 1 GB.

Memory Cards

Driven by appliances such as digital cameras, a number of removable data storage formats have been invented, some of which are also deployed in mobile phones. These memory cards or sticks can be used to store music, pictures, videos, and games, which might have been downloaded or, in the case of pictures and videos, captured by the user with the handset camera.

These memory devices are intentionally small, often a few tens of millimeters square, and weigh only a few grams. Their storage capability however might be up to several gigabytes. Typically, these devices are based on Flash technology and are housed within the handset in a manner similar to the SIM.

Memory cards or sticks are usually many-time programmable (MTP) but some are available in a one-time programmable (OTP) or read-only memory (ROM) format (Table 4.1).

An important aspect of content storage is the protection of that content against non-permitted copying and distribution. All the common card formats include some level of copyright protection. However, because there are a number of card

TABLE 4.1 Memory Card and Memory Stick Characteristics

	Multimedia Card (MMC)	Reduced Size (RS-MMC)	Secure Digital (SD)	MiniSD	Memory Stick Pro	Memory Stick Pro Duo
Capacity	Up to 256 MB	Up to 64 MB	Up to 1 G	Up to 256 MB	Up to 2 GB	Up to 512 MD
Size (mm)	32 × 24	24 × 18	32 × 24	21.5 × 20	50 × 21.5	31 × 20
Weight (g)	1.5	1	2	1	4	2
Connection pins	7	7	9	11	10	10
Copyright protection	Yes	Yes	Yes	Yes	Yes	Yes

formats, there are — not surprisingly — a number of copyright protection technologies. For example, the Multimedia Card Association (MMCA) is working on a secure digital format based on the Open Mobile Alliance (OMA) version 2 Digital Rights Management (DRM) framework, whereas the memory stick devices incorporate MagicGate protection technology.

Hard Disk Drives for Handsets

As handsets and computers converge, both in terms of capabilities and applications, it is not surprising to find that the architecture and components of a handset start to mirror those found in a computer. This trend is likely most evident in the memory components found in a handset, because these will need to include:

- Storage for *boot code*
- Storage for code, applications and data
- Execution space

As with a PC, the large volume of storage required for operating code, applications, and application and user data could be provided by means of a hard disk drive. A number of disk manufacturers have produced specialized, small format disks for the handset market, and these devices are now being incorporated into handsets. The disk drives are usually in a format of one inch or smaller and have storage capabilities ranging from 1 to 4 GB (although this will increase with time). Similar disk products have been used for some time in portable entertainment terminals such as MP3 players (see Figure 4.23).

One major consideration for integrating components such as disk drives into handsets is the difference in treatment by users between their PCs and mobile phones. For this reason, the manufacturers of the disk drives also offer proprietary protection techniques that protect the drive should the handset be dropped or suffer severe mechanical shock.

While converged devices now have total storage capabilities of many gigabytes (a figure that will continue to increase), it should be noted that voice-centric devices will still only require a few megabytes of storage — sufficient to support their non-memory-intensive applications and to provide some limited storage for user data.

Review Questions

Q1. Which memory type would you most likely find in a mobile terminal?
    a. Magnetic tape
    b. Removable Flash
    c. 128-pin DIMM
    d. Removable disk drive

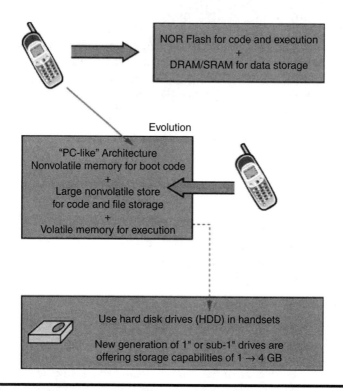

FIGURE 4.23 Hard disk drives for mobile terminals.

Q2. The type of memory used to store application code is _____.
   a. SRAM
   b. DRAM
   c. E2PROM
   d. HDD

Q3. Which of the following characteristics *cannot* be attributed to NAND flash memory?
   a. Fast write and erase time
   b. High density and lower cost
   c. Sequential access only
   d. True random access

Q4. The SIM/USIM stores information about the user and the network, and this information is used to influence the behavior of the mobile device.
   a. True
   b. False

Q5. Which of the following removable memory media types uses MagicGate technology to protect media content?
a. MMC
b. SD
c. MiniSD
d. Sony memory stick

4.2.3 Display Technologies

The displays used in handsets are based on liquid crystal diode (LCD) technology that has been used in consumer products for some time. Liquid crystal materials have some special properties that are exploited to create displays; most importantly, the crystals have a twisted structure and the amount of twist can be altered by applying a voltage to the crystal material.

LCD Display Structure

A basic LCD display is a sandwich of layers through which light passes (Figure 4.24). One of these layers is the liquid crystal that is situated between two layers of glass that contain electrical connections. By altering the signal to these connections, the crystals can be made to alter their twist, which has an effect on the polarization of the light and can be used to create the dark/light contrast necessary for a display.

The specific LCD technology found in many displays is super twisted nematic (STN), which relates to the special form of liquid crystal that is used. Displays can be characterized as being either reflective or transmissive. A reflective display relies on incident light from the front of the display, passing through all the layers to a final reflective layer where it is reflected back to the front of the display. It is possible to provide front-lighting or back-lighting to reflective displays. A transmissive display uses backlight from within the display. The use of back- or front-lighting will increase the energy consumption of the display; and when used in handsets, the lighting has an associated sleep circuit to switch off the light after a few seconds of *user inactivity*.

There are a number of variants of the twisted nematic (TN) display, although they all generally employ the same principles of operation.

Color Displays

Adding color to a display is relatively simple (Figure 4.25). Each pixel in the display has three separate filters associated with it: one red, one green, and one blue. Therefore, each pixel is effectively divided into three sub-pixels. The filters can be activated so that only light of a particular color can pass through for that pixel. The three sub-pixels can be manipulated to create a range of colors. The number of bits

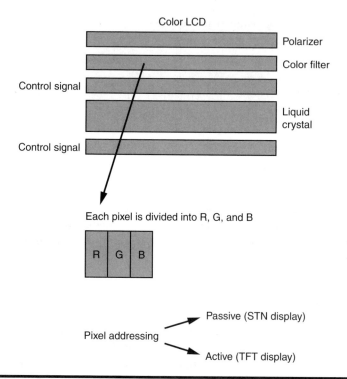

FIGURE 4.24 LCD structure. Reflective LCD device (may be improved using backlight). Transmissive devices do not rely on incedent light (i.e., they use backlight). Liquid crystals can be made to *twist* by applying a voltage. They can be used either to allow light through or to block it.

per pixel will determine how many different colors can be displayed and various resolutions are found in handsets:

- 256 colors; 8 bits per pixel
- 4096 colors; 12 bits per pixel
- 65,536 colors; 16 bits per pixel (65K)
- 262,144 colors; 18 bits per pixel (262K)

The backlight used in handset display can be a number of colors; however, the advent of white LED means that one choice is to illuminate with white light.

Display Types

The way pixels in a display are addressed, so that they can be switched between states, has led to two main display technologies, the so-called passive and active displays.

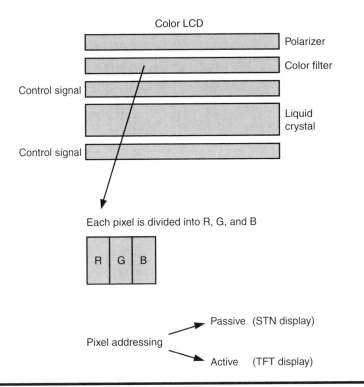

FIGURE 4.25 Color display structure.

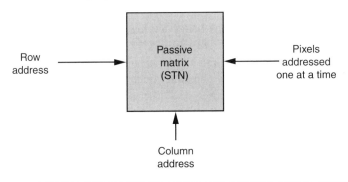

FIGURE 4.26 Super-Twisted Nematic (STN) display.

In a super-twisted nematic display, a passive display (often called STN, Figure 26), the individual pixels are addressed by row and by column signals, one pixel at a time, and thus the display is relatively slow because it takes time to build up an image pixel by pixel.

On the other hand, active displays add another component in the form of a transparent transistor at each pixel, hence these displays are referred to as thin film transistor (TFT). Using the active technology allows a whole row (or column) of

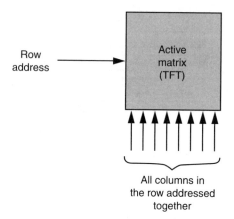

FIGURE 4.27 TFT display.

pixels to be addressed at once, which means that creating an image is much more rapid than in an STN display.

The disadvantages of STN displays are their relatively slow operation, and there are also issues relating to brightness and angle of view. However, they are cheap to manufacture and use less energy than a TFT display (Figure 4.27), which corrects the major problems of the STN format. In handsets with two displays, where a simple display is used for phone functions and a higher specification display is used for viewing videos and playing games, it is common to find both technologies deployed — STN for the simple display and TFT for the high-quality display.

Other display technologies are being developed; one in particular, the organic LED (OLED), is receiving a lot of interest. OLED are emissive devices and thus do not require a backlight, they can also be created on very thin layers of polymer (almost like printing), and they consume much less energy than LCD displays, which makes them ideal for handsets and other power-constrained devices.

OLEDs are being demonstrated in consumer products and much research is being conducted into these components to overcome some of their limitations, at which point widespread deployment would become a reality. For example, the lifespan of blue OLED elements is only a few thousand hours, which means their use in a TV or phone display is not yet commercially possible.

Review Questions

Q1. The term STN (Super Twisted Nematic) refers to _____.
   a. The layered structure of the display
   b. The nature of the liquid crystal itself
   c. The control signal applied to the display
   d. The application of display backlighting

Q2. A display that relies on incidental light falling on the display to make the information visible is known as _____.
   a. Transmissive
   b. Reflective

Q3. A display that supports 16 bits per pixel can display how many colors?
   a. 256
   b. 4096
   c. 65K
   d. 262K

Q4. This statement, "… the individual pixels are addressed by row and by column signals, one pixel at a time, and thus the display is relatively slow," describes what type of display?
   a. Passive — TFT
   b. Active — TFT
   c. Passive — STN
   d. Active — TFT

Q5. Which of the following is *not* a characteristic related to OLED technology?
   a. Can be created on thin layers
   b. Low power consumption
   c. No backlight required
   d. Long life span

4.2.4 Handset Sound Capabilities

Handsets need audio coders and decoders for a variety of reasons, the most fundamental being the encoding and decoding of human speech for telephony services. However, with a typical handset now supporting multimedia functions, the sound capabilities will support music, the audio that accompanies a video clip, and applications such as games and ringtones.

Ringtones

Ringtones have become big business as users attempt to differentiate their handset models from all identical models. The ringtone represents an easy route to phone personalization. There are numerous formats (Figure 4.28) in which to provide ringtones, and this represents the dramatic changes in capability from the very early phones, which could only support monophonic tones, to the devices of today, which are able to support complex music files coded with the same techniques used to record CDs.

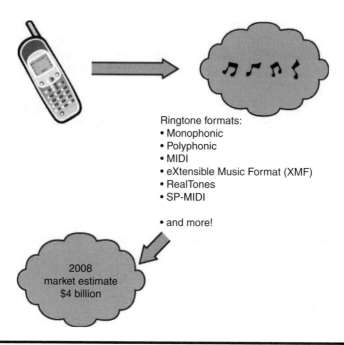

FIGURE 4.28   Ringtone formats.

Some of the ringtone formats that have appeared in the marketplace are proprietary and are perhaps only supported by a limited range of models from one supplier or a small number of suppliers.

Ringtone Formats

Many early handsets had very limited capability in terms of ringtone support — the tone output was monophonic as the sound elements could only play one note at any moment in time. The manufacturer would maybe supply a handful of built-in ringtones for the user to choose from, and there was no capability for downloading new tones. Monophonic tones have a very artificial sound to them.

Increased capability brought polyphonic ringtones to handsets and, combined with features and services that allowed users to download new tones on their phones, the market for ringtones was created. There are a number of polyphonic ringtone formats (see Table 4.2), including the Musical Instrument Digital Interface (MIDI). MIDI differs from the other tone formats in that the MIDI file does not actually contain coded music, but rather a set of instructions about the notes to be played, the *voice* to be used for each note, and the duration and depth of each note. The consequence is that MIDI files are very compact and therefore ideally suited to ringtone downloads.

An improvement on MIDI is Scalable Polyphonic MIDI, which allows the same content to be played on devices that differ in terms of their polyphonic capability.

TABLE 4.2 Ringtone Formats

Format	Characteristics
Monophonic	Simplest ringtone format; phone only plays one note at a time in sequence
Polyphonic	Phone can play several notes simultaneously; more realistic than monophonic ringtone
MIDI (Musical Instrument Digital Interface)	These files are not musical sequences but sets of instructions to synthesize sound
SP-MIDI (Scalable Polyphonic MIDI)	Supports phones with variable n-note polyphony
XMF	Overcomes the 128 instrument bank limitation of MIDI by using downloaded sounds
RealTones	An MP3-based format for real or true music recordings

A low-end phone might only have four-note polyphony while a high-end phone might have 32-note polyphony; the same file could play on both handsets through a process of scaling.

The eXtensible Music Format (XMF) was introduced to overcome the limitations of the 128 fixed instrument pallet of MIDI. XMF allows downloading of new sounds to replace the default MIDI sounds.

More recently, ringtones have been provided as MP3 files using the same audio coding techniques used for music distribution. These files offer much more realistic sound capabilities, and the availability of chart music as ringtones is evidence of the popularity of this format.

There are a number of manufacturer-specific ringtone formats in use and also formats devised by third parties; for example, the polyphonic Synthetic music Mobile Application Format (SMAF) from Yamaha is supported on a range of phones from different manufacturers.

Audio Coding

As with ringtones, handsets must be capable of supporting a wide range of audio formats if an end user wants to decode audio from a variety of sources. There are two distinct families of audio coder found in handsets. The first family is related to the need to code the human voice for telephony services, although some of the coders used are derivatives that support signals with a wider bandwidth than speech (e.g., music). The second family of coders consists of those that comprise the audio layer used in video coding techniques.

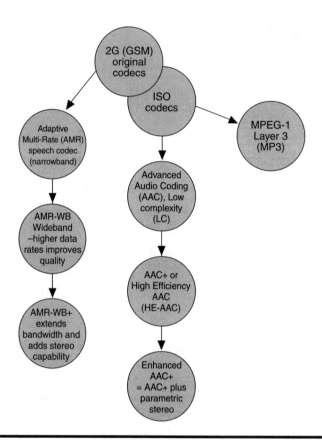

FIGURE 4.29 Codec types.

GSM handsets were originally built around the Full-Rate (FR) codec, which was later supplemented by the Half-Rate (HR) codec, the Enhanced Full-Rate (EFR) codec, and the Enhanced Half-Rate (EHR) codec (Figure 4.29). All these coding mechanisms are built around a model of the human voice and, therefore, while they offer good quality for speech, they are not optimized for non-speech signals such as music.

The GSM specifications moved on to an Adaptive Multi-Rate (AMR) codec that was also adopted as the standard by 3GPP for UMTS networks. This codec could switch rates according to needs and conditions, but was still speech oriented. However, recent improvements have been made to the codec, first by improving quality, and second by extending the audio bandwidth and adding stereo capability. Thus, the codec has evolved to support not only voice, but also high-quality audio, including stereo music.

The second family of codecs found in handsets is based on the Advanced Audio Codec (AAC) taken from the MPEG specifications. As with the AMR codec, the AAC codec has evolved to improve quality and support stereo signals.

Codec Mode	Codec Bit Rate	Notes
AMR_SID	1.8 kbps	
AMR_4.75	4.75 kbps	
AMR_5.15	5.15 kbps	
AMR_5.90	5.90 kbps	
AMR_6.70	6.70 kbps	PDC-EFR
AMR_7.40	7.40 kbps	IS-641
AMR_7.95	7.95 kbps	
AMR_10.20	10.20 kbps	
AMR_12.20	12.20 kbps	GSM-EFR

FIGURE 4.30 AMR codec data rates.

AMR Codecs

The AMR codec is specified for both GSM and UMTS networks, although its use has been driven by 3G rather than a new generation of GSM phones. The AMR codec contains coding and decoding functions, a voice activity detector (VAD) for discontinuous transmission (DTX), and a comfort noise (CN) generator. The input to the codec is 13-bit linear PCM (or transcoded 8-bit nonlinear PCM), and the signal is then coded using a multi-rate algebraic code excited linear predictor (ACELP). With evolutions of this coder now specified, it is retrospectively referred to as the AMR-narrowband (AMR-NB) codec.

The 3GPP has agreed to a modification to the AMR codec known as AMR-wideband (AMR-WB); this codec is still based on ACELP and supports nine output rates ranging from 6.6 to 23.85 kbps. The major difference between the narrowband and wideband codecs is that AMR-WB supports an audio input range from 50 Hz to 7 kHz, rather than the upper limit of the AMR-NB, which was 3.4 kHz. The AMR-WB codec still includes DTX, VAD, and CN functions (Figure 4.30).

The AMR-WB+ codec builds on the AMR-WB system by adding higher sampling rates and through the support of both mono and stereo signals (Figure 4.31). The AMR-WB+ input can extend up to 20 kHz, and the output is scalable between 6 and 36 kbps for mono and between 8 and 48 kbps for stereo. The AMR-WB+ codec has a wide bit-rate range (from 6 to 48 kbps). Mono rates are scalable from 6 to 36 kbps, and stereo rates are scalable from 8 to 48 kbps. The AMR-WB+ codec is backward compatible with the AMR-WB codec.

These newer codecs are liable to be applied for a range of services, including ringtones, music downloading, Packet Switched Stream (PSS) and the Multimedia Broadcast Multicast Service (MBMS).

Bluetooth Audio Coding

The Bluetooth wireless technology can be used to convey both voice and data signals, depending on the profile in use. When voice is carried over Bluetooth, the profile definition specifies the coding mechanism for the audio signal (Figure 4.32).

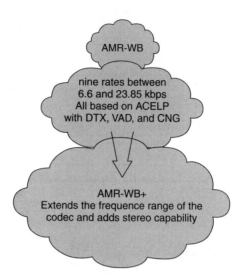

FIGURE 4.31 AMR evolution.

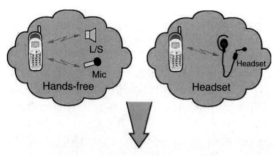

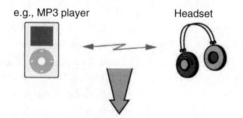

FIGURE 4.32 Bluetooth audio coding.

A number of profiles are defined that support voice, and a range of appropriate coders have been specified. For example, the hands-free and headset profiles both use Bluetooth to convey voice signals from a mobile handset to peripheral devices; in both cases, the speech coding can be either nonlinear 8-bit PCM (either A-law or μ-law encoded), or 8/16-bit linear PCM, or Continuously Variable Slope Delta (CVSD) modulation. These are both standard voice coding technologies that have been used in networks such as the PSTN or private PABX networks.

Another profile known as the Advanced Audio Distribution Profile (A2DP) supports the connection of devices such as MP3 players to wireless headsets, or the connection of a wireless microphone to a portable recorder. In this case, the profile specifies a range of coding options:

- Subband coding (SBC)
- MPEG-1, -2, or MPEG-4 (AAC)
- ATRAC

SBC is a low-complexity, royalty-free coder that uses psycho-acoustic analysis within frequency sub-bands to determine the threshold to apply in each band. The output of each sub-band is quantized and combined with all the other sub-bands in a frame structure.

The MPEG-1 video coding standard originating from the International Standards Organization (ISO) included audio coding options referred to as Layers 1, 2, and 3. ATRAC, which stands for Adaptive Transform Acoustic Coding, is a proprietary Sony format used in mini-disks and other systems.

Additional Audio Coding Types

In addition to the AMR and AAC coding techniques, there are numerous other coding formats employed in commercial systems; many of these are proprietary technologies, and some are open standards (Figure 4.33). Formats encountered in relation to mobile phones include:

- WAV (a Windows format for the PC)
- Windows Media Audio (WMA)
- ATRAC
- MP3pro

The WAV format (pronounced "wave"), a very common PC format for audio files, is based on PCM coding and can optionally support compression. The fact that it is a waveform coder means that the file sizes produced are relatively large in comparison to lossy coding techniques.

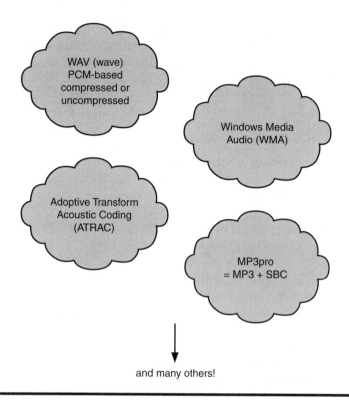

FIGURE 4.33 Additional audio coding types.

WMA is another Windows format originally intended to compete with MP3 but is now perhaps positioned against AAC. WMA files can be played on the Windows Media Player.

There are a number of versions of the Sony ATRAC coding system. The original ATRAC coder, used for minidisks, encoded audio at a rate of 292 kbps for full stereo. ATRAC3 and ATRAC3 plus are both aimed at Web applications and can give equivalent quality to MP3 at rates of only 132 kbps and 64 kbps, respectively.

Review Questions

Q1. Handsets require audio codec functions for a number of reasons. Which of the following is *not* related to audio processes?
   a. A/D voice conversion
   b. Copyright of musical downloads
   c. Support of ringtone services
   d. Support of multimedia applications

Q2. What type of audio codec does the following describe? "The file does not actually contain coded music, but rather a set of instructions about the notes to be played."
   a. Monophonic
   b. MIDI
   c. XMF
   d. MP3

Q3. The AMR-WB codec has the following characteristics.
   a. Input range 50 to 3400 Hz; output range 1.8 to 12.2 kbps
   b. Input range 50 to 3400 Hz; output range 6.6 to 23.85 kbps
   c. Input range 50 to 7000 Hz; output range 1.8 to 12.2 kbps
   d. Input range 50 to 7000 Hz; output range 6.6 to 23.85 kbps

Q4. The AMR-WB+ is *not* compatible with previous versions of AMR.
   a. True
   b. False

Q5. Which codec is most likely used for a Bluetooth hands-free kit?
   a. CVSD
   b. MPEG-3
   c. ATRAC 3 Plus
   d. SBC

4.2.5 Image and Video Capabilities

The inclusion of a camera device has become an expectation of today's handset users. Combined with the capabilities of the screen (or screens) on the handset, the user has a range of options supporting visual information:

1. Taking, storing, and forwarding photographs
2. Capturing, storing and forwarding video clips
3. Downloading photographs and movie clips
4. Streaming video reception
5. Video calling and video conferencing

Capturing the Image

Image capture by a handset is performed by a camera element that is functionally identical to those found in digital cameras. The optical chain consists of a lens, the detector device, and some associated processing electronics. The lens can be plastic,

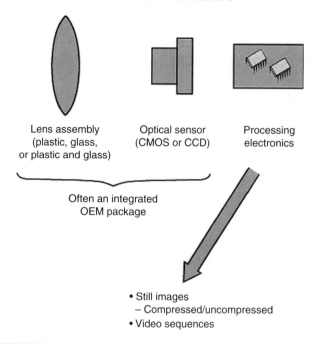

FIGURE 4.34   Image capture process.

glass, or a compound lens with both plastic and glass layers, and is normally in an integrated package that includes the optical sensor. There are two main sensor types: (1) complementary metal oxide semiconductor (CMOS) or (2) charge-coupled device (CCD). CCDs for a long time were viewed as high-end components, but handsets are now starting to incorporate these devices as the optical sensor. The number of pixels supported by phone cameras has increased rapidly; however, it should be noted that the quality of the lens and the manufacturing tolerances in the optical package are more important to final image quality.

The sensor chip is accompanied by a matched processing element that is able to take either still or moving images and process these — maybe by way of applying a compression technique to reduce the file size of an image or a video clip (Figure 4.34).

Features associated with the camera include digital zoom, which is an image manipulation technique that allows users to select a portion of an image and increase its size. In an attempt to increase image quality under a range of conditions, camera phones are appearing in the marketplace that include optical zoom capability, autofocusing, and flash units for illumination in low ambient light conditions.

Image Formats

Handsets typically support a range of formats (Figure 4.35) for both still images and video, although there may only be limited options for image encoding, particularly

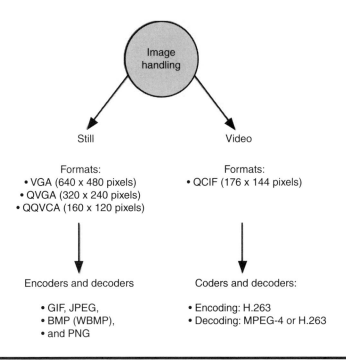

FIGURE 4.35 Image coding.

for video. This is a result of the technology licensing costs associated with many of the industry-standard coding formats.

For still images, the typical image format is based on the Video Graphics Array (VGA) family, which is widely used for CRT screen resolution settings. The base format VGA has a resolution of 640 × 480 pixels, a total of 307,200 pixels. The other VGA formats are mathematically related to this resolution and either have more or less pixels; at the top end, the resolutions extend into the megapixel range associated with digital cameras and high-end phones.

Image storage and manipulation options include the Graphics Interchange Format (GIF), Joint Photographic Experts Group (JPEG), bitmaps (BMP), and Portable Networks Graphics (PNG).

Moving images are often processed using an intermediate format from a family known as the Common Interchange Format (CIF), where the full CIF format has a resolution of 352 × 288 pixels. The Quarter CIF (QCIF) format, at 176 × 144 pixels, has a quarter of the pixels in the CIF format. The intermediate format is used to decrease the total number of pixels in an image and is a coarse method for reducing the bit rate of the final coded video signal. Further reduction in bit rate is achieved using coding/compression algorithms such as ITU-T H.263.

Even with the increase in storage capacity and techniques that are associated with mobile phones, the finite limit of storage means that image processing plays

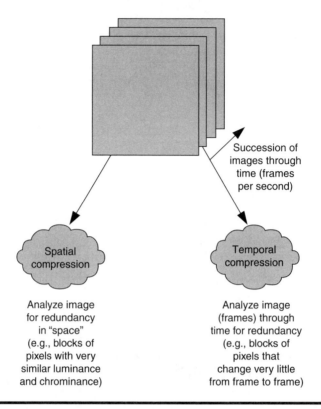

FIGURE 4.36 Video coding.

an important role in striking the balance between image quality and the file size or bit rate of the coder output.

Video Coding

Although reducing the number of pixels in an image reduces the bit rate of a video signal quite dramatically, further reductions are necessary if video is to be supported over the limited bandwidth channels of a cellular network. For most applications, the number of frames per second can also be reduced from the standard 50 or 60 frames per second used in broadcast systems to a figure between 10 and 30 frames per second. Frame rates as low as 10 frames per second will be acceptable for some applications, such as videoconferencing or videocalling.

Further reductions in bit rate are achieved by employing coding or compression techniques. Although there are a number of techniques on the market for processing video, they have similarities in terms of concept, using a combination of spatial and temporal compression (Figure 4.36). Spatial compression techniques analyze redundancy within a frame, produced for example by a large number of adjacent pixels all having the same or similar levels of brightness (luminance) and color (chrominance). This redundancy can then be removed by coding. In a similar fashion, temporal compression looks for redundancy between adjacent frames; this is often the result of an image background, for example, that does not change significantly between one frame and the next. Again, this redundancy can be removed by coding.

As the power of electronic processors, particularly digital signal processors (DSPs), has improved, video codecs have been designed that are able to offer equivalent quality to their predecessors but at reduced bit rates.

Coding for still images is based on the same spatial compression as used for video; there is no need to apply temporal compression to a single image. However, the different still image formats are better suited to one type of image or another. For example, JPEG works well with black and white or color *natural images* (such as photographs), whereas GIF works better for black and white images that contain lines and blocks (such as cartoons).

There is also a distinction between coding that is lossy and coding that is lossless. The video coding techniques described here and JPEG for still images are all classified as lossy in that they remove information through coding that cannot be regenerated later. On the other hand, GIF is a lossless coding technique and subsequently does not remove information through its coding.

Video Coding Standards

A number of video coding standards exist in commercial applications and two main groups have worked on these standards: (1) the International Standards Organization (ISO) and (2) the International Telecommunication Union – Telecommunications branch (ITU-T). The ISO is responsible for the Moving Pictures Expert Group (MPEG) that has produced a series of video coding systems (Figure 4.37).

The first MPEG standard, MPEG-1, was released in 1992 and was aimed at providing acceptable, but sub-broadcast, quality video that could be used for CDs and games. The video coding produced an output at 150 kbps. The audio coding portion of the MPEG-1 standard included three coding options offering progressively greater compression rates for the audio component. The third of these options, MPEG-1 Layer 3 or simply MP3, has become a dominant standard for the distribution of music and audio over the Internet.

In 1994, MPEG-2 was released, and offered improvements on the MPEG-1 standard. MPEG-2 can work at a variety of bit rates, but at 1 to 3 Mbps outperforms the quality of MPEG-1. MPEG-2 is used for digital versatile disks (DVDs) and digital video broadcast (DVB), and includes a range of audio coding options such as advanced audio coding (AAC).

The most recent addition to the MPEG family, MPEG-4 was originally designed for low bit services, with channels operating sub-64 kbps but can also be employed at high bit rates into the Mbps range. The design intention was to provide video coding for some of the "new" applications that were appearing, such as streaming services

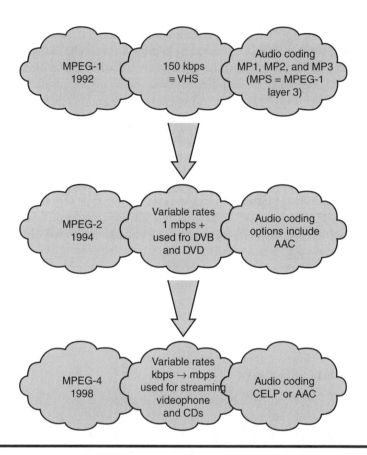

FIGURE 4.37   ISO video coding standards.

and videophones. MPEG-4 is common in mobile handsets as the coding technique was aimed at low rate channels typical of those found in mobile networks.

ITU Video Codecs

In parallel with ISO, the ITU-T has produced a series of video coding technologies that have been included in a number of umbrella standards that define videoconferencing services over a range of channel types: ISDN, telephone, and LAN/IP based (Figure 4.38). The three main codes defined by ITU-T are:

1. H.261 for n × 64-kbps channels
2. H.263 for telephone channels (with modems)
3. H.264 for LAN/IP applications

The final codec, H.264, is identical to that defined in MPEG-4 part 10 and is the result of collaboration between the ISO and ITU-T.

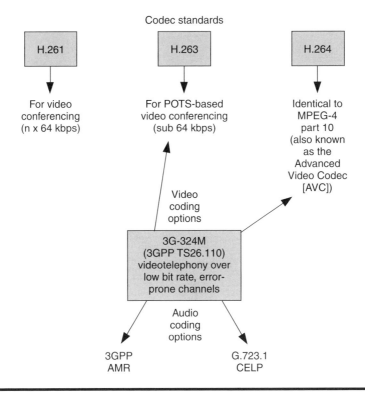

FIGURE 4.38 ITU video codec standards.

ITU-T H.324 is an umbrella standard for videoconferencing over plain old telephone services (POTS) and included the H.263 codec standard. A derivative of this recommendation, known as H.324M, has been adopted by the 3GPP as 3G-324M (video telephony over low-bit-rate, error-prone channels). The primary document for 3G-324M is 3GPP TS26.110.

3G-324M includes video coding using either the original H.324 mechanism (i.e., H.263) or the Advanced Video Coder (AVC) specified in H.264 (and MPEG-4). For audio coding, the options in 3G-324M are the adaptive multi-rate coder (AMR) or a code excited linear prediction technique defined in ITU-T G.723.1.

The 3G-324M standard aims at video delivery over circuit-switched connections. However, there are services that deliver video over packet connections using streaming techniques, and this is where another format known as 3GP can be applied.

The 3GP Format

3GP is a file format originally specified for the Multimedia Message Service (MMS) in 3G networks, but has subsequently been applied to media streams delivered by the Packet Stream Service (PSS). For the PSS service (Figure 4.39), the handset

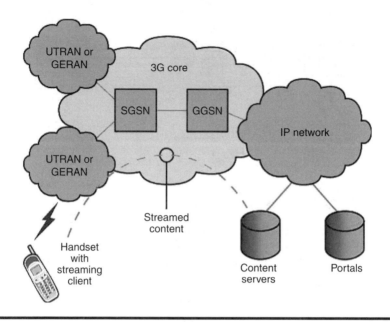

FIGURE 4.39 Packet-switched streaming.

requires a streaming client that is able to receive and decode streamed media, which could be audio or video, or indeed a mixture of media formats. The user would typically launch a browser and search for content, which could be available either live or on demand, and could then navigate to a content server that would be responsible for streaming the content.

The 3GP specification defines a file format that can be used to deliver multiple media elements to a device, and to identify how those media elements are coded and how they should be synchronized with respect to one another. The video coders supported by 3GP include:

- H.263
- MPEG-4/H.264

And the audio codecs supported include:

- AMR, narrow, wide and extended
- Enhanced AAC+
- MPEG-4 AAC

The 3GP file format is used to provide a means to organize and transport multimedia elements between devices and express the relationship between these elements. For the purposes of digital rights management (DRM), the 3GP format is able to convey encrypted content.

Review Questions

Q1. Which of the following components would you *not* expect to find in a mobile phone camera?
   a. Compound lens
   b. Optical sensor
   c. Mechanical shutter element
   d. Image processor

Q2. Which of the following is *not* an image storage format?
   a. VGA
   b. JPEG
   c. BMP
   d. PNG

Q3. When sending and receiving still and video images, which of the following can offer greater efficiency?
   a. Provide more radio bandwidth
   b. Send the data faster
   c. Adopt image compression techniques
   d. Send half as many pictures

Q4. Which of the following MPEG standards was designed to work at sub-64 kbps rates?
   a. MPEG–1
   b. MPEG–2
   c. MPEG–4
   d. MPEG 1 Layer 3

Q5. Which of the following ITU defined codecs is designed for use over LANs and IP networks?
   a. H.261
   b. H.263
   c. H.264
   d. H.324

4.2.6 Serial Interfaces

Handset Interfaces

Most handsets have a number of serial interfaces that permit the phone to connect to other devices, such as personal digital assistants (PDAs) or PCs for data synchronization, or headsets and other cordless technologies. Some of the connectivity

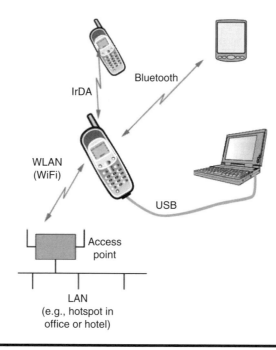

FIGURE 4.40 Mobile terminal interfaces.

options on the handset support connections to alternative networks. An example of this is the provision of wireless LAN (WLAN) connectivity, which allows the handset to connect via an access point (AP) to a LAN system. This connection acts as an alternative to the main cellular mode of operation of the handset and can be used for both data and voice connectivity.

As shown in Figure 4.40, the options for serial connections include:

- Infra-red Data Association (IrDA)
- Bluetooth
- Wireless LAN (WLAN or WiFi)
- Universal Serial Bus (USB)

It is not uncommon for handsets to support two, three, or even all four of these connection options. Other technologies that are coming to market may also require integration into handsets; for example, the personal area network (PAN) technology known as ZigBee offers the same type of connectivity as Bluetooth and IrDA.

Infra-red Data Association (IrDA)

IrDA connectivity was an early solution to the problem of the multiple cables that were required by users to interconnect their devices, in particular the connection

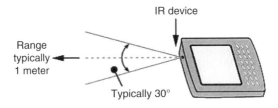

FIGURE 4.41 IrDA operation.

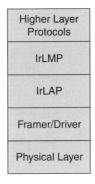

FIGURE 4.42 The IrDA protocol stack.

of a PC to a peripheral such as a printer. The Infra-red Data Association was established to produce standards for this new technology, which was named IrDA.

The IrDA standard is based on non-visible Infra-red directional connections that operate at a wavelength of around 875 nm and is intended to be a low cost solution for connecting devices. As shown in Figure 4.41, the range of IrDA is very limited and typically no more than one meter, although the optimal connection range is usually found between 5 and 60 cm. The IR devices are directional which means the connection is only effective within a cone, which typically has a half-angle of 15 degrees. Therefore, to connect two devices, they must be in close proximity, pointing at one another, and with no obstructions in between.

Although there are several versions of the physical layer in the IrDA standard, the baseline is called the Serial Infrared (SIR), which supports data rates equivalent to those found in RS-232 serial wired ports, that is, 9600 to 115,200 bps. It is the SIR version of IrDA that is implemented in many mobile handsets.

Figure 4.42 shows the IrDA protocol stack. Above the physical layer, the IrDA standard introduces layers that are responsible for data framing, for the discovery and negotiation process, which allows devices to pair, and that acts as an interface to the application protocols that will use the IR link.

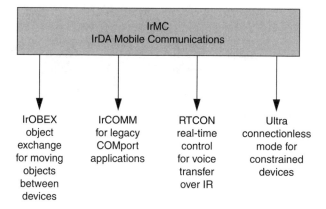

FIGURE 4.43 IrDA for mobile handsets.

IrDA for Mobile Handsets

The higher-layer functionality supported by IrDA, and therefore the applications that can make use of this connectivity, may vary from one handset to another. The Infrared Data Association created an option in the standard known as Infrared for Mobile Communications (IrMC). IrMC contains a number of higher-layer protocols aimed at supporting the types of tasks that would be performed by the users of mobile handsets (Figure 4.43).

IrMC consists of:

- Object Exchange (IrOBEX)
- Serial COM port emulation (IrCOMM)
- Real-time control (RTCON)
- Connectionless mode for constrained devices

Mobile handsets do not have to implement the full version of IrMC, and many include just the IrOBEX and IrCOMM functionality. This allows users to exchange objects such as business cards and contacts via the connection to use IrCOMM for serial port connectivity, perhaps for data synchronization.

It should be noted that IrDA only provides point-to-point connectivity, unlike some of the other connection options on a handset.

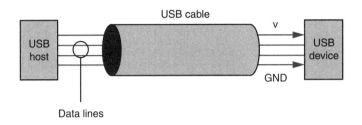

FIGURE 4.44 USB connections.

Universal Serial Bus (USB)

The USB initiative was an attempt by the IT industry to provide a simple, standardized interface that could support many applications and was capable of *plug-and-play* operation. The outcome of this was the definition of the USB connection. There are two major versions of USB (Figure 4.44): version 1.1, which supports data rates of 1.5 Mbps (low-speed) and 12 Mbps (full-speed), was supplemented recently by USB version 2.0, which supports 480 Mbps (high-speed).

USB is electrically and mechanically a very simple interface with data lines and power connections that allows a USB *host* device to provide power to *connected devices*. Although the USB standard has been modified over time to include the possibility of USB ports on mobile phones and similar devices, many handset vendors depart from the USB standard when it comes to the physical connection of USB on the phone.

The standard USB connection is too large for mobile devices; and although a small form-factor version is now in the standard, many handset vendors choose to use a proprietary physical connection for the USB on their handset ranges. This means that end users will require a manufacturer-specific cable to connect their handsets to other USB devices.

Commonly, the USB port found on handsets operates at full-speed (12 Mbps), with the handset acting as a USB *device*; some handsets support version 1.1 whereas more recent handsets support version 2.0.

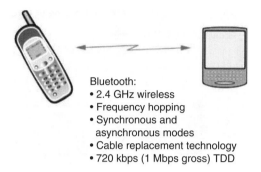

FIGURE 4.45 Bluetooth connections.

Bluetooth

Bluetooth is a radio-based connection option that aims to solve some of the issues addressed by IrDA, in particular the number of different cables that users require to interconnect the multitude of terminals they own.

To overcome some of the limitations of IrDA, Bluetooth operates in the 2.4-GHz, Industrial, Scientific and Medical (ISM) band (Figure 4.45). The advantage of this band is that it is license exempt, which means radio equipment operating in this band can do so without users requiring operating licenses. However, to support the coexistence of many radio applications in the band, it is regulated in terms of usable power levels and spectrum parameters.

Bluetooth was developed by the telecommunications industry, so the initial focus of the standard was to provide a means for mobile handsets to interconnect with associated devices, such as headsets, PDAs, and laptop computers. However, because of its ready availability, Bluetooth is finding its way into other consumer products.

The Bluetooth radio component operates across up to 79 channels in the 2.4-GHz band and, to mitigate interference, frequency hops around these channels at a rate of 1600 hops per second. It should be noted that the full range of Bluetooth channels is not available in all countries because of local regulatory constraints. The power classes defined for Bluetooth devices support typical ranges up to 10 meters, although the class 1 devices at 20 dBm can achieve ranges greater than 100 meters.

- Frequency band: 2.400 to 2.4835 GHz
- 70 channels (in most, but not all, countries)
- Power classes:
    - 1 = 2 dBm (100 mw)
    - 2 = 3 dBm (2.5 mw)
    - 3 = 0 dBm (1 mw)

Bluetooth is able to support both synchronous connection-oriented and asynchronous connectionless modes between devices, with the total gross transmission rate being 1 Mbps, and the net useable rate available to applications being around 720 kbps. Unlike IrDA, Bluetooth is not limited to point-to-point operation and can be used to create networks of devices known as piconets. A Bluetooth piconet can contain up to seven slave devices connected to one master device. The Bluetooth protocols support discovery and negotiation processes that allow the establishment of connections in an ad hoc manner. The specifications also include security functions, which include authentication and encryption.

Bluetooth Profiles

As the Bluetooth specifications were written, it became obvious that there were numerous *real-world* applications for the technology and that it would be unrealistic to include each and every one of these in the standards. Therefore, Bluetooth is based on the concept of a series of defined profiles, where a profile specifies how the Bluetooth protocols should operate to provide a set of functions. The profiles can be viewed as a series of building blocks from which real applications can be constructed. New profiles can be added to the Bluetooth standard after completing an agreement process.

An example of the relationship between a usage model and profiles is the 3-in-1 phone. The 3-in-1 phone has three operational modes: (1) it is able to act as a normal mobile handset and access the cellular network; (2) it can use Bluetooth to access a gateway device attached to a landline (therefore acting as a cordless phone); and (3) it is able to connect directly to other handsets using Bluetooth (therefore acting as an intercom device). This usage model is based on two Bluetooth profiles: (1) the Cordless Telephony Profile (CTP) and (2) the Intercom Profile (IP) (see Table 4.3).

Outside the profiles that support applications, there are two generic Bluetooth profiles, known as the Generic Access Profile (GAP) and the Service Discovery Application Profile (SDAP).

GAP concerns the discovery procedures that allow Bluetooth devices to find one another, and includes functions for establishing a Bluetooth connection and optionally adding security. SDAP is used by one Bluetooth device to discover what services are offered by a remote Bluetooth device.

Wireless LANs (WLANs)

Wireless LAN (WLAN) technology was developed to replace the wired LAN connection, usually Ethernet, which tethers users to their desks. WLANs can be deployed in a corporate environment to provide end users with a degree of freedom about where they use the LAN. The same wireless technology is also being deployed

TABLE 4.3 Bluetooth Profiles

Profile	Profile Name	Comments and Applications
GAP	Generic Access Profile	Covers discovery and link management
SDAP	Service Discovery Application Profile	Allows a device to discover the services registered on other devices
CTP	Cordless Telephony Profile	Supports cordless telephony and a 3-in-1 phone
IP	Intercom Profile	Supports intercom mode in the 3-in-1 phone
SPP	Serial Port Profile	Emulates serial port communications
HS	Headset Profile	Supports cordless connection between handset and handset
OPP	Object Push Profile	Supports the push (and pull) of objects such as business cards
DNP	Dial-up Networking Profile	Allows a mobile device to act as dial-up modem for connected Bluetooth devices

to support public hotspots in places such as hotels, conference centers, shops, and airports. Also, low-cost WLAN components are being deployed by many residential users to provide private WLAN capabilities at home.

Although there are a number of WLAN standards on the market, the dominant series are those produced by the Institute of Electrical and Electronic Engineers (IEEE). The IEEE project 802 oversees standards for all LAN technologies, both wired and wireless, and the working group 802.11 is responsible for WLAN standards. IEEE 802.11 has defined three radio technologies for WLAN, known as 802.11b, 802.11a, and 802.11g, and Draft 802.11n. These differ in terms of the data rates they support and the spectrum band in which they operate. Two of the WLAN standards are designed to operate in the 2.4-GHz ISM band, while the third, 802.11a, operates in bands around 5 GHz.

The preexistence of radar systems at 5 GHz in Europe means that 802.11a systems cannot be deployed in this region without suitable modification. Additional interference mitigation techniques were added by the 802.11h standard, which adapted 802.11a radio for use under European regulations.

The term "WiFi" (wireless fidelity) is often applied to 802.11 systems. WiFi is actually a brand name that belongs to an industry association, the WiFi Alliance, whose role is to test interoperability of WLAN products (Figure 4.46). Any device that carries the WiFi mark will have been tested against a baseline implementation of the standards and has been demonstrated to operate with products from other manufacturers.

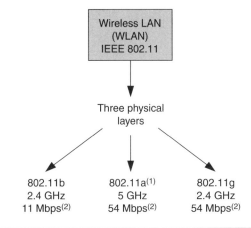

FIGURE 4.46 802.11 (WiFi) standards.

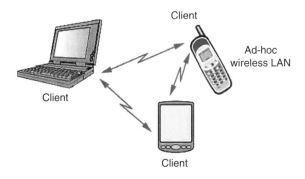

FIGURE 4.47 WLAN ad-hoc mode.

Integration of WLAN radio modules is already evident in the PDA market, and a number of handset manufacturers have announced plans to include 802.11 in their handset products.

WLAN Configuration

The wireless LAN technologies specified by the IEEE will work in one of two modes. The first mode is known as an ad-hoc WLAN (Figure 4.47) and is similar in concept to a Bluetooth piconet. A number of wireless LAN clients that are in close proximity come together in a temporary way to establish a network between them — there will be varying degrees of end-user intervention to create such a network. An ad-hoc WLAN is relatively simple to establish and requires no infrastructure and a minimal amount of administration.

The second mode of operation is known as an infrastructure WLAN, which uses *access points* (APs) to connect clients to a backbone LAN infrastructure

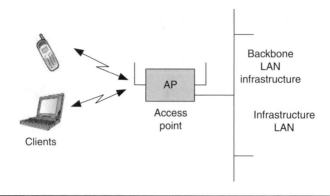

FIGURE 4.48 WLAN infrastructure mode.

(Figure 4.48). It is this configuration that is used in corporate WLAN, hotspots, and residential applications. An infrastructure WLAN requires more administration than an ad hoc WLAN.

IEEE 802 is not only responsible for the radio technology definitions for WLAN: its remit covers other aspects of LAN operation. For example, in response to concerns over WLAN security, which was quite basic in early systems, the IEEE has recently defined 802.11i, which is a suite of security measures for WLANs (it is also known as WiFi Protected Access 2 [WPA2] or the Robust Security Network [RSN]).

The growing use of WLANs to convey real-time traffic such as video and Voice-over-IP has raised concerns about the Quality of Service (QoS) for these services. This has been addressed by IEEE 802.11e, which introduces a range of QoS mechanisms for WLANs.

Another standard that is expected soon is IEEE 802.11n, which is the next generation of radio technology and will take transmission rates to 108 Mbps and perhaps beyond (some vendors already offer pre-802.11n equipment).

Review Questions

Q1. Which of the following is *not* a data interface you would commonly find on a mobile handset?
   a. WiFi
   b. IrDA
   c. RS 232
   d. Bluetooth

Q2. What is the typical data rate for IrDA SIR implementation found in most mobile handsets?
   a. 300–9600 bps
   b. 9600–14.4 kbps
   c. 9600–115.2 kbps
   d. 1–4 Mbps

Q3. Which of the following is *not* a Bluetooth profile supported by mobile devices?
   a. Headset profile
   b. Serial port profile
   c. Public access profile
   d. Object push profile

Q4. The interface characteristics of the IEEE 802.11g or wireless g are _____.
   a. 5 MHz band, 11 Mbps
   b. 2.4 GHz band, 54 Mbps
   c. 5 GHz band, 54 Mbps
   d. 2.4 GHz band, 11 Mbps

Q5. The IEEE standard concerning aspects of security including WPA and TKIP is known as _____.
   a. 802.11e
   b. 802.11f
   c. 802.11g
   d. 802.11i

4.2.7 Section 2 Review Questions

Q1. Complete the following diagram of a typical mobile radio circuit.

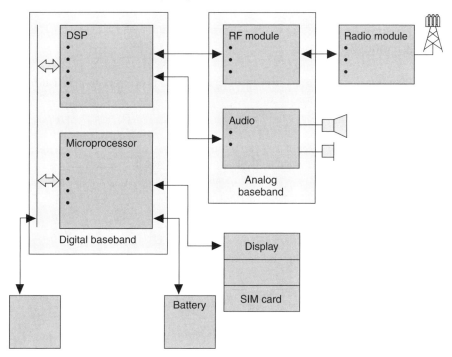

Q2. Mobile equipment contains different types of memory. List these memory types, and state for each its main feature and possible advantage and disadvantage to the mobile manufacturer or user.

Q3. Choose a mobile phone, preferably a feature phone, and identify the following features that it supports.
 a. Make and model of phone:
 b. Radio bands, radio modes supported:
 c. Type of display, resolution, number of supported colors:
 d. Video and image codecs supported:
 e. Audio codecs supported:
 f. Serial interfaces supported (include available data rates):

Q4. Which element within a mobile terminal is responsible for signal processing, voice coding, and message processing?
 a. RF circuitry
 b. Audio circuits
 c. Processors
 d. Memory

Q5. A typical 3G mobile device requires how many instructions per second to operate?
 a. 7 to 10 MIPS
 b. 40 MIPS
 c. 500 MIPS
 d. 1000 MIPS

Q6. Handset manufacturers may include additional processors to perform functions such as multimedia support.
 a. True
 b. False

Q7. The type of memory used to store application code is _____.
 a. SRAM
 b. DRAM
 c. E2PROM
 d. HDD

Q8. Which of the following characteristics *cannot* be attributed to NAND flash memory?
 a. Fast write and erase time
 b. High density and lower cost
 c. Sequential access only
 d. True random access

Q9. Which of the following removable memory media types uses MagicGate technology to protect media content?
   a. MMC
   b. SD
   c. MiniSD
   d. Sony memory stick

Q10. The term "STN" (Super Twisted Nematic) refers to _____.
   a. The layered structure of the display
   b. The nature of the liquid crystal itself
   c. The control signal applied to the display
   d. The application of display backlighting

Q11. A display that supports 16 bits per pixel can display how many colors?
   a. 256
   b. 4096
   c. 65K
   d. 262K

Q12. The statement below describes what type of display?

   " ...the individual pixels are addressed by row and by column signals, one pixel at a time, thus the display is relatively slow."

   a. Passive — TFT
   b. Active — TFT
   c. Passive — STN
   d. Active — TFT

Q13. Which of the following is *not* a characteristic related to OLED technology?
   a. Can be created on thin layers
   b. Low power consumption
   c. No backlight required
   d. Long lifespan

Q14. What type of audio codec is being described below?

   "The file does not actually contain coded music, but rather a set of instructions about the notes to be played."

   a. Monophonic
   b. MIDI
   c. XMF
   d. MP3

382 ■ *Introduction to Mobile Communications*

Q15. The AMR-WB codec has the following characteristics:
   a. Input range 50 to 3400 Hz; output range 1.8 to 12.2 kbps
   b. Input range 50 to 3400 Hz; output range 6.6 to 23.85 kbps
   c. Input range 50 to 7000 Hz; output range 1.8 to 12.2 kbps
   d. Input range 50 to 7000 Hz; output range 6.6 to 23.85 kbps

Q16. Which codec is most likely used for a Bluetooth hands-free kit?
   a. CVSD
   b. MPEG–3
   c. ATRAC 3 Plus
   d. SBC

Q17. Which of the following is *not* an image storage format?
   a. VGA
   b. JPEG
   c. BMP
   d. PNG

Q18. When sending and receiving still and video images, which of the following can offer greater efficiency?
   a. Provide more radio bandwidth
   b. Send the data faster
   c. Adopt image compression techniques
   d. Send half as many pictures

Q19. Which of the following MPEG standards was designed to work at sub 64-kbps rates?
   a. MPEG–1
   b. MPEG–2
   c. MPEG–4
   d. MPEG 1 Layer 3

Q20. Which of the following is not a data interface one would commonly find on a mobile handset?
   a. WiFi
   b. IrDA
   c. PCI
   d. Bluetooth

Q21. What is the typical data rate for IrDA SIR implementation found in most mobile handsets?
   a. 300 to 9600 bps
   b. 9600 to 14.4 kbps
   c. 9600 to 115.2 kbps
   d. 1 to 4 Mbps

Q22. The IEEE standard concerning aspects of security, including WPA and TKIP, is known as _____.
   a. 802.11e
   b. 802.11f
   c. 802.11g
   d. 802.11i

4.3 Mobile Operating Systems

4.3.1 Introduction to Mobile OS

The role of an operating system (OS) within a handset or handheld device is no different than the OS deployed in computing terminals; the major differences in the OS between the two environments are the result of handset constraints.

The OS is responsible for a range of tasks, which include management of the processor, memory, and devices (Figure 4.49). Processor management determines when an application can use the central processor and how to manage the resources when multiple processes have to operate simultaneously. Memory management allocates memory to processes so that they do not overlap and controls the reading and writing of data to memory locations. Additionally, the OS will look after storage of data, perhaps on a card or even a disk, and will also manage devices, or the input/output (I/O) capabilities. The user interface (UI) is considered part of the OS although not all OS include a UI that allows licensees to customize a UI to their own design.

Operating above the OS will in most cases be a series of applications; and to support these, the OS will have an application programming interface (API), which abstracts the functionality of the OS for application developers.

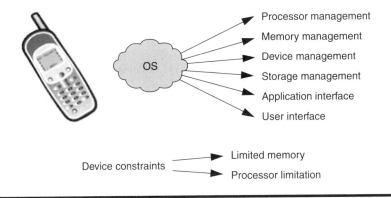

FIGURE 4.49 The OS is responsible for a range of tasks, which include management of the processor, memory, and devices.

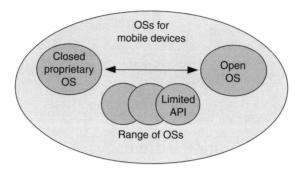

FIGURE 4.50 Types of mobile operating systems.

In the context of a mobile handset, an OS has a set of limitations placed on it that are the result of the processor capabilities and limited memory. Therefore, the OS in these cases needs a very small footprint, which means very efficient code writing. There are broadly two approaches to writing an OS for mobiles. The first is to develop an OS from the ground up, specifically for the mobile environment. The second is to take an OS that is perhaps used in desktop devices and produce a compact version.

One critical area for mobile OS coders is reliability. The end user of a mobile device would not tolerate systems crashes and lockups. This means not only reliability, but also robustness as the underlying connectivity between the device and the network is error-prone.

The range of handset OSs in today's market includes completely closed or proprietary systems through to open platforms (Figure 4.50). There are many variants in between these two ends of the scale, where developers are able to create content for a particular OS without having to know its technical details.

Proprietary Operating Systems

Many handset products have an OS that is proprietary, and in many cases the details of the OS are unavailable, even perhaps to developers. However, the popularity of smart phones, with comprehensive operating systems, and a recognition that content developers can generate revenues for carriers, has meant that even proprietary OS have a degree of *openness* associated with them.

For example, a proprietary OS will often include Java functionality, which offers developers a route to content production through standardized, well-published APIs; meanwhile, the core of the OS that drives the phone functionality remains hidden. Handset vendors that have moved along this path will generally offer developer platforms, software development kits (SDKs) and training, documentation, and support options.

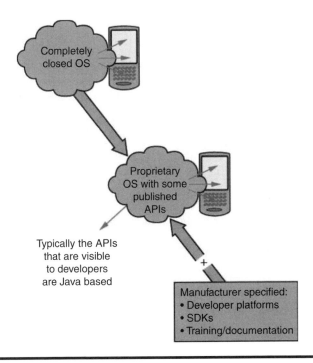

FIGURE 4.51 Proprietary operating system.

In the market for converged devices, or smart phones, which combine the features and functions of both mobile phones and PDAs, there are a number of OS options. A manufacturer could, of course, develop its own OS (Figure 4.51), although this can be a lengthy and costly process, or they could choose to license an OS from a third party (e.g., Symbian, Microsoft CE, Palm OS, or various Linux-based platforms). Although the origins of these OS options may be very different, they generally share a lot of common features, because it is necessary to support common functionality in converged devices.

4.3.2 Example Mobile Operating Systems

Java

Java is an object-oriented programming language that is popular in many environments, including mobile handsets. Java is not an OS, but OSs often include Java functions so that Java applications can run on a phone. In fact, content produced for one Java phone should be able to run on other Java-enabled devices. This is the *write-once, run-anywhere* promise of Java.

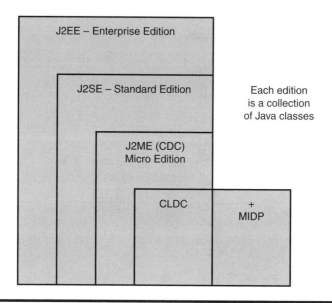

FIGURE 4.52 Java architecture.

As an object-oriented language, Java consists of a series of classes. A class can be thought of as a blueprint for creating instances of objects. The various editions of Java differ by virtue of the classes they contain. For example, the Java 2 Micro Edition (J2ME) is a subset of the classes in the Java 2 Standard Edition (J2SE) (Figure 4.52).

J2ME is of particular interest for mobile and handheld devices with two defined configurations: (1) the Connected Device Configuration (CDC), which is suited to devices such as PDA; and (2) the Connected Limited Device Configuration (CLDC) for handsets, although with convergence between PDAs and handset functions, it is likely that CDC will be implemented eventually in handsets.

CLDC is a very small subset of Java classes and does not even include UI functions; therefore, when it is used in mobile handsets, it is supplemented by additional classes that are part of the Mobile Information Device Profile (MIDP).

The range of Java-based content available for handsets includes games, news and other content, interactive applications such as mapping services, and screensavers.

The Java Mobile Information Device Profile (MIDP)

The Java CLDC includes a limited set of classes from three Java families: (1) Java.io, (2) Java.util, and (3) Java.lang. These classes support limited I/O functionality, basic types (integer, string, etc.), and a few utilities (e.g., calendar), respectively. The CLDC does not include any UI functions and is therefore augmented by adding classes with the Mobile Information Device Profile (MIDP) (Figure 4.53).

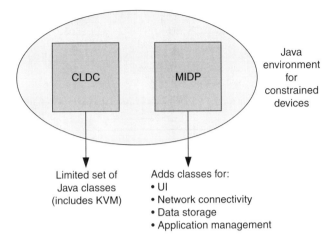

FIGURE 4.53 The Mobile Information Device Profile (MIDP). MIDP 1.0 has limitations (e.g., has no support for audio). MIDP 2.0 adds gaming and multimedia functions: multimedia playback, two-dimensional game-oriented API, and security features.

However, the CLDC does contain a small-footprint virtual machine, known as the KVM. MIDP includes classes for UI, network connectivity, data storage, and application management. There are two significant versions of MIDP found in handsets: version 1.0 and, more recently, version 2.0. MIDP 2.0 adds functionality aimed at the support of gaming and multimedia applications.

Microsoft Windows

Microsoft, responsible for the Windows family of operating systems for desktop and server applications, has also produced a number of OS versions specifically for handheld devices, handsets, and smart phones. At the heart of all of these products is Windows CE (Consumer Edition or Compact Edition). Windows CE was designed for the embedded OS market, but over time, like all operating systems, it has evolved and has been used as the base for a number of products such as Handheld PC, Windows Mobile, and Windows Smart phone (Figure 4.54).

The Handheld PC software was aimed at PDA-like devices and competes with other operating systems such as the Palm OS. Handheld PC has evolved into the Pocket PC product.

Windows Mobile, again based on Windows CE, is targeted at a range of devices such as pocket PCs, handsets, and tablet devices. Another OS, Windows Smart phone, is designed specifically with the smart phone market in mind, and there are a number of commercially available smart phone devices that are based on this OS. In this market, the main competitors are the Symbian OS and operating systems based on Linux.

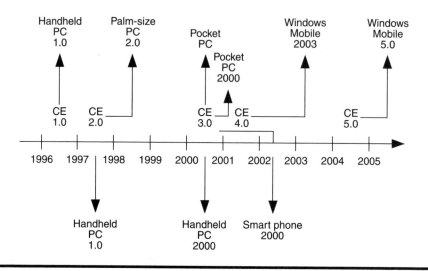

FIGURE 4.54 Windows Mobile OS evolution.

As with other systems, Microsoft is keen to encourage developers to create content for Windows-based phones, and offer a range of services such as software development kits (SDKs) and emulators to developers of compatible content.

With the market for smart phones predicted to grow significantly in the near future, there will be intense competition among the OS providers for these devices.

Windows Mobile 5.0

Launched in early 2005, Windows Mobile 5.0 is representative of the features expected in an OS for mobile devices (Figure 4.55). Initially code-named Magneto, Windows Mobile 5.0 requires 32 MB of ROM and 64 MB of RAM, and can operate on processors operating at speeds of 200 MHz or above. The software is based on CE version 5.1 and introduces a group of new features, including:

- Persistent memory storage (contents saved even when the battery drains)
- Support for high-bandwidth networks (including 3G and WiFi)
- Single-handed navigation and operation
- Integrated Windows Media Player 10
- Support of new technologies (PoC and video calling).

Windows Mobile is able to support the mobile versions of popular Microsoft software such as Outlook Mobile, Office Mobile, and Internet Explorer. The OS interfaces to the phone hardware and offers speed dial and call history features. Through the graphical user interface (GUI), a user is able to access call functions and control the Bluetooth and WiFi connectivity. The first device to use Windows Mobile 5.0 was launched in late 2005.

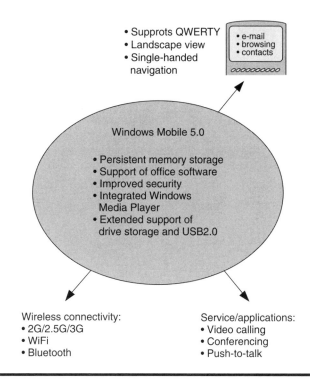

FIGURE 4.55   Windows Mobile 5.0.

Symbian OS

Symbian was created as a company in 1998 by Ericsson, Nokia, Motorola, and Psion with the intention of building an operating system for mobile devices that would be based on the Psion EPOC platform and would be freely available to handset manufacturers through licensing agreements. A total of more than 40 phones from eight manufacturers are presently available with the Symbian OS, and in 2004 a total of 25 million Symbian devices were shipped.

The structure of this OS includes basic telephony, security, communications, and networking modules on a kernel, or base (Figure 4.56). An application framework supports applications and provides the functionality to support the UI, although Symbian does not include its own UI (Figure 4.57). The messaging capabilities of Symbian include SMS, EMS, and MMS based on 3GPP network standards, and also include IETF-based e-mail capability.

Symbian's core applications engines support the granular requirements of applications such as contacts, agendas, help functions, and spreadsheet and chart handling. In addition, a Web engine, which is based on the Opera browser, provides HTML and XML support and is also able to browse local files. Designed for mobile handsets from the outset, Symbian is able to fit into the memory of a mobile and

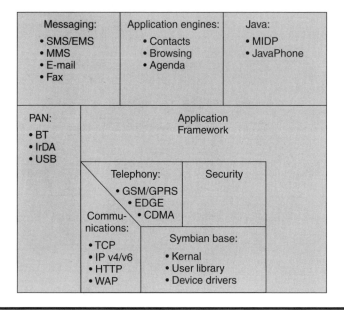

FIGURE 4.56 Symbian OS v7.0 architecture.

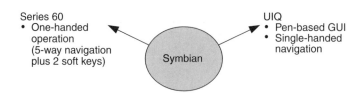

FIGURE 4.57 Symbian user interfaces (UIs).

can be supported on the processor architectures that are commonly encountered in these devices.

Symbian Versions

As with all operating systems, Symbian has evolved (Figure 4.58) to keep pace with the changing standards that define mobile networks and also to track the improvements in handset capabilities — for example, more powerful processors and greater memory availability.

The fist significant release of Symbian was release 5 in 1999, which was followed by release version 6 in 2000 and 7.1 in 2001. Release 5 was aimed at a single reference design, whereas release version 6 appeared in a number of different products. The evolution has continued, with the most recent release v9.1 launching in early 2005.

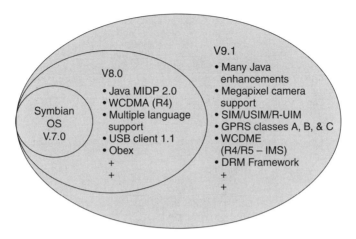

FIGURE 4.58   The evolution of Symbian.

Symbian OS v9 enhancements included:

- *Improved support of multimedia devices* (e.g., Bluetooth stereo, multi-megapixel camera support, three-dimensional graphics, and graphics acceleration).
- *Enhanced device management:* allows remote access to the Symbian device over the air so that new features can be provisioned or problems diagnosed.
- *Enterprise support:* integration of PIM applications based on either Lotus or Microsoft systems and improved Java capabilities targeted at mobile devices.
- *Improved security:* aimed at m-commerce and DRM-based services and also to protect sensitive information such as contacts from access by malware.
- *Performance:* Symbian is able to run on the latest and fastest processor architectures. Also, the structure of Symbian, with a hard real-time kernel, allows a single processor to be used, rather than the combination of main and multimedia processors found in many devices.
- *Reduced costs:* several reference designs are available from chip manufacturers that pre-integrate the Symbian OS. This can reduce time to market and the attendant costs.

Palm

Palm, Inc. (formerly palmOne, Inc.) was formed in 1992 and focused on the production of handheld computing devices and the operating software required by these devices. The first products, known as *Pilots,* launched in the mid-1990s and were based on the first version of the Palm OS. The OS continued to develop, adding functionality that supported color screens, new processor architectures, and telephony functionality.

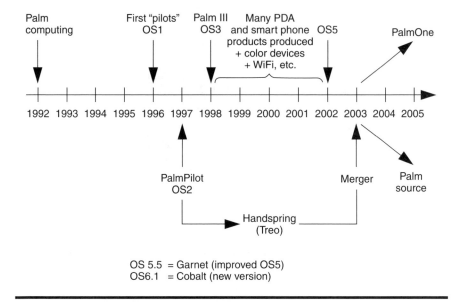

FIGURE 4.59 The evolution of Palm OS.

Palm manufactured the handheld devices and developed the software, although other companies were encouraged to license the software for their products. Support of telephony meant that the Palm OS could be used in converged products, and a wide range of PDAs and smart phones have appeared on the market powered by the Palm OS. Also, a significant developer community has been established, writing very diverse applications for Palm OS devices.

In 1998, some of the original founders of Palm left and set up Handspring to produce PDA devices based on the licensed Palm OS. In 2003, Palm and Handspring merged and, from the merger, two companies were created: PalmOne (now known as Palm), which produces hardware, and Palm Source, which is responsible for the Palm OS. Palm OS is still available under license to other manufacturers.

Figure 4.59 shows the evolution of the Palm OS. The most recent versions of the Palm OS are known as Garnet (OS5.5) and Cobalt (OS6). Garnet is seen as a *low-end* product while Cobalt is targeted at high-end, multimedia converged devices.

For many years, Palm hardware and software had a clear lead in the market, but increased competition means that other hardware manufacturers such as HP and software companies such as Microsoft are gaining market share.

Research in Motion (RIM)

To many, the term "BlackBerry" relates to the handheld device that has proved very popular with business users. However, there are actually four components in the BlackBerry system: (1) the wireless handheld terminal, (2) an enterprise server, (3) a browser, and (4) a development environment (Figure 4.60).

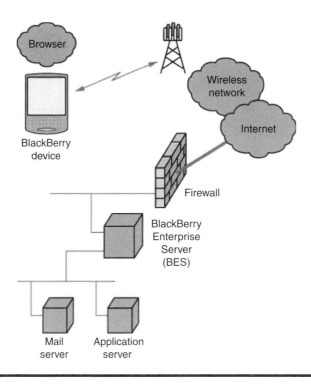

FIGURE 4.60   RIM BlackBerry architecture.

Initially driven by e-mail applications, the BlackBerry Enterprise Server (BES) can integrate with Lotus Domino and Microsoft Exchange servers, allowing users to have secure remote access to corporate e-mail systems. New e-mail messages are automatically sent to a user's handheld terminal. The BlackBerry system is able to support more than just e-mail. Because it is based on HTTP and XML standards, it can be integrated with a whole range of office systems supporting field-force automation across many industry sectors.

BlackBerry will operate over any of the major cellular network technologies, as evidenced by the fact that it has been employed in TDMA, CDMA, and GSM networks.

The BES provides a gateway between the mobile user and the corporate intranet, and integrates to the back-end business servers (see Figure 4.60). The link to the handheld device is optimized for wireless networks using efficient data coding schemes and is secured using industry-standard encryption mechanisms. The data coding, or transcoding, at the BES can be customized so that all data is delivered in the most efficient format for the handhelds. Delivery is over standard HTTP and TCP/IP protocols, which open up the system to a wide range of data types.

The BlackBerry system makes use of the wireless network's always-on capability (for example, using GPRS), which allows the support of push applications where data can be delivered to a mobile without the user having to initiate a session.

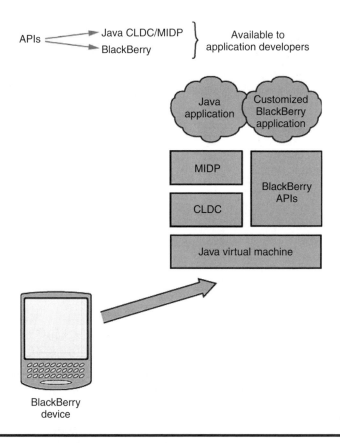

FIGURE 4.61 RIM application development.

The browser application on the handheld is able to render content in any of the standard Web formats, such as HTML, but also supports wireless-optimized formats such as Wireless Markup Language (WML) and compact-HTML (c-HTML).

Developing BlackBerry Applications

The fourth component of the BlackBerry system is the development environment, which allows developers to create custom applications for BlackBerry devices. There are two sets of APIs that are available to developers: (1) a standard set of Java APIs (CLDC/MIDP) and (2) the BlackBerry-specific APIs (Figure 4.61).

RIM (Figure 4.61) supports a BlackBerry Java Development Environment (JDE), which is available to content developers. This includes tools for editing and debugging, handheld simulation, APIs, and example applications.

The system hides the details of the wireless network from developers who can use standard HTTP and TCP/IP connectivity for their applications. Examples of customized applications include database access and sales order tracking.

As an example of system capabilities, a user could use a handheld to browse for content. However, using the push model, content could be pushed to the browser, pushed to the device cache, or pushed using a *channel*. In a third option, the user will have a specific icon on the device behind which is the latest version of the pushed channel information. This could be used to provide up-to-date prices of commodities or updated news stories.

As RIM has gained significant market share with the BlackBerry system, other manufacturers have looked to support the functionality. In 2003, Palm announced that a joint project with RIM would allow Palm OS devices to connect to the BES. The Palm OS devices would also support the BlackBerry browser function.

Linux-Based Handset Operating Systems

Mirroring developments in desktop OS, a number of companies are now producing versions of Linux (Figure 4.62) that target consumer devices, and handsets in particular. The organization Tux Mobile keeps track of companies that have produced mobile variants of the Linux operating system. Motorola, NEC, and Panasonic have released a number of phones to market based on Linux.

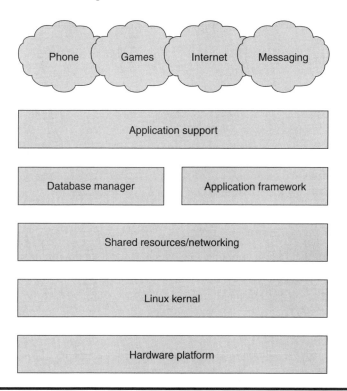

FIGURE 4.62 The Linux operating system.

In addition to Linux-based OS, there are also several operating systems that are based completely on Java, and handset manufacturers have stated that they are scheduling releases of these products.

Review Questions

Q1. Which of the following is *not* a function of a typical operating system?
   a. Processor management
   b. Memory management
   c. Signal processing
   d. User interface

Q2. From a user perspective, which is the most important aspect of the OS?
   a. Efficient code
   b. OS stability
   c. Open OS platforms
   d. Support of MP3

Q3. In today's mobile market, the open OS is an important tool to support the convergent services increasingly seen on smart phone-style devices.
   a. True
   b. False

Q4. Java is an operating system.
   a. True
   b. False

Q5. Which of the following Java Editions is most suited to the mobile processing environment?
   a. J2ME
   b. J2SE
   c. J2EE

Q6. The handheld PC is supported by which of the following mobile OS?
   a. Symbian
   b. Palm
   c. Windows CE
   d. Linux

Q7. Which of the following is *not* one of the four elements of the BlackBerry system?
   a. Enterprise server
   b. Handheld device

c. Development environment
   d. Application software

Q8. The Tux Mobile group tracks the use of which of the following OS?
   a. Symbian
   b. Palm
   c. Windows CE
   d. Linux

4.3.3 Section 3 Review Questions

Q1. Choose one of the four mobile OS (Symbian, Microsoft CE, Linux, or Palm OS) and answer the following questions.
   a. What handsets support your chosen OS? Find at least three handsets: a low-feature, feature, and smart phone device?
   b. What are the specific OS features supported on the smart phone device?
   c. How does the chosen OS differ from the other systems?

Q2. What Java applications are available for your chosen OS and mobile device?

Q3. What are the specific conditions required for Java to run on your OS (i.e., compatibility, versions, virtual machines, etc.)?

Q5. Which of the following is *not* a function of a typical operating system?
   a. Processor management
   b. Memory management
   c. Signal processing
   d. User interface

Q6. From a user perspective, which is the most important aspect of the OS?
   a. Efficient code
   b. OS stability
   c. Open OS platforms
   d. Support of MP3

Q7. In today's mobile market, the open OS is an important tool to support the convergent services increasingly seen on smart phone style devices.
   a. True
   b. False

Q8. Java is an operating system.
   a. True
   b. False

Q9. Which of the following Java Editions is most suited to the mobile processing environment?
   a. J2ME
   b. J2SE
   c. J2EE

Q10. The handheld PC is supported by which of the following mobile OS?
   a. Symbian
   b. Palm
   c. Windows CE
   d. Linux

Q11. Which of the following is *not* one of the four elements of the BlackBerry System?
   a. Enterprise Server
   b. Handheld device
   c. Development environment
   d. Application software

Q12. The Tux Mobile group track the use of which of the following OS?
   a. Symbian
   b. Palm
   c. Windows CE
   d. Linux

4.4 Services and Security

4.4.1 Mobile Handset Services

Given the range of mobile device capability in today's market, there are many types of service that may be available to the user. Some of these services depend on handset capability; others rely on the services supported by the network.

Until recently, many services relied on proprietary techniques in the handset or network, which means that services are often limited to just a certain handset model or are only available with a particular mobile operator. The process of defining services by organizations such as 3GPP has led to the standardization of many services, or the development of standard environments where services can be developed, deployed, and executed.

Described below are some of the services that can be seen on handsets that are available today. All of these services have benefited from the process of standardization.

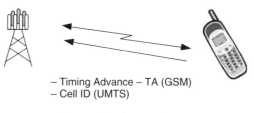

FIGURE 4.63 Cell-based location.

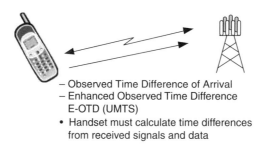

FIGURE 4.64 Network-based location system.

Support for Location Technologies

There are a number of techniques for providing location information for handsets. Some of these require involvement of the handset in the location process; others use data that is already collected by the handset during normal modes and operation and do not place additional processing requirements on the handset.

The least accurate location technique uses either Timing Advance (TA) in GSM or Cell Identity (Cell ID) in UMTS. When a mobile connects to the network (in dedicated mode), the TA or Cell ID is known and can be reported to a location server and used to deliver some form of location service to the user. More accurate location information can be obtained using a triangulation process, wherein the handset measures information from multiple base stations and uses this, along with data obtained from the network, to calculate the true time differences of the signals it has received (Figure 4.63).

In GSM, this technique is known as Observed Time Difference of Arrival (OTDOA) (Figure 4.64). In UMTS, the equivalent process is Enhanced-Observed Time Difference (E-OTD). In both cases, the handset measures the time difference of signals between base stations; it also receives information about the real-time difference (either sent by broadcast messages or to one specific mobile) and it calculates from these a geometric time difference. This geometric time difference is the result of location and can be reported to the network to deliver location services. This calculation requires some additional processing in the handset.

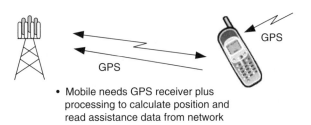

FIGURE 4.65 Assisted GPS location technique.

The final location mechanism is potentially the most accurate, and involves integrating a GPS receiver in the handset (Figure 4.65). When required, the handset can take GPS measurements and report these to the location servers in the network. To speed the satellite acquisition process, the network can broadcast assistance data, which tells handsets in a particular area which satellites are visible. GPS-based location has the largest impact on the mobile because of the GPS receiver and the processing that is required to resolve position information. It is possible to combine location techniques; thus, for example, when GPS fails to give a fix because the handset is inside a building, the location information could be obtained from the time-difference techniques.

Mobile TV Reception

The multimedia capabilities of handsets have raised the possibility of delivering television to mobile or handheld devices, and indeed a number of services have been launched that allow users to download short video clips and even video ringtones onto their phones.

However, these services are using the bearers or channels in the 2.5G or 3G network, which in many cases may not have the bandwidth or the format to support high-quality video transmission. There is also an impact on all other services. Because video has a relatively high bandwidth, it will limit the capacity for all the other services delivered by the network.

A possible solution is the use of a technology known as Digital Video Broadcast-Handheld (DVB-H). DVB-H is a derivative of the main DVB Terrestrial (DVB-T) format, which includes a number of features targeted at delivering content to mobile devices (Figure 4.66).

To minimize power consumption in a DVB-H receiver, the information sent to the handheld device is time-sliced. That is, it is delivered in concentrated bursts so the receiver is not switched on all the time. This reduces power consumption by up to 95 percent compared to DVB-T. In addition, the DVB-H standard includes a number of features aimed at supporting user mobility, such as seamless handovers between DVB transmitters. The signal format is able to accommodate users moving at several hundred kilometers per hour.

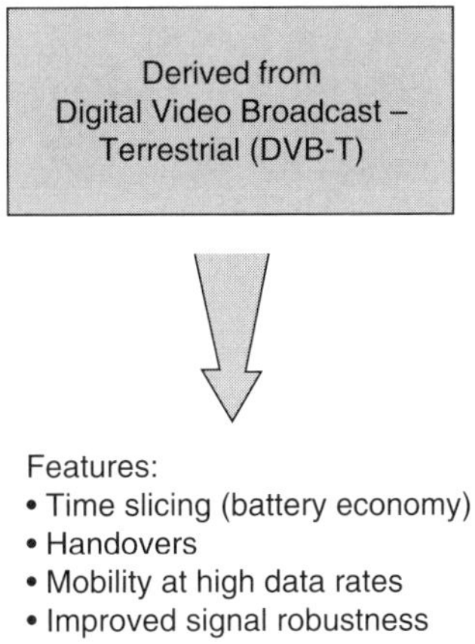

Derived from
Digital Video Broadcast –
Terrestrial (DVB-T)

Features:
- Time slicing (battery economy)
- Handovers
- Mobility at high data rates
- Improved signal robustness

FIGURE 4.66 DVB Terrestrial (DVB-T) format.

Some other changes have been made to the DVB standards to improve the robustness of the signals, compensating for the variances in signal witnessed by mobile receivers. This is provided by adding additional *forward error correction* to the DVB signal.

The DVB-H content can be conveyed using IP-based protocols, which means that the system can be built around standard components that are already being used to store and manipulate content. The DVB-H system can therefore be used to deliver audio, video, and any other IP-based content.

DVB-H and UMTS

Studies have been conducted in relation to the roll-out of DVB-H and its possible integration with cellular networks. DVB-H is backward-compatible with DVB-T, and the two formats can be mixed in the same digital multiplex. While one multiplex can convey six to eight DVB-T channels, it can be used for up to fifty DVB-H channels.

The DVB technology can be kept separate and used to deliver content to DVB-H-enabled terminals, or a cellular technology, such as UMTS, could be used to provide a reverse channel that allows users to browse for content and, when content selection is made, have that content delivered over a DVB-T channel (Figure 4.67). This would mean that some degree of cooperation would be required between the UMTS network operator and a broadcaster. The convergence of telecommunications and entertainment services means that such cooperation is very likely.

Short and Enhanced Message Services

The architectures for Short Message Service (SMS) and Enhanced Message Service (EMS) are identical because EMS is really an extension of SMS (Figure 4.68).

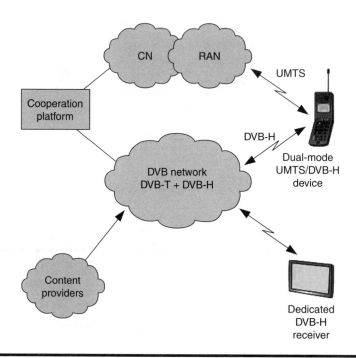

FIGURE 4.67   DVB-H and UMTS.

SMS is a store-and-forward mechanism, which for an SMS from one mobile to another is really a two-part procedure. The first part is SMS-Mobile Originated (SMS-MO) between the originating mobile and the Short Message Service Center (SMSC). The SMSC will then attempt to deliver this message to the terminating mobile. The second procedure is SMS-Mobile Terminated (SMS-MT).

It is important to note that the SMSC that receives and then delivers the message is in the originator's network. There is no SMSC involved in the terminating network. However, as with an incoming call, a gateway function is used in the terminating network to locate the subscriber for message delivery. The store-and-forward will occur if the terminating mobile is not available to receive the message. It will be stored at the SMSC, which will attempt delivery later. It is possible by means of flags in the location registers to indicate that a mobile has a message waiting so that when it next connects to the network, delivery can be organized.

The SMS allows for the transfer of text messages of up to 140 octets. EMS adds the ability to apply simple text formatting, such as underlining, bold, italics, and text alignment, and it is possible to include simple graphics, cartoons, and tunes within the message body.

EMS never really became very popular because support was not complete throughout the industry and the much more advanced Multimedia Messaging Service (MMS) was being defined. SMS and EMS are typically transferred over signaling channels in GSM networks but can be conveyed on packet channels when appropriate.

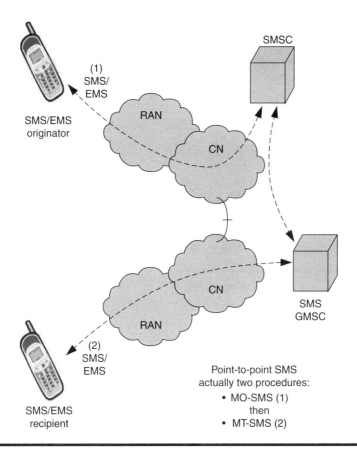

FIGURE 4.68 SMS/EMS message transfer.

Multimedia Messaging Service (MMS)

The MMS allows a user to send complex messages containing numerous media components to other MMS-equipped phones or to other terminal devices such as computers and PDAs (Figure 4.69). If a terminal does not support MMS, then it is possible to send an SMS that contains a URL where the MMS content can be viewed. The major difference between the SMS and MMS architectures is that with MMS between networks, a Multimedia Message Service Center (MMSC) is required in both the originator's network and the termination network.

MMS can be conveyed over a number of bearer types but is best suited to packet mode channels, and it is usual to employ the Wireless Access Protocol (WAP) to support MMS.

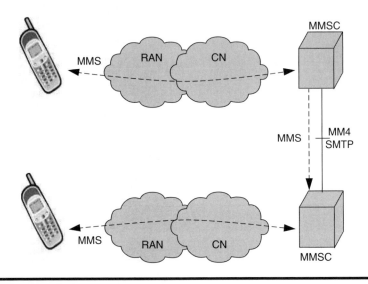

FIGURE 4.69 MMS message exchange.

Unified Messaging

Unified messaging systems are systems capable of adapting messages of various formats and media types into formats chosen by the user (Figure 4.70). This allows for a greater degree of freedom and choice for the user. Users may, for example, choose to receive a fax or e-mail in a spoken form; text-to-voice technology is used to support this type of service.

Unified messaging platforms or devices normally reside in the network and may not require any specific application or standard interaction to be defined at the mobile end. However, with mobile devices supporting an increasing number of messaging technologies, it is possible for platform developers to take advantage of these delivery techniques. Fax-to-text and MMS-to-e-mail are other examples of how this technology can be deployed.

Instant Messaging (IM)

Instant messaging (IM) applications on the Internet have been experiencing rising popularity over the past few years and have been touted as the next big thing or killer application for use within wireless networks (Figure 4.71).

IM is a point-to-point or point-to-multipoint exchange of text information in real-time; that is, it does not use store-and-forward techniques like the SMS. IM normally requires users to log in before they begin to use the service, which gives them a greater degree of control over when they will receive messages and who is

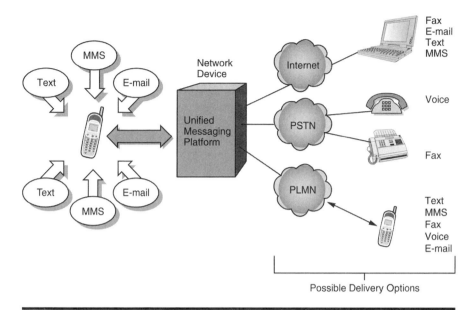

FIGURE 4.70 Unified messaging.

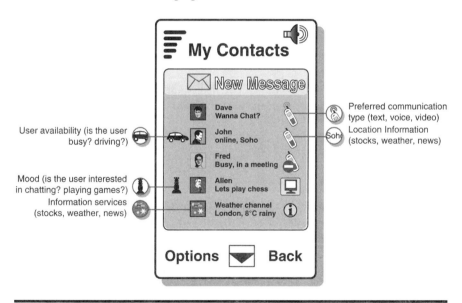

FIGURE 4.71 Wireless IM features.

able to send them messages. IM also brings the concept of chat rooms, buddy lists, and user presence to wireless networks.

The implementation of IM over wireless networks requires the support of software in the phone device and platforms in the network to manage the presence information, buddy lists, etc. There are many solutions on the market that can offer these services but they generally suffer from a lack of compatibility with each other. This is a common issue with IM software used on the Internet.

In the future, mechanisms will be standardized through common architectures and protocols. Release 5 of the UMTS specifications describes an IP Multimedia Sub-system (IMS) that makes use of the Internet Engineering Task Force (IETF)-defined Session Initiation Protocol (SIP). IM is one of the possible applications that may be supported by this.

Press-to-Talk

Press-to-Talk (PTT) service is a real-time, point-to-point and point-to-multipoint, voice-based instant messaging application. This has proved a popular service in the United States on the Nextel network, based on Motorola's iDEN architecture. Many of the features are the same as text-based IM, with buddy lists and chat rooms, etc. Once again, the biggest potential problem affecting the widespread adoption of the service will be compatibility between handsets and across networks.

There are, at present, four PTT solutions on the market.

1. *Motorola iDEN.* This is a well-established system that has a proven track record in the United States with good performance. Motorola also supplies a wide range of handsets to support the service. However, only Motorola manufactures handsets for this service. iDEN does not support a presence service.
2. *Kodiak.* This system uses circuit-switched connections, which results in good performance but misses out on the efficiency of packet-based delivery. The system is also technology-agnostic. It has been deployed on GSM, CDMA, and analog networks. However, there is a limited range of handsets available for this service. Kodiak has a thin-client available to allow vendors to implement the service.
3. *Qualcom Qchat.* This service is proprietary to Qualcom and is only supported on cdmaOne 1X only. In this case, the performance of the 1X network may not be sufficient to ensure adequate performance. The QChat service is implemented on Qualcomm's BREW platform.
4. *PTT over Cellular (PoC).* The final method is a proposed standard mechanism from Ericsson, Nokia, Motorola, and Siemens for a PTT over Cellular (PoC) system. This has been presented to the Open Mobile Alliance (OMA), which is developing this proposal as a standard in line with the 3G UMTS and IMS specification from the 3GPP. It is based, generally, on existing protocols and methods: IP and SIP. This system has the advantage that it will run over any network. There will be products available in GSM and GPRS, UMTS, and cdma2000. It is anticipated that it will be possible to interoperate a PTT service across network boundaries.

Review Questions

Q1. Which of the following is not a location-based service proposed for 2.5G and 3G networks?
   a. Cell ID based
   b. Assisted GPS
   c. Cellular triangulation
   d. Observed Time Difference of Arrival (OTDOA)

Q2. One potential problem of GPS systems is _____.
   a. Poor accuracy
   b. Does not work inside buildings
   c. Poor battery life
   d. Expensive

Q3. DVB-H can achieve up to 95 percent power saving over DVB-T using which method?
   a. Seamless handover
   b. Time slicing
   c. Channel coding
   d. UMTS return channel

Q4. SMS is a real-time delivery service for short text messages on a point-to-point basis.
   a. True
   b. False

Q5. Which of one the following PTT methods was chosen by the 3GPP for use in UMTS?
   a. QChat
   b. OMA PoC
   c. iDEN
   d. Kodiak

4.4.2 Security Issues for Mobile Handsets

Security Issues for Mobile Handsets

The wireless connectivity and mobility of subscribers in a mobile network introduce a range of security issues that are not present in fixed networks (Figure 4.72). The major vulnerabilities include authentication and authorization, that is, making sure that subscribers are valid and only access services they are entitled to, such as interception and eavesdropping of radio signals and integrity of information and identities.

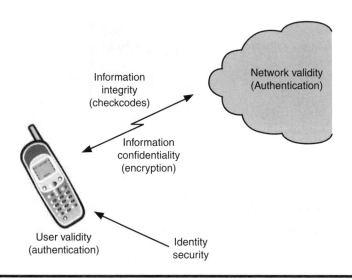

FIGURE 4.72 Aspects of mobile security.

Most wireless standards include a number of features aimed at providing security measures, and these typically include authentication mechanisms, encryption, and identity security. In the recent 3G standards, weaknesses that existed in 2G security have been acknowledged, and aspects such as mutual authentication and message integrity have been added. In addition, the trend has been to move away from secret nonpublished algorithms to mechanisms that are published and open to scrutiny and checking by the cryptographic community. Other major security issues relate to denial-of-service (DoS) attacks and fraudulent use of services (e.g., roaming fraud).

Handset Validation

In systems such as GSM and UMTS, where there is a distinct separation between the mobile device and the user's subscription, which is represented by the SIM or the USIM card, it is normal to provide a means of validating the handset hardware. For this purpose, a GSM or UMTS handset contains a unique identity known as the International Mobile Equipment Identity (IMEI) (Figure 4.73). The IMEI has a structure that identifies the:

- Manufacturer
- Country of manufacture
- Serial number
- (Optionally) the software version

Handset, Services, Media, and Content Distribution ■ 409

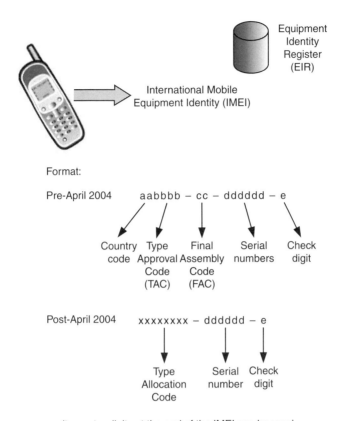

FIGURE 4.73 Mobile hardware identity.

The IMEI can be requested by the network using a request message, to which the handset will respond with a message that contains the IMEI. The intention is that the IMEI can then be checked against a database of IMEI numbers to check that this mobile is type-approved and that it has not been reported as stolen. The database is held at the Equipment Identity Register (EIR).

While the concept of IMEI checking is well established, the practical implementation is more of a problem because the database must be able to check mobiles that belong to this network, but also any roamers (from any other network in the world). Therefore, maintaining the contents of the EIR in an up-to-date state is very difficult.

The GSM Association does hold a Central Equipment Identity Register (CEIR) in Dublin, Ireland, which can be accessed by networks. The CEIR holds two lists of IMEI: (1) the so-called whitelist of approved handsets and (2) a so-called blacklist of mobiles that have been reported stolen or are otherwise unsuitable for use.

410 ■ *Introduction to Mobile Communications*

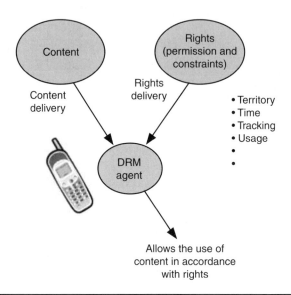

FIGURE 4.74 Digital rights management (DRM).

Cooperating networks download their own blacklists to the CEIR on a daily basis, and this data can then be used to update the national EIR of other networks.

In other cellular technologies, the role played by the IMEI is provided by means of an *electronic security number* (ESN), which is preprogrammed into handsets. As with the IMEI, the ESN can be checked against databases for validity.

Digital Rights Management (DRM)

The use of mobile and portable devices to download and use content raises the issues of rights management. The originator of the content, whether music, images, movies or games, will have spent a lot of time and money developing that content and will want to exert some control over the future use of that content. This is where DRM systems can be applied to limit the use, and therefore prevent misuse of content (Figure 4.74).

There are a large number of proprietary DRM systems at present, which reflects the growing availability of digital content over the Internet and other networks, and is indicative of the lack of standards in this area. The Open Mobile Alliance (OMA) has worked on enablers for DRM systems aimed at mobile handsets and other portable devices, and many handset manufacturers now support OMA DRM but may also support proprietary techniques as the market dictates. DRM systems are built around a trusted entity known as a DRM agent (or client), which in the case of a mobile network resides in the handset.

The DRM client is able to download content and the rights to use that content. The content and rights can be kept separate or can be delivered in a combined format. Rights are a mixture of permissions and constraints that indicate what can be done with the associated content.

For example, content may be valid for 30 days, or until the end of the month; the number of *plays* may be limited; etc. It is the responsibility of the DRM agent to apply the rights to the content and to track its usage, so that if, for example, it is a song that can only be played 30 times, then the content cannot be used unless new rights are obtained.

OMA release 1 DRM included a number of basic features, which were forward lock (preventing content forwarding), combined delivery, and separate delivery. Separate delivery, where the content and rights are kept separate, supports something known as super-distribution, whereby users are able to distribute content at will, but not the rights for that content. This can be very useful where the content has some sort of *preview* mode that can be used to encourage recipients to acquire their own rights to that content.

Open Mobile Alliance (OMA) Digital Rights Management

After release 1 of the DRM enabler, OMA followed up with release 2, which added new functionality to the DRM model, including security and charging mechanisms. The architecture for OMA DRM includes a Content Issuer (CI), a Rights Issuer (RI), and the DRM Agent (Figure 4.75).

Content is formatted in what is known as DRM Content Format (DCF), which consists of a non-encrypted header and a payload that is encrypted or protected content. The header will contain metadata that describes the content and includes the Uniform Resource Indicator (URI) for the RI — so the agent knows where to acquire the rights for this content.

The RI is responsible for generating a *rights object* (RO) for agents on request (and after payment if appropriate); the RO contains the rights in terms of permissions and constraints and also includes the *content encryption key* (CEK), which can be used to decrypt the protected content. The RO is usually encrypted using a *rights encryption key* (REK) and is digitally signed using certificates; this process is used to cryptographically bind the rights object to a particular agent.

This prevents an RO acquired from one agent from having any validity to another agent. This DRM model allows users to store content on internal and external storage elements and also to transfer content to other devices deemed to belong to a trusted domain. Additionally, protected content can be forwarded to other agents by means of super-distribution; those agents can then use the header data to acquire the rights for that content.

Digital Rights Management Models

There are numerous proprietary and open DRM standards in existence. These are driven by concerns over activities such as file sharing and also as a result of the

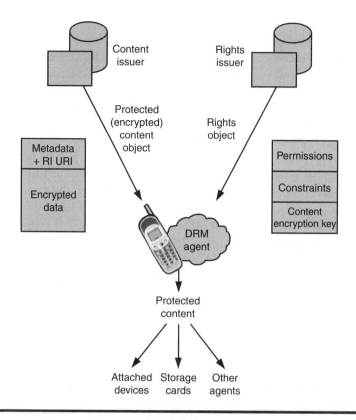

FIGURE 4.75 OMA architecture for DRM.

variety of content coding formats employed in the market. Although many handset manufacturers are deploying OMA DRM, or maybe have indicated that they will, handsets will in some cases also support other DRM models.

OMA DRM supports content that is *pushed* to a device or *pulled* to a device, for example, after browsing; it also supports streamed content. The core protocol used to manage rights is the Rights Object Acquisition Protocol (ROAP). The standard includes a Rights Expression Language (REL), which is a standardized way of expressing rights. Microsoft is an example of a company that has produced a proprietary DRM technology, which is part of Windows Media. This is specifically aimed at mobile devices and as such does not require significant amounts of memory or processing power.

Although similar in concept to OMA DRM, Windows Media does support a mode called *indirect license acquisition;* this allows the rights to be obtained on one device (connected) yet transferred to another device (non-connected). The increasing prevalence of Windows as an operating system (OS) for smart phones means that the use of Windows Media DRM may become more widespread.

In addition to the Windows DRM technology, there are other DRM systems used in mobile devices, such as those based on Java technology. Again, Java is widespread in mobile devices and may spur the adoption of associated DRM techniques.

Summary of Digital Rights Management Models

OMA DRM (rel 2):

- Supports Push, Pull, and streaming
- Separates content and rights objects
- Uses Rights Object Acquisition Protocol (ROAP)

Proprietary DRM Models:

- Windows Media DRM
    - Aimed at mobile devices
    - Supports indirect license acquisition
- Java-based DRM systems
    - Content packaged in JAR (Java Archive) file format
    - Based on Mobile Information Device Profile (MiDP)

Security of Attached Devices

It is now very common for users to attach other devices to their mobile handsets. These include storage devices, to overcome the storage limitations of the handset, and other information devices such as PCs, PDAs, and headsets. The physical means of connecting these devices include Bluetooth and other personal area networking (PAN) technologies, USB ports, and proprietary connections.

If the connection is wireless in particular, then security will become an issue. Bluetooth and ZigBee, which are both 2.4-GHz wireless systems, have built-in security, whereas IrDA (Infra-red Data Association) connectivity does not, by default, include security. It is assumed that the limited range of IrDA, at around one meter, means that security is not a major concern, although it could be applied at the application layer.

Using a physical connection such as a USB also means that security is not a paramount design consideration. However, most DRM systems do address the security of information when transferred from a handset to an external element such as a storage device. In this case, by storing data in its encrypted format without the rights object, it renders the content unusable by anyone else.

Bluetooth Security

Bluetooth includes an authentication and encryption mechanism that aims to provide security on Bluetooth radio links (Figure 4.76). The security sequence for

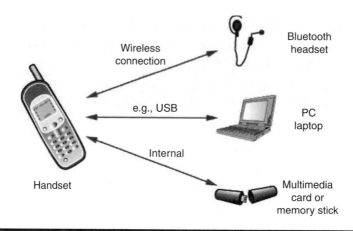

FIGURE 4.76 Bluetooth radio connection.

Bluetooth is part of the pairing procedure between two Bluetooth devices and commences with the generation of an initialization key. This key is generated from a personal identification number (PIN) that is either preprogrammed in some Bluetooth components or entered manually by the user. The length of the PIN can be up to sixteen digits, although often only four digits are used, and it is well documented that attacks on Bluetooth security are greatly aided by the selection of short PIN codes. After generation, the initialization key is used to create a link key via random number exchanges between the two Bluetooth devices. Subsequently, this key can be used to generate keys for the encryption process. After generating the link key, the two devices proceed to stage of mutual authentication, which is based on the challenge-response model (Figure 4.77).

These three stages are all based on an algorithm known as the SAFER+ algorithm. This is also used during encryption but generates only the keys; it is not used for the actual generation of encryption sequences, which is done by a series of shift registers.

There have been a number of well-documented attacks on Bluetooth technology, and handset vendors have at times had to introduce patches to fix their implementation of security. In addition, the evolution of the Bluetooth specifications has addressed some of the security issues. The short range of Bluetooth devices was once considered to limit security risks. However, devices capable of picking up Bluetooth signals over 100 meters are available and thus this is no longer a justifiable view.

Viruses and Malware

There is much debate in the telecommunications sector at present about the extent and possible impact of viruses on handsets and mobile devices. There is no doubt that as smart phones, complete with an operating system (OS) and a host of applications become more prevalent, there may be more viruses appearing. Figure 4.78 illustrates threats to mobile devices.

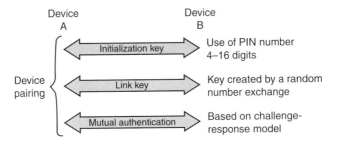

FIGURE 4.77 Bluetooth security model.

There is an issue of critical mass and therefore device density. This why most viruses found in the PC sector are aimed at Microsoft Windows, the most widespread OS. There are viruses in circulation that are aimed at mobile devices, most notably at Symbian OS-based handsets.

Many of these viruses are thought to be *proof of concept* viruses and the result of infection is often no more serious than reduced battery life. This symptom is the result of viruses that spread via Bluetooth and keep the Bluetooth radio switched on in an attempt to find other devices to infect. The biggest concerns at present for mobile devices is that either a mobile device is used as a vehicle to upload a virus onto a corporate network, or that devices are rendered non-usable — not because of a virus, but because the user has downloaded software from a non-trusted source.

Not withstanding this, the traditional anti-virus software companies are starting to produce versions of their AV software that are targeted at handsets and handheld devices. Additional protection for mobile devices is also available in the form of firewall software that has equivalent functionality to personal firewall software packaged with desktop OS. Some commercial mobile firewall products include additional features such as authentication and encryption technology to provide mobile employees with a complete security suite on their handheld devices.

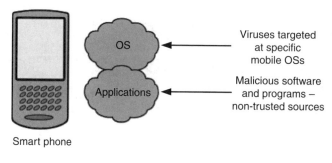

- Anti-virus software for mobile devices
- Firewall software
- Security enhancements:
  - Authentication
  - End-to-end encryption

FIGURE 4.78 Potential threats from mobile viruses.

Review Questions

Q1. Which of the following is currently considered a weak element of mobile network security?
  a. User authentication
  b. Data encryption
  c. Network or mutual authentication
  d. User identity protection

Q2. Which numbers appear in the International Mobile Equipment Identity (IMEI)?
  a. Mobile country code (MCC)
  b. Mobile network code (MNC)
  c. Equipment serial number
  d. None of the above

Q3. Which of the following is a feature of OMA DRM release 2.0?
  a. Forward lock
  b. Security and charging model
  c. Combined distribution
  d. Super distribution

Handset, Services, Media, and Content Distribution ■ 417

Q4. Windows Media DRM solution is based on the OMA open DRM model.
   a. True
   b. False

Q5. The SAFER+ algorithm is used to provide security in which wireless technology?
   a. ZigBee
   b. UWB
   c. Bluetooth
   d. WiMAX

4.4.3 Section 4 Review Questions

Q1. Select a network operator, investigate the location-based services it provides, and answer the following questions:
   a. What location-based technologies are used by the network?
   b. What services are available that are supported by the LBS?
   c. How accurate and useful are these services?

Q2. Many mobile devices available today are capable of supporting MP3 playback and downloading often associated with a music downloading Web site, such as iTunes. Select one of these mobiles and answer the following questions:
   a. What musical formats does your handset support?
   b. What DRM solution does it support?
   c. What are the limitations of the DRM solution?

Q3. Which of the following is *not* a location-based service proposed for 2.5G and 3G networks?
   a. Cell ID based
   b. Assisted GPS
   c. Cellular triangulation
   d. Observed Time Difference of Arrival (OTDOA)

Q4. One potential problem of GPS systems is _____.
   a. Poor accuracy
   b. Does not work inside buildings
   c. Poor battery life
   d. Expensive

Q5. DVB-H can achieve up to 95 percent power saving over DVB-T using which method?
   a. Seamless handover
   b. Time slicing

c. Channel coding
d. UMTS return channel

Q6. SMS is a real-time delivery service for short text messages on a point-to-point basis.
   a. True
   b. False

Q7. Which of the following PTT methods has been chosen by the 3GPP for use in UMTS?
   a. QChat
   b. OMA PoC
   c. iDEN
   d. Kodiak

Q8. Which of the following is currently considered a weak element of mobile network security?
   a. User authentication
   b. Data encryption
   c. Network or mutual authentication
   d. User identity protection

Q9. Which numbers appear in the International Mobile Equipment Identity (IMEI)?
   a. Mobile country code (MCC)
   b. Mobile network code (MNC)
   c. Equipment serial number
   d. None of the above

Q10. Which of the following is a feature of OMA DRM release 2.0?
   a. Forward lock
   b. Security and charging model
   c. Combined distribution
   d. Super distribution

Q11. Windows Media DRM solution is based on the OMA open DRM model.
   a. True
   b. False

Q12. The SAFER+ algorithm is used to provide security in which wireless technology?
   a. ZigBee
   b. UWB
   c. Bluetooth
   d. WiMAX

Index

A

Access network, 15–16
Access signaling, 136–137
Access to network services, 21
Aggregators, samples of, 48
Alarm call, 24
American National Standards Institute, 94
AMR codecs, 357
Analog frequency modulation, 159–160
Analog mobile networks, 13–14
Analog mobile phone technologies, 14
Analog signals, 151–152
Analog telephony connection, 178
Analog television, 190
Analog to digital, 10
  subscriber loop, 178–180
Announcements, control of, 298
ANSI. *See* American National Standards Institute
Answer phone services, 25
Antennas, 186
ARIB. *See* Association of Radio Industries and Businesses
ARPU forecasts, 76–78
Association of Radio Industries and Businesses, 95–96
Attached device handsets, security, 413
Attenuation, 155
Audio circuits, 231, 336
Audio coding, handsets, 355–356, 359–360
Authentication center, 244
Automatic call forwarding, 103

B

Bandwidth, data rates, relationship between, 165
Batteries, 232, 337
  handsets, 316–317
    Battery technologies, 316
Bearer services, teleservices, 112–113
Benefits of regulations, 99–100
Billing, 34–35, 103, 107, 109–110, 138–143, 302–307
  how services billed, 139
  in network, 142–143
  what can be billed, 141–142
Binary environment for wireless, 295
BlackBerry
  architecture, 393
  developing applications, 394–395
BlackBerry PDA mobile, 42
Bluetooth, 191, 374–375
  audio coding, 357–359
  connections, 374
  profiles, 375–376
  radio connection, 414
  security, 413–415
Branding, 28–31
  mobile phone branding, 29–30
  network branding, 30–31
BREW. *See* Binary environment for wireless
Broadband data connections, 25
BSS. *See* Global System for Mobile Communications, radio network elements

419

BT Bluephone, 87–88
Business support systems, 302–307
    billing, 306–307
    content, 306
    data transfer rate, 306
    operations system, 303–304
    post-paid, 306
    prepaid, 306
    requirements, 302–303
    scale of problem, 303
    throughput, 306
    time connected, 306
    total data, 306
    type of service/quality of service, 306

C

Call barring, 25
Call blocking, 25
Call diversion/forwarding, 25
Call waiting, 24
Caller line identity, 24
CAMEL. *See* Customized applications for mobile networks enhanced logic
Carrier-to-interference ratio, 193–195
CDMA2000, 215–217
CDMAOne, 215–217
    worldwide distribution, 62
Cell-based location, 399
Cell planning, 199–203
    seven-call reuse pattern, 200
Cellular principles, 192–204
Cellular radio, 12–13
Cellular spectrum use, 195–197
CEPT. *See* European Conference of Postal and Telecommunications Administration
Changing nature of services, 21–22
Channels, 163–170
    bandwidth, 164–165
    error protection, 166–167
    errors, 165–167
Characteristics of waveform, 150–151
China, 65–68
Circuit-switched core networks/packet-switched, 240–249
    core network requirements, 240–244
        circuit-switched domain, 241
        gateway mobile switching center, 241
    packet-switched domain, 244–245
        serving GPRS support node, 244–245

signaling/control in core network, 245–248
    connecting core network elements, 246–248
Circuit switching, 130–133
    principles, 128–129
Co-location, facility sharing, 101
Coaxial cable, 122, 172
Code division multiple access, 198, 200–202
    cellular planning, 201
    systems based on, 211–218
    technologies, 61–62
Code Division Multiple Access Development Group, 95
Color displays, handsets, 349–350
Comparison of mobile technologies, 59
Competition, 100
Conference call, 24
Confidentiality, 103, 154
Connections to fixed networks, types, 8
Content providers, 47–48
Copper coaxial cable, 122
Copper twisted-pair cable, 121–122
Coprocessors, 341
Core network vendors, 45
Core networks, 17–18
Corporate users, 42
Current mobile technologies, 54–65
    code division multiple access technologies, 61–62
    Enhanced Data for Global Evolution, 59–60
    Global System for Mobile Communications, 55–56
        development, 56–57
        phases, 58–59
        statistics, 59
    GPRS, 59–60
    mobile generations, 54–59
    3G systems, evolving to, 62–64
Customized applications for mobile networks enhanced logic, 298–301

D

Data protection regulations, 102–104
Data rates, bandwidth, relationship between, 165
Dedicated mode handovers, 257
Defining areas of network, 15–20
    access network, 15–16

core networks, 17–18
radio access network, 17
Definition of telecommunications, 106–110
Dense Wavelength Division Multiplexing, 177
Dial-up lines, 9–10
Diffraction, 188
Digital, analog subscriber loop, 178–180
Digital mobile networks, 14–15
Digital mobile phone technologies, 14
Digital rights management, 410
    9DRM, 410–411
    models, 411–413
Digital signals, 152–154
    noise immunity, 153
Digital subscriber line, 179–180
Digital terrestrial television, 190
Direct broadcasting satellite, 190
Directories of subscribers, 103
DRM. *See* Digital rights management
DSL. *See* Digital subscriber line
DVB-H, handset, 401
DWDM. *See* Dense wavelength division multiplexing

E

E-mail via message switching, 133
E1 transmission, 172
Early sample, mobile phone, 14
EDGE. *See* Enhanced Data for Global Evolution
Effects of multipath propagation, 189
Electromagnetic radio wave, 185–186
Emergency call, 23
Enhanced Data for Global Evolution, 59–60, 204–211
Enhanced logic, intelligent networks, 296–302
    customized applications, 298–301
    service control platforms, 297–298
Enhanced message services, 401–403
Entertainment, handsets, 332
Equipment ID Register, 244
Equipment manufacturers, 315
Error protection, 166–167
Errors in radio systems, 189
Errors in received signal, sample, 167
ETSI. *See* European Telecommunications Standards Institute
European Conference of Postal and Telecommunications Administration, 93–94

European Telecommunications Standards Institute, 91
Evolution of telecommunications, 3–5
ExtremeMob, screenshot from, 41

F

Factors in evolution of mobile services, 80
Fading, 189
Feature phones, 332
Fiber-optic cable, 172
Fixed access network, 17
Fixed communication networks, 7–11
    analog to digital, 10
    dial-up lines, 9–10
    integrated services digital networks, 10
    connections, 10–11
    public switched telephony network, 8–9
    connections to, 9
Free phone numbers, 25
Frequency, graphical representation, 151
Frequency division multiplexing, 168
Frequency modulation, 159
Frequency spectrum, 154–155
Functional elements of handset, 336
Future technologies, 74–76, 82–88

G

Gateway mobile switching center, circuit-switched core networks/packet-switched, 241
Global mobile markets, 65–73
    China, 65–68
    United States, 69–72
    Western Europe, 68–69
Global revenues, voice, data, 77
Global System for Mobile Communications, 55–56, 93, 204–211
    architecture, 225–226
    band spectrum, 206
    base station subsystem, 233
    core network connections, 247
    dedicated mode, 256
    development, 56–57
        significant events in, 57
    Enhanced Data for Global Evolution, 268–272
    frequency allocation, 205–206
    GPRS
        connections, 265–266

mobility management, 261–264
network elements, 226–227
packet data in, 207–208
radio resource procedures for,
    268–272
handovers, 256, 271
identities, 227–228
idle mode, 255–256
location, 256–258
location registers, 243–244
location update, 257–259
making calls, 261
    calls from mobile networks, 261
    calls to mobile networks, 261
mobility management states,
    253–255
network diagram, 250
network elements, 226
packet data protocol contexts, 264
paging, 269–270
phase timeline, 58
phases, 58–59
    releases, 204
power control, 272
predictions, 205
procedure sequences, 267–268
procedures, 224–228, 253–261
    categorizing, 224–225
radio network connections, 237
radio network elements, 49–50
radio resource, 268–273
    connection establishment, 269
    in idle mode, 268–269
    procedures, 270
radio site, 234
railway industry, 43
routing area updates, 256–259
sample procedures, 261–267
sequence of procedures for, 267
specifications, 204
states modes, 253
statistics, 59
subscribers, 205
success, 204–205
timeslots, 207
GPRS, 59–60, 204–211
    connections, 265–266
    mobility management, 261–264
    network elements, 226–227
    packet data in, 207–208
    procedures, 261–267

radio resource procedures for, 268–272
serving support node, 244–245
Ground waves, 123
Growth, penetration forecast, distinguished,
    73
GSM. *See* Global System for Mobile
    Communications

H

Handset market, 50–53, 330–333
    handset complexity, features, 52–53
    low-cost handsets, 51
    samples of mobile devices, 53
Handset-user interface, elements of, 320
Handsets, 44, 313–418
    architecture, 337–340
    batteries, 316–317
    capturing image, 361–362
    components, 336–383
    coprocessors, 340
    device manufacturers, 315–316
    display technologies, 349–353
        color displays, 349–350
        display types, 350–352
        LCD display structure, 349
    features of, 52
    form factors, 321–322
    functional blocks, 231, 336–337
    functions, 336–342
    human interfaces, 318–320
    image capabilities, 361–369
    image formats, 362–364
    inter-network roaming, 317–318
    International Telecommunication
        Union, 366–367
    memory storage
        hard disk drives for, 347
        hardware memory, 343–344
        memory cards, 346–347
        memory growth, 344
        requirements, 342–349
        terminal memory types, mobile, 342
        universal subscriber identity mobile
            cards, 344–346
    mobile operating systems, 383–398
    original equipment, 315–316
    processing, 338–339
    processor architectures, 339–340
    requirements, 314
    security, 407–417

attached devices, 413
Bluetooth security, 413–414
digital rights management, 410–413
handset validation, 408–410
malware, 414–416
Open Mobile Alliance, 411
viruses, 414–416
serial interfaces, 369–379
Bluetooth, 374–375
handset interfaces, 369–370
infra-red data association, 370–371
IrDA for mobile handsets, 372
universal serial bus, 373
wireless LANs, 375–378
services, 29, 398–418
DVB-H, 401
enhanced message services, 401–403
instant messaging, 404–405
location technologies, support for, 399–400
mobile TV reception, 400–401
multimedia messaging system, 403–404
Press-to-Talk, 406
short message services, 401–403
UMTS, 401
unified messaging, 404
sound capabilities, 353–361
additional audio coding types, 359–360
AMR codecs, 357
audio coding, 355–356
Bluetooth audio codings, 357–359
ringtones, 353–354
standards bodies, 323–330
Open Mobile Alliance, 325–327
open service access, 327
testing, 324–325
type approval, 324–325
Web services, 327–329
technologies, 313–335
testing, 325
3GP format, 367–368
types, 330–333
entertainment, 332
feature phones, 332
low-feature phones, 331–332
music, 332
segmentation, 330–331
smart phones, 332–333
user interfaces, 318

video capabilities, 361–369
video coding, 364–365
standards, 365–366
Hard disk drives, mobile terminals, 348
Hardware identity, mobile, 409
Hardware memory, 343
handsets, 343–344
Highs-speed downlink packet access, 82–83

I

i-mode, 293–295
Idle mode, 253–255, 263
Image capabilities, handsets, 361–369
Image formats, handsets, 362–364
IMEI. *See* International mobile equipment identity
IMSI. *See* International mobile subscriber identity
IMT2000 family, 214–215
Increasing service needs, 22–23
Incumbent telecom operators, 33–34
Information transfer, 150–185
channels, 163–170
bandwidth, 164–165
error protection, 166–167
errors, 165–167
interleaving, 167
frequency division multiplexing, 168
multiplexing channels, 167–169
time division multiplexing, 168–169
modulation, 158–163
analog frequency modulation, 159–160
modulation techniques, 158–159
pulse code modulation, 160–162
representing information, 150–157
analog signals, 151–152
characteristics of waveform, 150–151
digital signals, 152–154
frequency spectrum, 154–155
noise, 155–157
propagation attenuation, 155–157
sample transmission systems, 170–181
analog, digital subscriber loop, 178–180
analog telephony connection, 178
Dense Wavelength Division Multiplexing, 177
Digital Subscriber Line, 179–180
E1, T1 transmission, 172
integrated services digital network, connections, 178–179

local loop, 178
plesiochronous digital hierarchy, 173, 176
primary multiplexing, 170–171
synchronous digital hierarchy, 174–176
Infra-red data association, 370–371
Infrastructure, mobile networks, 224–253
circuit-switched/packet-switched core networks, 240–249
core network requirements, 240–244
gateway mobile switching center, 241
Global System for Mobile Communications, 243–244
architecture, 225–226
base station controller, 233
base transceiver station, 233
connecting radio access network elements, 236–238
elements of UMTS Terrestrial Radio Access Network, 235
Enhanced Data for Global Evolution, 234
GPRS network elements, 226–227
high-speed downlink packet access, 236
identities, 227–228
node B, 235–236
procedures, 224–228
radio network controller, 236
radio network elements, 232–240
wireless LANs, 238–239
handset types, 229–230
mobile switching center, 241
packet-switched domain, 244–245
signaling/control in core network, 245–248
smart cards, 232
subscriber identity mobile cards, 232
system overview, 224–229
user equipment/radio access network, 229–240
requirements, 230–232
visitor locator register, 241–242
Instant messaging, 404–405
Integrated services digital networks, 10
connections, 10–11, 178–179
Intelligent networks
enhanced logic, 296–302
customized applications, 298–301
service control platforms, 297–298
service provision, 296
Intelligent routing, 298
Inter-network roaming, handsets, 317–318

Inter-symbol interference, 189
Interleaving, 167
frequency division multiplexing, 168
multiplexing channels, 167–169
time division multiplexing, 168–169
Internal market development, 100
International, long-distance connections, 20
International direct dial, 24
International mobile equipment identity, 228
International mobile subscriber identity, 228
International Telecommunication Union, 91, 366–367
Internet, 5
access, 292–295
binary environment for wireless, 295
i-mode, 293–295
methods, 292–295
wireless application protocol, 293
influence of, 6
protocol address, 228
IrDA for mobile handsets, 372
IrDA protocol stack, 371
ISDN. *See* Integrated services digital networks
ITU. *See* International Telecommunication Union

J

Japanese plesiochronous digital hierarchy system, 174
Java, 385–387

K

Kodiak, 406

L

LCD display structure, handsets, 349
Line of sight, 124, 186
Linux-based handset operating systems, 395–396
Linux operating system, 395
Local call
sample, 137
signaling for, 138
Local loop, 178, 180
Location-based services, 82
Location clients, 283
Location registers, 243
Location technologies, support for, 399–400

Long-distance connections, 20
Loop/disconnect signaling, 136
Low-end phones, 331
Low-feature phones, 331–332
Low switching barriers, 100

M

Malware, 414–416
Market, mobile telecommunications, 1–104
    ARPU forecasts, 76–78
    BT Bluephone, 87–88
    evolution of mobile services, 80
    forecasts, 73–79, 81–82
    future technology, 74–76, 82–88
    global, 65–73
        China, 65–68
        United States, 69–72
        Western Europe, 68–69
    handset market, 50–53, 330–333
        complexity, features, 52–53
        low-cost handsets, 51
        samples of mobile devices, 53
    high-speed downlink packet access, 82–83
    location-based services, 82
    mobile Internet in 2G, 81
    penetration forecasts, 73–74
    segmenting market, 33–50
        content providers, 47–48
        core network vendors, 45
        Global System for Mobile
            Communications, 49–50
        IP multimedia sub-system, 46–47
        messaging platforms, 45–46
        mobile market sectors, 33
        operators in market, 33–44
        radio access networks, 44
        voicemail platforms, 46
    subscription forecasts, 73–74
    third-party services, 82
    trends in telecommunications, 79
    2G forecasts, 79
    voice-over-IP, 86–87
    WiFi, 83
    WiMAX, 83–86
Media, available, comparison of, 125–126
Media types, comparison of, 126
Memory, 232, 337
    mobile terminals, 342
Memory cards, 346–347
Memory growth, handsets, 344
Memory requirements, growth in, 345
Memory stick characteristics, 346
Message switching, 130–133
Messaging platforms, 45–46, 287–292
    voicemail platforms, 287–291
        enhanced messaging service, 289
        mobile originated point-to-point, 289
        mobile terminated point-to-point, 288
        multimedia messaging service, 289–291
        short message service SMS, 288
        SMS cell broadcast, 289
Microsoft Windows, 387–388
Microwave links, 190
Microwave radio, 172
Mobile communication networks, 11–15
    analog mobile networks, 13–14
    background, 11–12
    cellular radio, 12–13
    digital mobile networks, 14–15
    private mobile radio, 12
Mobile data card, from Orange, 84
Mobile generations, 54–59
Mobile operating systems, 383–398
    sample, 385–397
Modulation, 158–163
    analog frequency modulation, 159–160
    modulation techniques, 158–159
    pulse code modulation, 160–162
    techniques, 158–159
Motorola iDEN, 406
Multimedia capability, 154
Multimedia messaging system, 403–404
Multimedia service support, 212
Multinational carriers, 35
Multipath propagation, 188
Multiplexing channels, 167–169
Music, handsets, 332

N

Network-based location system, 399
Network branding, 30–31
Network operator branded mobile phones, 30
Network provided value-added services,
        115–116
    samples, 116
Network roaming, 317
Network signaling, 137–138
Networking principles, 110, 143–144
    bus, 143
    hierarchic, 143

mesh, 143
ring, 143
star, 143
Networks, 192–199
Noise, 155–156
Nokia 9500 communicator, with WiFi, 84
North American plesiochronous digital
    hierarchy system, 174
Number translation services, 297
Numbering, control of, 101

O

Ofcom regulatory principles, 102
Open Mobile Alliance, 325–327
    digital rights management, 411
Open service access, 328
    handsets, 327
Open services architecture, 281
Operational/business support systems,
        302–307
    billing, 306–307
    business support, 304–305
    content, 306
    data transfer rate, 306
    operations system, 303–304
    post-paid, 306
    prepaid, 306
    requirements, 302–303
    scale of problem, 303
    throughput, 306
    time connected, 306
    total data, 306
    type of service/quality of service, 306
Operators in telecommunications market,
        33–44
    fixed telecom operators, 33–35
    handsets, 44
    mobile telecom operators, 35–39
    subscriber markets, 39–44
Optical fiber, 124–125
    propagation in, 125
Orange, mobile data card from, 84

P

Packet-switched/circuit-switched core networks,
        240–249
    core network requirements, 240–244
        circuit-switched domain, 241
        gateway mobile switching center, 241

Global System for Mobile
    Communications, 243–244
mobile switching center, 241
visitor locator register, 241–242
packet-switched domain, 244–245
    serving GPRS support node, 244–245
signaling/control in core network, 245–248
    connecting core network elements,
        246–248
Packet switching, 130–133
Palm, 391–392
    evolution of, 392
PAMR. *See* Public access mobile radio
PDH. *See* Plesiochronous digital hierarchy
Penetration forecasts, 73–74
Phone branding, mobile, 29–30
Plesiochronous digital hierarchy, 173–174, 176
Pre-intelligent network service provision, 296
Premium rate numbers, 25
Press-to-Talk, 406
Primary multiplexing, 170–171
Privacy regulations, 102–104
Private mobile radio, 12
Procedures, mobile network, 253–276
Product branding, 28–31
Propagation attenuation, 155–157
Proprietary operating system, 385
Providers, samples of, 48
PSTN. *See* Public switched telephony network
Public access mobile radio, 190
Public switched telephony network, 8–9
    connections to, 9
Pulse code modulation, 160–162
    need for, 160

Q

Qualcom Qchat, 406
Quality of location information, 284

R

Radio access networks, 17, 44
    Global System for Mobile Communications,
        232–240
    base station controller, 233
    base transceiver station, 233
    connecting radio access network
        elements, 236–238
    elements of UMTS Terrestrial Radio
        Access Network, 235

Enhanced Data for Global Evolution, 234
high-speed downlink packet access, 236
node B, 235–236
radio network controller, 236
wireless LANs, 238–239
handset types, 229–230
requirements, 230–232
schemes, 197
smart cards, 232
subscriber identity mobile cards, 232
technologies, 198–199
user equipment, 229–240
Radio frequencies, management of, 101
Radio in fixed network, 8
Radio propagation, 187–188
Radio resource management, 269
Radio spectrum, 186
Radio systems, 122–124, 185–192
  effects of multipath propagation, 189
  electromagnetic radio wave, 185–186
  errors in radio systems, 189
  inter-symbol interference, 189
  radio propagation, 187–188
  radio spectrum, 186
  sample radio systems, 190–191
  UHF band, 186–187
  wireless connections, 191
Railway industry, Global System for Mobile Communications, 43
Redundancy, 281
Regulations, 89–104
  American National Standards Institute, 94
  Association of Radio Industries and Businesses, 95–96
  benefits, 90, 99–100
  case study, 101–102
  Code Division Multiple Access Development Group, 95
  data protection, 102–104
  European Conference of Postal and Telecommunications Administration, 93–94
  European Telecommunications Standards Institute, 91
  Global System for Mobile Communications Association, 93
  International Telecommunication Union, 91
  need for, 98–99
  objectives, 100–101
  Ofcom regulatory principles, 102
  regulators of world, 102
  standards bodies, 89–98
  Telecommunication Technology Committee, 96
  Telecommunications Industry Association, 94
  Telecommunications Technology Association, 96
  Third Generation Partnership Project, 91–92
  UMTS Forum, 92
  user privacy, 102–104
  Wi-Fi Alliance, 96
  WiMAX Forum, 97
Representation of information
  analog signals, 151–152
  digital signals, 152–154
  frequency spectrum, 154–155
  noise, 155–157
  propagation attenuation, 155–157
Requirements of telecommunications, 106–107
Resilience, 281
Rights of way, 101
Ring back when free, 24
Ringtones, 353–354
  formats, 354–355

S

Sample radio systems, 190–191
Sample transmission systems, 170–181
  analog, digital subscriber loop, 178–180
  analog telephony connection, 178
  Dense Wavelength Division Multiplexing, 177
  Digital Subscriber Line, 179–180
  E1, T1 transmission, 172
  integrated services digital network, connections, 178–179
  local loop, 178
  Plesiochronous Digital Hierarchy, 173, 176
  primary multiplexing, 170–171
  Synchronous Digital Hierarchy, 174–176
Satellite systems, 190
Sectored site, 201
Sectorization, 200
Security, 103, 391
  handsets, 407–417
  mobile, 408
Segmenting market, 33–50
  content providers, 47–48

core network vendors, 45
Global System for Mobile Communications, 49–50
IP multimedia sub-system, 46–47
messaging platforms, 45–46
mobile market sectors, 33
operators in telecommunications market, 33–44
    fixed telecom operators, 33–35
    handsets, 44
    mobile telecom operators, 35–39
    subscriber markets, 39–44
    radio access networks, 44
    voicemail platforms, 46
Serial interfaces, handsets, 369–379
Service creation tools, 119
Service platforms, 277–287
    capacity, 278–279
    IP multimedia sub-system, 285–286
    location platforms, 283–284
    location techniques, 284–285
    logical view, 279
    open service architecture, 280–282
    redundancy, 279–280
    requirements, 277–278
    software requirements, 277–278
    vendors, 278
    virtual home environment, 282–283
Services, 21–32, 107, 111–115
    access to, 21
    bearer services, teleservices, 112–113
    branding, 28–31
        mobile phone branding, 29–30
        network branding, 30–31
    changing nature of services, 21–22
    fixed network, 23–25
    on handset, 28
    increasing service needs, 22–23
    mobile network services, 25–28
    provisioning, 117–118
    quality of service, 114
    service requirements, 114
    short message service, 27
    supplementary services, 115
Seven-call reuse pattern, 200
Seven-cell repeating cluster, 194
Short message service, 26–27, 401–403
    applications, 27
Short message service center, 288
Signaling systems, 134–138
    access signaling, 136–137
    control requirements, 134
    functions, 135–136
    importance of, 134
    local call sample, 137
    network signaling, 137–138
    tones, 137
Sky waves, 123
Small antennas, 186
Small regional operators, 33–35
Smart cards, 337
Smart phones, 230, 332–333
Sound capabilities, handsets, 353–361
SS7, 137–138
Standards, 89–104, 323
    American National Standards Institute, 94
    Association of Radio Industries and Businesses, 95–96
    benefits, 90, 99–100
    case study, 101–102
    Code Division Multiple Access Development Group, 95
    data protection, 102–104
    European Conference of Postal and Telecommunications Administration, 93–94
    European Telecommunications Standards Institute, 91
    Global System for Mobile Communications Association, 93
    handsets, 323–330
    International Telecommunication Union, 91
    need for, 98–99
    objectives, 100–101
    Ofcom regulatory principles, 102
    regulators of world, 102
    standards bodies, 89–98
    Telecommunication Technology Committee, 96
    Telecommunications Industry Association, 94
    Telecommunications Technology Association, 96
    Third Generation Partnership Project, 91–92
    UMTS Forum, 92
    user privacy, 102–104
    Wi-Fi Alliance, 96
    WiMAX Forum, 97
STN display. *See* Super-twisted nematic display
Subscriber markets, 39–44
Subscribers by technology, United States, 71

Subscription forecasts, 73–74
Super-twisted nematic display, 351
Supporting systems, mobile networks, 277–312
    accessing Internet, 292–295
        binary environment for wireless, 295
        i-mode, 293–295
        methods, 292–295
        wireless application protocol, 293
    billing, 302–307
    enhanced logic, intelligent networks, 296–302
        customized applications, 298–301
        service control platforms, 297–298
    messaging platforms, 287–292
    operational/business support systems, 302–307
        billing, 306–307
        business support, 304–305
        content, 306
        data transfer rate, 306
        operations system, 303–304
        post-paid, 306
        prepaid, 306
        requirements, 302–303
        scale of problem, 303
        throughput, 306
        time connected, 306
        total data, 306
        type of service/quality of service, 306
    service platforms, 277–287
        capacity, 278–279
        IP multimedia sub-system, 285–286
        location platforms, 283–284
        location techniques, 284–285
        open service architecture, 280–282
        redundancy, 279–280
        requirements, 277–278
        software requirements, 277–278
        vendors, 278
        virtual home environment, 282–283
    voicemail platforms, 287–291
        enhanced messaging service, 289
        mobile originated point-to-point, 289
        mobile terminated point-to-point, 288
        multimedia messaging service, 289–291
        short message service SMS, 288
        SMS cell broadcast, 289
Switching, 109, 128–133
    circuit switching, 130–133
    principles of circuit switching, 128–129

Symbian, 389–390
    architecture, 390
    evolution, 391
    user interfaces, 390
    versions, 390–391
Synchronous digital hierarchy, 174–175
    multiplexing, 175
    ring structures, 176

T

T1 transmission, 172
TDMA. *See* Time division multiple access
Technology coverage areas, 202–203
    samples, 202
Technology principles, 105–222
    cellular concept, 192–204
    channels, 163–170
    code division multiple access-based systems, 211–218
    Enhanced Data for Global Evolution, 204–211
    Global System for Mobile Communications, 204–211
    GPRS, 204–211
    modulation, 158–163
    radio systems, 185–192
    representing information, 150–157
    sample transmission systems, 170–181
    telecommunications basics, 106–111
Telecom operators, mobile, 35–39
Telecommunication Technology Committee, 96
Telecommunications Industry Association, 94
Telecommunications services
    in fixed network, 23–25
    on handset, 28
    mobile network services, 25–28
Telecommunications Technology Association, 96
Terminal memory types, handsets, mobile, 342
Third generation network requirements, 212
Third Generation Partnership Project, 91–92
Third generation technologies, 64
Third-party services, 82, 116–117
    providers, 278
    samples, 117
3G systems, 62–64, 211–213
3GP format, 367–368
Tier 1 carriers, 35
Time connected, 141

Time division multiple access, 198, 206–207
Time division multiplexing, 168–169
Tones, 137
Traffic data, 103
Transit exchanges, need for, 130
Transmission systems, 107–108, 120–121, 126–127, 170–181
Trends in telecommunications, 79
Trunk circuits, need for, 129
TV reception, mobile, 400–401
Twisted-pair cable, 172
2G systems, 79, 81
Types of networks, 7–21
    analog mobile phone technologies, 14
    defining areas of network, 15–20
        access network, 15–16
        core networks, 17–18
        radio access network, 17
    digital mobile phone technologies, 14
    fixed communication networks, 7–11
        analog to digital, 10
        dial-up lines, 9–10
        integrated services digital networks, 10–11
        public switched telephony network, 8–9
    mobile communication networks, 11–15
        analog mobile networks, 13–14
        background, 11–12
        cellular radio, 12–13
        digital mobile networks, 14–15
        private mobile radio, 12
    radio in fixed network, 8

U

Ultra high frequency band, 186–187
UMTS, 401
    license allocations, 213–214
UMTS Forum, 92
UMTS Terrestrial Radio Access Network, 235
Unified messaging, 404–405
United States market, 69
United States operators, 72
Universal serial bus, 373
Universal subscriber identity mobile cards, 344–346
Unsolicited communications, 103–104
User equipment/radio access network, 229–240
    Global System for Mobile Communications, 232–240

base station controller, 233
base transceiver station, 233
connecting radio access network elements, 236–238
elements of UMTS Terrestrial Radio Access Network, 235
Enhanced Data for Global Evolution, 234
high-speed downlink packet access, 236
node B, 235–236
radio network controller, 236
wireless LANs, 238–239
handset types, 229–230
requirements, 230–232
smart cards, 232
subscriber identity mobile cards, 232
User perspectives, 21–32
    access to network services, 21
    branding, 28–31
        mobile phone branding, 29–30
        network branding, 30–31
    changing nature of services, 21–22
    fixed network services, 24
    increasing service needs, 22–23
    mobile network services, 26
    short message service, applications, 27
    telecommunications services, 23–
        mobile network services, 25–28
        services in fixed network, 23–25
        services on handset, 28
USIM. *See* Universal subscriber identity mobile
UTRAN. *See* UMTS Terrestrial Radio Access Network

V

Validation, handset, 408–410
Value-added services, 115–118
    copper coaxial cable, 122
    copper twisted-pair cable, 121–122
    media, available, comparison, 125–126
    network provided samples, 116
    optical fiber, 124–125
    radio, 122–124
    service provisioning, 117–118
    third-party provided services, 116–117
    transmission systems, 120–121, 126–127
    user experience, 117

Vendor markets, 43–44
Vendor spaces in mobile telecommunications, 43
Video capabilities, handsets, 361–369
Video coding, handsets, 364–365
    standards, 365–366
Viruses, 414–416
Vodafone live
    sample screenshot, 30
    web page, 31
Voice, data core network connections, 19
Voice over Internet, 87
Voice-over-IP, 86–87
Voicemail platforms, 46, 287–291
    enhanced messaging service, 289
    mobile originated point-to-point, 289
    mobile terminated point-to-point, 288
    multimedia messaging service, 289–291
    short message service SMS, 288
    SMS cell broadcast, 289
VoIP mobile telephone call, 87
VoIP. *See* Voice-over-IP

W

W-CDMA system, 213–214
Waveform characteristics, 150–151
Web services, handsets, 327–329
Western Europe, global market, 68–69
Wi-Fi Alliance, 96
WiFi, 83–85, 191
WiMAX, 83–86, 191
WiMAX Forum, 97
Windows mobile 5.0, 388–389
Windows mobile OS evolution, 388
Wireless instant messaging, 405
Wireless LANs, 375–377
    ad-hoc mode, 377
    configuration, 377–378
    infrastructure mode, 378

Y

Youth market, 39–41
Youth services, 40